Genomics-Driven Drug Discovery Through Pharmacogenomics

Arli Aditya Parikesit
Indonesia International Institute for Life Sciences, Indonesia

Mehrdad Naderi
University of Latvia, Latvia

Anupam Singh
Bansal Institute of Engineering and Technology, India

Rajesh G. Konnur
Eudoxia Research Centre, India

Saptarshi Mukherjee
GoldSmith Institute of Advanced Research, India

IGI Global
Scientific Publishing
Publishing Tomorrow's Research Today

Vice President of Editorial	Melissa Wagner
Managing Editor of Acquisitions	Mikaela Felty
Managing Editor of Book Development	Jocelynn Hessler
Production Manager	Mike Brehm
Cover Design	Phillip Shickler

Published in the United States of America by
IGI Global Scientific Publishing
701 East Chocolate Avenue
Hershey, PA, 17033, USA
Tel: 717-533-8845
Fax: 717-533-8661
Website: https://www.igi-global.com E-mail: cust@igi-global.com

Library of Congress Cataloging-in-Publication Data

Names: Parikesit, Arli Aditya, editor.
Title: Genomics-driven drug discovery through pharmacogenomics / edited by Arli Parikesit, Mehrdad Naderi, Anupam Singh, Rajesh Konnur, Saptarshi Mukherjee.
Description: Hershey, PA : IGI Global Scientific Publishing, [2025] | Includes bibliographical references and index. | Summary: "This book aims to provide a comprehensive overview of how genetic information is revolutionizing the identification of drug response markers and the personalization of medicine"-- Provided by publisher.
Identifiers: LCCN 2024050863 (print) | LCCN 2024050864 (ebook) | ISBN 9798369365977 (h/c) | ISBN 9798369365984 (s/c) | ISBN 9798369365991 (eISBN)
Subjects: LCSH: Pharmacogenomics.
Classification: LCC RM301.3.G45 G47 2025 (print) | LCC RM301.3.G45 (ebook) | DDC 615.7--dc23/eng/20250107
LC record available at https://lccn.loc.gov/2024050863
LC ebook record available at https://lccn.loc.gov/2024050864

British Cataloguing in Publication Data
A Cataloguing in Publication record for this book is available from the British Library.

This book is dedicated to humanity and the pursuit of global peace.

Table of Contents

Detailed Table of Contents

Over the years, the field of drug discovery has seen significant progress, with technological advancements proving crucial in the identification of new therapeutic candidates. High-Throughput Screening (HTS) which combines large-scale data analysis with automated micro-quantitative experiments, and can rapidly analyse a large number of chemical compounds against the drug targets. This chapter explores the recent advancements in high-throughput screening platforms that significantly enhanced drug discovery process, and also includes the incorporation of miniaturization, automation, and image-based analysis to increase throughput and data quality. The significance of these advanced HTS platforms on accelerating the discovery and development of new therapeutic drugs is discussed, along with the challenges. This chapter, taken as a whole, will offer a comprehensive overview of the most advanced high-throughput screening (HTS) platforms available for use in the drug development industry and will address how they might completely transform the process.

Genomics-driven drug discovery in precision medicine tailors treatments to individuals' genetic profiles, yet raises complex ethical, legal, and social issues (ELSI). Concerns include patient privacy, consent, and genetic discrimination. Legal challenges involve intellectual property, data ownership, and regulatory frameworks. Socially, attitudes toward genetic testing and access to technology influence perception. Addressing these ELSI requires collaboration across stakeholders to establish robust governance systems ensuring individual rights, fair access, and trust in genomic research and healthcare. This chapter explores these implications and offers strategies for navigating ELSI in genomics-driven drug discovery.

The discovery and validation of genomic biomarkers are crucial for advancing drug development and personalized medicine. Genomic biomarkers, including DNA and RNA characteristics, help predict therapeutic responses, drug efficacy, and potential adverse effects, enabling precise treatment strategies. This chapter explores the identification and validation of genomic biomarkers, their role in drug development, and the high-throughput sequencing technologies and bioinformatics tools driving their discovery. It examines validation techniques, biomarker integration into drug design and clinical trials, and relevant regulatory frameworks. Through case studies, the chapter illustrates the transformative impact of genomic biomarkers on drug efficacy and patient safety, emphasizing the need to overcome technical, ethical, and legal challenges to optimize patient outcomes and advance medical science.

Pharmacogenomics, the study of how genes influence an individual's response to drugs, is changing the scene of medication disclosure and improvement by joining pharmacology and genomics to make powerful, custom-made to an individual's hereditary profile. This arising field is vital in customized medication, where medicines are upgraded in light of individual hereditary varieties. This chapter offers an in-depth exploration of the global impact of pharmacogenomics, underscoring key examination foundations and undertakings, moral contemplations, difficulties and future headings. The joint efforts exhibit the capability of joint endeavor to defeat difficulties and advance the field by fitting worldwide guidelines and creating strategies that help even-handed admittance to pharmacogenomic progressions are significant for the worldwide reconciliation of customized medication.

This chapter provides a comprehensive overview of the role of genomics in accelerating the drug development process. It highlights the significant advancements in large-scale population collections and whole genome sequencing approaches, which have significantly influenced multiple stages across the lifespan of a drug discovery program. The chapter emphasizes the increasing importance of genomic data in target identification, target prioritization, and tractability, as well as predicting outcomes from pharmacological perturbations. It also discusses integrating genomic technologies with other omics approaches, such as proteomics and metabolomics, to understand disease mechanisms better and identify potential therapeutic targets. The integration of bioinformatics is pivotal in harnessing the power of genomic data for drug discovery endeavors. This chapter delves into the computational approaches and analytical frameworks that enable extracting meaningful insights from vast genomic datasets. Furthermore, it examines the contributions of bioinformatics in the field of pharmacogenomics.

Abhidha Kohli, Department of Biotechnology, Noida Institute of Engineering and Technology, Greater Noida, India
Sonia Verma, Department of Biotechnology, Noida Institute of Engineering and Technology, Greater Noida, India
Neeraj Agarwal, Department of Biotechnology, Noida Institute of Engineering and Technology, Greater Noida, India
Aryan Kushwaha, Department of Biotechnology, Noida Institute of Engineering and Technology, Greater Noida, India
Arpit Gupta, Department of Biotechnology, Noida Institute of Engineering and Technology, Greater Noida, India
Ashwini Kumar, School of Basic Sciences and Research, Sharda University, Greater Noida, India

Precision Medicine encompasses decision making about prevention, diagnosis, and treatment of a particular disease in a patient tailored based on the information derived from a patient's genetic profile, environment, and lifestyle. It is patient-specific based on their specific clinical and biological characteristics with minimum side effects. With advancements in DNA sequencing technology, this concept gained attention among scientists. The projects like Human Genome Project and International HapMap Project provided deep insight into human genomics and information about specific disease-associated gene variants. Moreover, pharmacogenetics and pharmacogenomics have helped in understanding molecular basis of diseases and corresponding therapeutic strategy designed for individuals for a particular drug and the effect of multiple variations in genome on drug treatment, helping biomarker identification, influencing pathogen and drug effects on body, establishing the need for genomic biomarkers discovery for designing patient-specific treatment modalities for a specific disease.

Foreword

The landscape of drug discovery is experiencing an unprecedented evolution, driven by breakthroughs in genomics and the expanding field of pharmacogenomics. As the world moves toward precision medicine, the ability to tailor treatments based on an individual's genetic makeup is not just a goal but an imperative. This book, "Genomics-Driven Drug Discovery Through Pharmacogenomics", is a timely and critical contribution to this transformation, offering a deep dive into the pivotal role of genomics in reshaping pharmaceutical sciences and healthcare.

Through all its expertly crafted chapters, this book covers a wide spectrum of emerging topics, from the fundamentals of genomics-driven drug discovery to the sophisticated integration of pharmacogenomics into personalized medicine. The book's focus on next-generation sequencing technologies, functional genomics, and the impact of artificial intelligence in drug development highlights the interplay between biological data and computational tools, demonstrating how these innovations are streamlining drug discovery and advancing therapeutic strategies.

Additionally, the book's emphasis on drug repurposing, genomic biomarkers, and global perspectives offers readers a comprehensive view of the current landscape and future directions of pharmacogenomics. The integration of advanced high-throughput screening platforms and systems medicine provides cutting-edge methodologies that accelerate the transition from discovery to clinical application. What truly distinguishes this work is its careful examination of the ethical, legal, and social implications of these advancements, ensuring a responsible and balanced approach to innovation.

At a time when the demand for personalized and precision healthcare is intensifying, "Genomics-Driven Drug Discovery Through Pharmacogenomics" stands as an indispensable resource. It offers profound insights for academic researchers, industry professionals, clinicians, and policymakers alike, equipping them with the knowledge to pioneer new frontiers in medicine.

I applaud the editors and contributors for their dedication to delivering a work of both depth and breadth, making complex concepts accessible while driving forward the narrative of genomics-based drug discovery. This publication will not only serve as a vital reference but also inspire the next wave of advancements in precision medicine and pharmaceutical sciences.

Arif Nur Muhammad Ansori
Universitas Airlangga, Indonesia & Uttaranchal University, India

Preface

CHALLENGES AND SOLUTIONS

We merge genomics and pharmacogenomics into drug discovery for a new era of personalized medicine. This book, "Genomics-Driven Drug Discovery Through Pharmacogenomics," addresses the question of how much scientific development might clinically be translated into medical treatment tailored to the individual's genetic profile. We will closely examine pharmacogenomics's focus on how the advent of genomic data translates into drug development and therapeutic strategies (Spreafico et al., 2020).

Pharmacogenomics gives insight into genes' role in determining the response an individual might give to a drug and can be used as a cornerstone in drug discovery and development. This approach uses global gene expression patterns that enable biologically relevant endpoints of drugs' efficiency. For instance, in the cases of cystic fibrosis, pharmacogenomic approaches take advantage of gene expression profiles corrected by gene therapy to formulate decisions in drug development(Hassanzad et al., 2017; Parikesit, 2021; Pollard et al., 2001; Sakon & Tillman, 2022). This postulates that efficient drug and gene therapies share common points. This incorporation of genomics in drug discovery is two-dimensional: one that increases our understanding of the disease process and the other that allows us to design drugs aimed at the known genetic mutations (Mahgoub et al., 2024).

This is further supported by the fact that pharmacogenomics also contributes to predicting the safety and efficacy of the drug with more precision. It studies the effects of thousands of genes via genome-wide studies and provides information on the target polymorphism of drugs and their changes in drug response. Besides that, this capability is important in stratifying patients by genotype in clinical trials, leading to fewer adverse reactions and better trial outcomes. The FDA recently started recognizing pharmacogenetic data in drug labeling, showing its increasing impact on regulatory processes (Guin et al., 2019; Patel, 2014).

Also, pharmacogenomics finds its applications in biomarker discovery, which plays a vital role in the development of targeted therapeutics. These biomarkers are key determinants in defining the treatment regime and avoiding adverse effects for better treatment outcomes. Rapid regulatory approval of biomarker-based personalized therapeutics underlines the broadening role of pharmacogenomics at the forefront of multiple therapeutic fields, ranging from molecularly targeted cancer therapies to wider applications in personalized medicine(Sadee et al., 2023; Scandolara et al., 2022).

Despite the promise, there are problems with translating pharmacogenomics into the clinic, including those related to data validity across diverse populations and ethical considerations concerning genetic information. This obviously calls for massive industry-academia collaboration to systematically collect and validate pharmacogenetic data, overriding these challenges. Indeed, this will also be important in developing genetic profiles that predict drug response, accelerating the development of novel drugs and diagnostic tests (Gurwitz et al., 2009; Otsubo et al., 2013).

This book describes those recent developments and explains methodologies, applications, and implications for genomics-driven drug discovery. It will interest the researcher, clinician, student, and policymaker who wishes to understand how genomics is reconfiguring medicine toward a future of ever-more personalized and effective treatments. This book endeavors to shed light on the way forward by integrating scientific insights with practical applications in leveraging the full power of genomics in health care.

ORGANIZATION OF THE BOOK

The radical development in genomics and personalized medicine accelerates the pace of change in drug discovery and development. This book examines these tectonic changes in integrating genomics, pharmacogenomics, artificial intelligence, and high-throughput screening technologies in drug discovery. Each chapter details one aspect of this evolution and assesses methodology, applications, and implications.

Chapter 1. A Paradigm Shift on the Emergence of Pharmacogenomics in the Era of Personalized Medicine: Transforming Health Care-Role of Genetic Profiling in Personalized Treatment Approaches. This chapter deals with the translation of pharmacogenomics into the transformation of health care by enabling genetic profiling for personalized treatments. This chapter addresses integrating genetic data into clinical practice to enhance therapeutic precision. The growing effect of pharmacogenomics is evident through increasing FDA approvals of personalized therapeutics involving biomarkers.

Chapter 2. Advanced High Throughput Screening Platforms in Drug Discovery touches on HTS technologies, which have become integral to modern drug discovery. High-throughput screening platforms offer fast screening of a large compound library, fast-tracking the process of identifying potential drug candidates. The platforms also fit into integrating genomic data into drug discovery workflows to identify promising therapeutic compounds efficiently.

Chapter 3. Ethical, Legal, and Social Implications of Genomics-Driven Drug Discovery completes the volume by considering some ethical, legal, and social challenges associated with genomics-driven drug discovery. Data privacy, consent, and equity feature in questions integral to the complex discussion of integrating genomics into healthcare.

Chapter 4. Genomic Biomarkers Discovery and Validation in Drug Development covers methodologies related to discovering genomic biomarkers and their validation, underlining the importance of biomarkers for enhancing the effectiveness and safety of drugs through targeted therapeutic intervention. Guidelines through regulatory agencies such as the FDA and EMA and involvement in drug development pinpoint the pivotal roles that genomic biomarkers will play for successful regulatory approval.

Chapter 5. Global Perspective on Pharmacogenomics and Drug Discovery details the adoption of pharmacogenomics into drug development across different regions. The chapter also covers various global initiatives related to the use of genetic information in developing novel therapeutic options. International collaborations will henceforth be vital in ensuring that there is overcoming the obstacles toward pharmacogenomic implementation and a guarantee of equal access to genomic technologies.

Chapter 6. Introduction to Genomic-Driven Drug Discovery by highlighting how the introduction of genomics in drug discovery is a field with tremendous changes. This chapter reviews the transition from traditional approaches to more precise, data-driven, and informed by genetics to identify novel therapeutic targets. Genomics-driven drug discovery has become key to accelerating the development of new drugs, a trend also underscored in recent guidelines and frameworks supporting cross-population meta-analyses and GWAS.

Chapter 7. Precision Medicine in View of Genomic Biomarker Discovery deals with discovering and applying genomic biomarkers associated with precision medicine. Genomic biomarkers offer important possibilities for predicting drug efficiency and safety and the diagnosis and prognosis of pathological conditions. These biomarkers accelerate the pace of precision medicine by developing molecularly targeted drugs and optimizing clinical trial designs.

Chapter 8. Revolutionizing Pharmaceutical Sciences-A Comprehensive Review of Artificial Intelligence's Impact on Drug Discovery and Development: The chapter throws light on how artificial intelligence has been reshaping the face of pharmaceutical sciences. AI has become important in enhancing drug design, predicting drug interaction, and giving personalized treatment regimens. Despite this promise, AI faces challenges regarding data quality, generalizability, and ethical considerations.

Chapter 9. Trends of Pharmacogenomics in Personalized Medicine largely reviews pharmacogenomics's important role in developing treatment in medicine based on an individual's genetic background. Pharmacogenomics provides the basis for precision medicine in healthcare, enabling providers to select drugs and their dosing based on patient genetics. While promising, its clinical translation must be improved due to a limited uptake. This is partly because most studies have used traditional investigational approaches to explain clinical variability in therapeutic outcomes.

This book will interest any researcher, clinician, student, or policy maker eager to learn about the new dynamics in drug discovery opened by the genomic era. The scientific insights and practical applications evoke a holistic view of how genomics is reshaping medicine toward a future of treatments that will be increasingly personalized and effective. Moving even further into this frontier, we must be able to address challenges head-on and seize opportunities developed by the advantage of these advanced technologies.

REFERENCES

Guin, D., Rani, J., Singh, P., Grover, S., Bora, S., Talwar, P., Karthikeyan, M., Satyamoorthy, K., Adithan, C., Ramachandran, S., Saso, L., Hasija, Y., & Kukreti, R. (2019). Global text mining and development of pharmacogenomic knowledge resource for precision medicine. *Frontiers in Pharmacology*, 10(JULY), 839. Advance online publication. DOI: 10.3389/fphar.2019.00839 PMID: 31447668

Gurwitz, D., Zika, E., Hopkins, M. M., Gaisser, S., & Ibarreta, D. (2009). Pharmacogenetics in Europe: Barriers and opportunities. In *Public Health Genomics* (Vol. 12, Issue 3). DOI: 10.1159/000189625

Hassanzad, M., Orooji, A., Valinejadi, A., & Velayati, A. (2017). A fuzzy rule-based expert system for diagnosing cystic fibrosis. *Electronic Physician*, 9(12), 5974–5984. DOI: 10.19082/5974 PMID: 29560150

Mahgoub, E. O., Cho, W. C., Sharifi, M., Falahati, M., Zeinabad, H. A., Mare, H. E., & Hasan, A. (2024). Role of functional genomics in identifying cancer drug resistance and overcoming cancer relapse. In *Heliyon* (Vol. 10, Issue 1). DOI: 10.1016/j.heliyon.2023.e22095

Otsubo, Y., Ishiguro, A., & Uyama, Y. (2013). Regulatory perspective on remaining challenges for utilization of pharmacogenomics-guided drug developments. In *Pharmacogenomics* (Vol. 14, Issue 2). DOI: 10.2217/pgs.12.194

Parikesit, A. A. (2021). Introductory Chapter: The Emerging Corner of the Omics Studies for Rational Drug Design. In Parikesit, A. A. (Ed.), *Drug Design: Novel Advances in the Omics Field and Applications* (p. 4). IntechOpen., DOI: 10.5772/intechopen.95544

Patel, J. N. (2014). Application of genotype-guided cancer therapy in solid tumors. In *Pharmacogenomics* (Vol. 15, Issue 1). DOI: 10.2217/pgs.13.227

Pollard, H. B., Eidelman, O., Jacobson, K. A., & Srivastava, M. (2001). Pharmacogenomics of cystic fibrosis. *Molecular Interventions*, 1(1), •••. PMID: 14993338

Sadee, W., Wang, D., Hartmann, K., & Toland, A. E. (2023). Pharmacogenomics: Driving Personalized Medicine. In *Pharmacological Reviews* (Vol. 75, Issue 4). DOI: 10.1124/pharmrev.122.000810

Sakon, C. M., & Tillman, E. M. (2022). Pharmacogenomics: a tool to improve medication safety and efficacy in patients with cystic fibrosis. In *Pharmacogenomics* (Vol. 23, Issue 9). DOI: 10.2217/pgs-2022-0025

Scandolara, T. B., Barreto Pires, B. R., Vacario, B., de Amorim, I. S. S., Siqueira, P. B., Serpeloni, J. M., Mencalha, A. L., Bonvicino, C. R., & Panis, C. (2022). An Overview Regarding Pharmacogenomics and Biomarkers Discovery: Focus on Breast Cancer. *Current Topics in Medicinal Chemistry*, 22(20), 1654–1673. Advance online publication. DOI: 10.2174/1568026622666220801115040 PMID: 35927918

Spreafico, R., Soriaga, L. B., Grosse, J., Virgin, H. W., & Telenti, A. (2020). Advances in Genomics for Drug Development. *Genes*, 11(8), 1–19. DOI: 10.3390/genes11080942 PMID: 32824125

Acknowledgment

The editors would like to acknowledge the help of all the people involved in this project and, more specifically, to the authors and reviewers that took part in the review process. Without their support, this book would not have become a reality.

First, the editors would like to thank each one of the authors for their contributions. Our sincere gratitude goes to the chapters authors who contributed their time and expertise to this book.

Second, the editors wish to acknowledge the valuable contributions of the reviewers regarding the improvement of quality, coherence, and content presentation of chapters. Most of the authors also served as referees; we highly appreciate their double task.

Third, the editors would like to acknowledge Department of Research and Community Services of Indonesia International Institute for Life Sciences (LPPM i3L) and Mr. Brijesh Soni, President and CEO of GoldSmith Institute of Advanced Research [GIAR], India, for their heartfelt support toward this book project.

Chapter 1

A Paradigm Shift on Emergence of Pharmacogenomics in the Era of Personalized Medicine

Neha Kukreti
https://orcid.org/0000-0003-0802-4616
Chitkara College of Pharmacy, Chitkara University, Rajpura, India

Khushi Lamba
Chitkara College of Pharmacy, Chitkara University, Rajpura, India

Pankaj Pant
School of Pharmacy, DIT University, India

Sharad Visht
3Tishk International University, Erbil, Iraq

Shilpa Rana
Uttaranchal Institute of Pharmaceutical Sciences, Uttaranchal University, India

ABSTRACT

Pharmacogenomics is the study of genetic modifications in patients that influence their potential to metabolize medications, encouraging their tolerability and desired effect. Genetic (like genome-wide association & candidate gene) methods may help in choosing the most suitable drug and dosage, by impeding adverse drug reactions

DOI: 10.4018/979-8-3693-6597-7.ch001

(ADRs). Pharmacogenomics are usually documented as key steps in personalized medicine. The significance of personalized medicine in healthcare is developing due to its defined genetic and molecular basis. Personalized medicine anticipates the beneficial or serious side effects of a medication by using specific individual's genetic profile. In the recent year's pharmacogenomics approaches are being widely used in personalizing therapies for diseases like Cancer, Rheumatoid arthritis, Asthma, Type-1 diabetes and many psychiatric disorders (depression, psychosis, schizophrenia etc.).This chapter provides an overview of the present importance of pharmacogenomics and identifies the challenges that need to be conquered for effective personalized medicine.

1. INTRODUCTION

With the evolution of studies regarding drugs, genes and variation of diseases and factors in each individual, the rationales for tailoring different therapy for each individual are becoming popular. 'Personalized medicine' refers to treating, monitoring and preventing diseases in an individual with unique approaches and therapies considering different physiology, metabolism, behaviour, environmental exposure and genetic profile. In recent years, the clinical adoption of personalized medicine has seen a widespread rise. But the total adoption of the practice by all the key stakeholders is still a long way to go, as with its implementation there are greater challenges (Goetz, & Schork, 2018).

Every individual has differences in thoughts, behaviour, environmental exposure, food habits and of course genetic makeup of each one of us is unique. There are different approaches to treat the same disease in two different individuals. For example, in one individual diuretics can be started as a first line drug to treat hypertension whereas another individual whose family history shows severe kidney disorders, diuretics are not chosen for the treatment. Understanding of family history can be a crucial tool to understand genetic make-up and possibility of certain mutations in a person (Krzyszczyk, *et al.,* 2018). This information will lead to optimized interventions by healthcare professionals. In recent years personalized therapies are used for the treatment of many diseases like cancer, cardiovascular conditions, Parkinson's disease, asthma etc.

In personalized medicine, genes are critical because they allow medical professionals to customize therapies based on each patient's unique genetic profile. Genetic testing determines the inclination for a disease, enabling tailored medicines and early intervention

Numerous genes add up to various phenotypic expressions, while some may have more prominence than others. Millions of genetic variations are possible in the whole human population. Thus, a large portion of the current focus on personalized medicine is based on these genetic findings (Sugandh, *et al.,* 2023). It is important to note that while genetic research forms the basis of personalized medicine, it is generally acknowledged that other factors, such as developmental processes and epigenetic modifications, environmental exposures, and behavioural factors, must also be considered when deciding how best to treat a given patient (Carlsten, *et al.,* 2014).

This book chapter emphasizes on the emerging role of pharmacogenomics in personalized therapy for the patients. It delves into pharmacogenomics, its history and clinical adoption of pharmacogenomics in recent years. Along with this, the chapter discusses the challenges faces and future implications of pharmacogenomics in personalized medicine.

2. PHARMACOGENOMICS

Pharmacogenomics is the study of various genetic variations that exist among populations and affect the drug response. It integrates the fields of pharmacology, the study of medications with genomics (Meenu, *et al.,* 2023). The main goal of pharmacogenomics is to identify underlying genetic mechanisms and achieve better efficacy and decreased toxicity. The advancements in pharmacogenomics further contribute in expanding the use of current drugs for new conditions and will serve as an instrument for the development of new drugs. The drug response is attributed to Pharmacokinetics and Pharmacodynamics. In many cases, genetics largely define the ways how drug responses. These genetic variants alter its absorption, distribution, metabolism, excretion and pharmacodynamics (Auwerx, *et al.,* 2022). Sequence variation in medication target genes can modify the pharmacodynamics of a drug, and variation in sequences of medication mediation nature alters out the drug's pharmacodynamics properties (Mirsadeghi & Larijani, 2017).

Pharmacogenomics evolved from pharmacogenetics with time and advancements in technology. Researches and data collection on human genome acted as a stepping stone in development of modern pharmacogenomics .Pharmacogenomics refers to the whole set of genes involved in determining the safety and efficacy of drugs, whereas pharmacogenetics refers to monogenetic differences that impact medicine response (Meenu *et al.,* 2023).

Pharmacogenomics finds its usage in various clinical procedures, from diagnostics to treatments. While the personalization of therapies using pharmacogenomics is discussed further in the chapter in detail, the diagnosis using pharmacogenomics still remains an underutilized tool. Pharmacogenomic based diagnosis are being used for various disorders like major cardiac and brain disorders. In cardiac patients, CYP2C19 genotyping is implemented on a large scale (Duarte & Cavallari, 2021). Even for cancer detection pharmacogenomics finds widespread application. Tests like Endo-Predict and PAM-50 have been released for diagnosis of breast cancer in latest years, these tests based on gene analysis assist in identification and treatment of breast cancer (Rodríguez-Vicente, *et al.,* 2016).

With the advancements in the field of pharmacogenomics with time many databases have been generated to facilitate research in the domain. Pharm-GKB is a web public repository for genotypic and phenotypic information. CBio-Portal, Cell-Miner, Connectivity Map (CMAP 2.0), SPHINX are the major pharmacogenomics databases developed by renowned institutions. These tools serve as a helping hand in implementation of pharmacogenomics in personalized medicine (Gong, *et al.,* 2021).

3. HISTORY AND TYPES OF PHARMACOGENOMICS

3.1 History

Pharmacogenomics did not come into existence suddenly; the correlation of diseases with genes has been a topic of study for a long period of time. It all begun with a thought that different people respond different to the same drug and in the year 1931, Archibald Garrod predicted the possibility of what we call pharmacogenomics today (Garrod, 1931).Following this in the year 1949, JBS Haldane observed some atypical reactions to medications on the basis of biological uniqueness of an individual (Haldane, 1949). Fredrich Vogel was the first one to use the term 'Pharmacogenetics' in 1959 (Vogel, 1959). As the time passed, many more studies were published in attempt to understand the effect of genes on drug response. In 1962, "Pharmacogenetics: Heredity and Response to Drugs" was published by Werner Kalow (Grant & Tyndale, 2008).The emergence of Pharmacogenetics to pharmacogenomics was gradual with the advancements of technology in the late 20th century. The major stepping stone can be considered the description of the debrisoquine hydroxylase genetic deficiency, hereafter referred to as CYP2D6, cytochrome P450 enzyme (Gonzalez, *et al.,* 1988). In 1990, Human Genome Project was started with the major goal to decode the molecular arrangement of the whole human genetic material, or genome (Collins & Fink, 1995) and this ambitious project was completed in 2003. Meanwhile in 1999, Pharm-GKB, the Pharmacogenomics Knowledge Base, a web resource was launched by National Institutes of Health (NIH) that till date offers descriptions of pharmacogenomics interactions together with the supporting data. This information is useful for pharmacogenetics and pharmacogenomics research (Thorn *et al.,* 2013) and year 2002 marked the beginning of International Hap-Map Project to identify the prevalent DNA sequence variation patterns in the human genome ("The International HapMap Project," 2003). In 2004, the FDA authorized the Amplichip CYP450, the first pharmacogenetic test to look for 30 common variants in CYP2D6 and CYP2C19 (Moaddeb & Haga, 2013). The Translational Pharmacogenetics Program was launched by PGRN i.e. Pharmacogenomics Research Network in 2011 in an effort to evaluate the further application of pharmacogenomics in regular treatment by identifying obstacles and creating solutions (Krebs & Milani, 2019). Precision medicine initiative started in 2015 by NIH, to collect data in order to advance personalized medicine (Collins & Varmus, 2015). FDA has also taken measures to implement pharmacogenomics in an ethical way, in 2019, a list regarding pharmacogenomics biomarkers in drug labelling was released (FDA, 2019) and in 2023 FDA issued latest guidelines for pharmacogenomics data submissions. New updates are constantly coming in the pharmacogenomics domain as it forms the basis of personalized therapies for the future **(Figure 1).**

Figure 1. Timeline of pharmacogenomics.

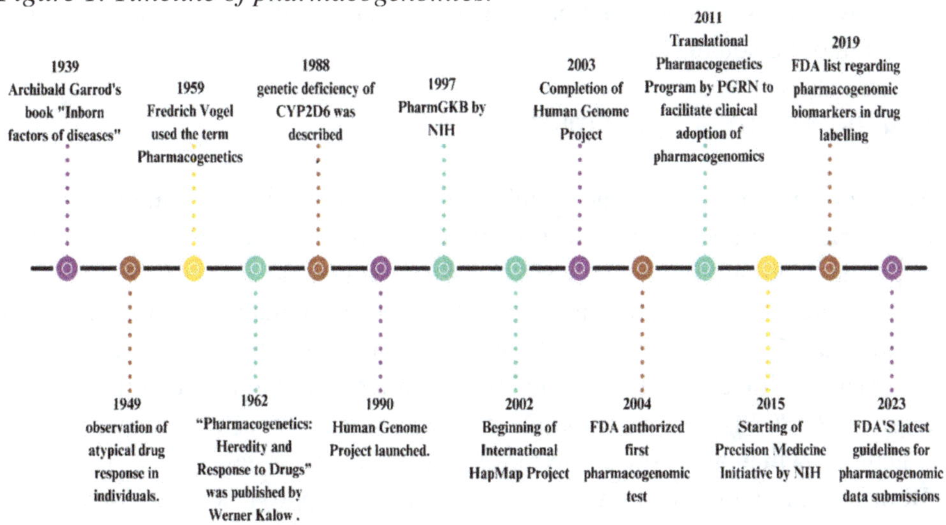

3.2 Types of Pharmacogenomics

Pharmacogenomics being an extensive area of research finds many applications in healthcare. It is currently being employed in drug development process as following the completion of the human genome project; over 3000 genes linked to illness have been identified and the majority of them remain underused. Genetic polymorphism affects how much and to what degree genes are expressed, which can lead to inter-individual differences in how drugs respond to them and how they behave inside the body, which can change phenotypic traits and cause disease vulnerability (Gupta & Jhawat, 2017).

The pharmacogenomics studies can be categorized into two approaches:-

1. Structural Pharmacogenomics
2. Functional Pharmacogenomics

Structural Pharmacogenomics

Structural pharmacogenomics identifies the basic structural differences among the individual genes which are involved in transportation of drug, interaction with target and metabolism of drug. This involves gene mapping, chromosome mapping, DNA sequencing and single nucleotide polymorphism (Jhawat *et al.,* 2020 and Gupta & Jhawat, 2017).The knowledge of structural changes in these genes become crucial for the development of new drugs with better efficacy and reduced toxicity. It also provides insights of variations in genetic makeup of an individual and aids in selection of personalized therapies.

Functional Pharmacogenomics

Functional pharmacogenomics studies the changes in function brought about by genetic variation. Functioning process within the body can be affected by modification of the kind of protein synthesized because of genetic changes. Due to these functional alterations of genes, any ailment or deformity may potentially develop (Chawla, *et al.,* 2021).The functional pharmacogenomics include understanding of disease markers, linkage analysis, gene expression and gene-target interaction studies (Jhawat *et al.,* 2020). It focuses on underlying molecular mechanisms associated with genetics which are responsible for effectiveness of a drug.

Both structural and functional pharmacogenomics approaches are essential for tailoring therapies for individuals and understanding the treatment response. These two are capable of anticipating, identifying, and arranging and optimizing medication therapy for the treatment of genetically impacted conditions.

4. WORKFLOW OF PHARMACOGENOMICS

Pharmacogenetics and pharmacogenomics play a significant role in personalized medicine, especially in chronic therapies and treatments with narrow therapeutic indexes. Genetic variations can impact drug metabolism, efficacy, and toxicity, as seen in cancer treatments where both host and tumor genome polymorphisms influence outcomes. An example of successful implementation is the upfront DPYD genotyping in patients receiving fluoropyrimidine chemotherapy, which helps reduce toxicity and improve treatment response. Ethical considerations are crucial in the integration of AI and digital health into precision medicine (Pant, *et al.,* 2023). Issues like patient privacy, data security, and equitable access need thorough examination to ensure responsible use (Nath, *et al.,* 2017). Patient-centricity is key, focusing on tailoring treatments to individual needs and involving patients actively in decision-making processes. This aligns treatments with personal preferences and values (Rao, *et al.,* 2023; Johnson *et al.*, 2020). See how patient genetic testing is used in personalized medicine with this pharmacogenomics workflow chart (**Figure 2**).

Figure 2. The workflow chart of pharmacogenomics for personalized medicine using genetic testing of the patients

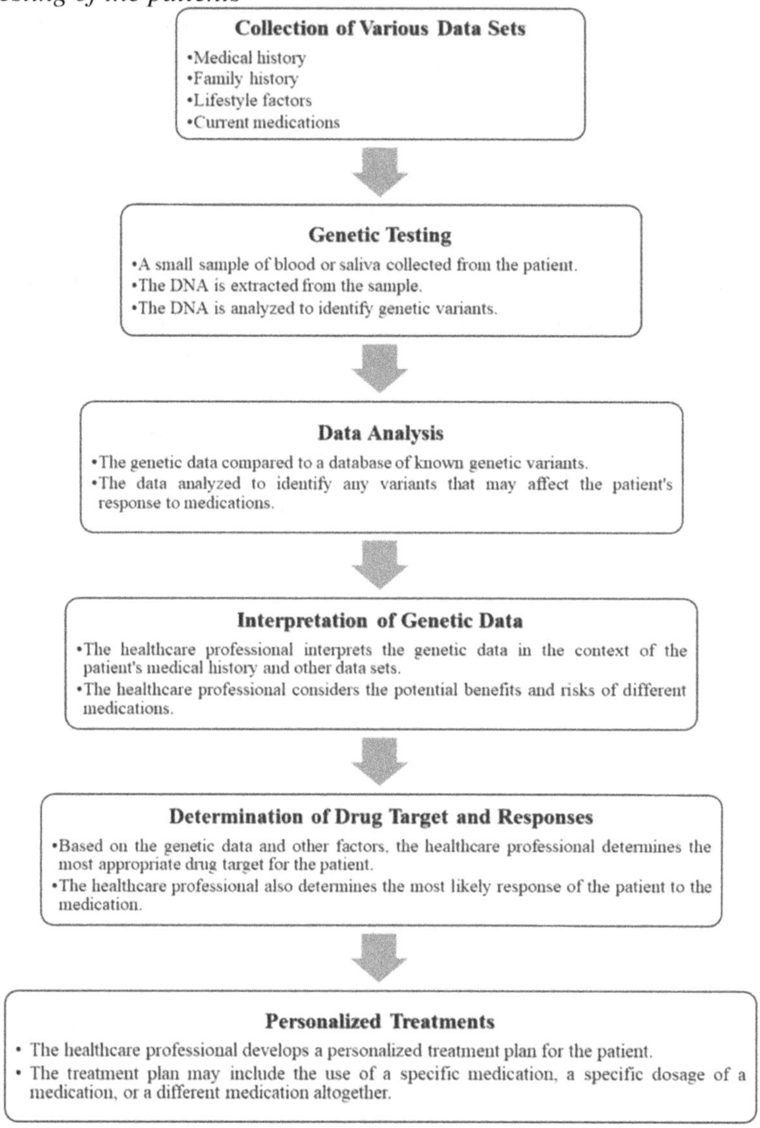

On a global scale, the adoption of precision medicine varies, with comparative analyses highlighting the challenges and successes in different healthcare systems. The long-term impact of AI and digital health is analysed through factors like cost-effectiveness, scalability, and sustainability, providing strategies to overcome barriers for widespread adoption. Integration with public health initiatives emphasizes the role of AI-fuelled precision medicine in disease detection, prevention, and management at a population level (Giansanti, 2024).

Education and training for healthcare professionals are essential, prompting the inclusion of relevant skills in medical and allied health curricula to prepare a future-ready workforce. Addressing health disparities is another critical focus, ensuring AI-driven precision medicine benefits diverse populations equitably. Regulatory frameworks and policies are necessary for the ethical use of AI in precision medicine, with international collaborations establishing common guidelines (Xu *et al.,* 2024). Continuous monitoring and feedback are vital, advocating for real-time data integration into precision medicine models. This allows for the adaptation and improvement of treatment strategies over time, ensuring ongoing effectiveness (Marques *et al*., 2024).

Specific studies exemplify the application of AI and pharmacogenomics in clinical settings. For instance, Leung et al. explore the use of planning computed tomography-based radiomics for prognosticating high-risk localized prostate cancer, showing promising accuracy. Authors assessed the impact of the COVID-19 pandemic on pulmonary tuberculosis using AI, revealing a significant decrease in TB diagnoses during the pandemic's initial year (Tapper *et al*., 2024; Behl, *et al*., 2020). Issa et al. investigate the impact of CBCT viewing parameters on identifying the inferior alveolar nerve, finding optimal settings to improve visibility and accuracy (Issa, *et al*., 2023).

In the context of cardiovascular diseases, Elvas et al. use machine learning models to predict myocardial ischemia incidents, demonstrating high accuracy and enabling proactive interventions. Personalized structural heart disease interventions consider predefined variables to tailor treatments to individual patient needs, enhancing the accuracy and efficacy of interventions (Chen *et al*., 2024; Elvas *et al*., 2023).

Overall, integrating the AI and pharmacogenomics into clinical practices holds the promise of more effective, personalized healthcare, improving patient outcomes and addressing health disparities across diverse populations.

5. PERSONALIZED MEDICINES AND ITS EVOLUTION IN RECENT YEARS (ORDER CHANGED)

In this modern era, personalized medicine is an emerging practice that leverages an individual's genetic profile to guide decisions regarding the prevention, diagnosis, and treatment of diseases. By understanding a patient's unique genetic makeup, doctors can select the appropriate medication or therapy and administer it in the correct dose or regimen (Goetz & Schork, 2018). The advancements in personalized medicine owe much to the Human Genome Project, which has been continuously refined and expanded through ongoing research and data collection (Delpierre & Lefèvre, 2023)

It is well understood that at different stages of life, there are variations in body mass, volume, behavioural levels, environmental exposure, biochemical and hormonal levels, and other physiological parameters (Vandenberg *et al.*, 2012). Not everyone shares the same baseline levels for these parameters. Therefore, it is unrealistic to expect a one-size-fits-all approach to be effective for every individual. Medical and scientific experts are increasingly focusing on understanding how different people develop diseases and how their bodies respond to treatments (Williams, 2008).

The traditional practice of medicine, which relies on broad population averages, often misses the mark because each person's genetic makeup is unique, leading to significant differences in how individuals experience health and respond to medical interventions (Johnson *et al.*, 2020). Precision medicine, also known as individualized, personalized, or genomic medicine, is moving us closer to more precise, predictable, and powerful healthcare tailored to the individual patient (Horwitz *et al.*, 2021; Marques *et al.*, 2024).

Our growing understanding of genetics and genomics—and how they influence health, disease, and drug responses—is enabling doctors to provide better-quality disease prevention, more precise diagnoses, safer remedy prescriptions, and more effective management. Tailoring healthcare for each person's distinctive genetic profile is the promising knowledge behind precision medicine, pointing toward improved overall health outcomes by acknowledging and addressing individual genetic differences (Strianese, *et al.*, 2020; Mathur & Sutton, 2017). Emerging technologies such as DNA sequencing, proteomics, imaging protocols, and wireless health monitoring devices have revealed significant inter-individual variation in disease processes, confirming the potential of personalized medicine. Personalized medicine, based on each patient’s unique genetic makeup, is starting to overcome the limitations of traditional medicine. Its benefits over traditional systems include shifting the focus from reaction to prevention, prescribing more effective drugs, predicting disease susceptibility, customizing prevention strategies, avoiding drugs with predictable side effects, pre-empting disease progression, improving disease detection, and reducing the time, cost, and failure rate of clinical trials(Goetz & Schork, 2018; Malik, *et al.,*2022)

However, challenges remain in obtaining regulatory approval and broad acceptance among healthcare stakeholders, including physicians, executives, insurance companies, and patients (Stefanicka & Kurpas, 2023). These challenges often revolve around proving that personalized medicine outperforms traditional methods, especially given the high costs of tailored therapies like CAR-T cell transplants for cancer and mutation-specific drugs like invocator for cystic fibrosis (Knowles, *et al.*, 2017)

The concept of personalized medicine dates back over a century to Archibald Garrod, who studied inborn errors of metabolism such as alkaptonuria. He observed that individuals with alkaptonuria had distinct biochemical profiles compared to unaffected family members, leading him to conclude that metabolic differences underlie phenotypic variations. This insight hinted at the broader idea that humans vary widely in their metabolism, which can influence disease susceptibility and manifestation (Blau, *et al.,* 2022; Arnold, 2018).

6. ROLE OF PHARMACOGENOMICS IN DRUG DISCOVERY AND DEVELOPMENT

The most vital phase is the evidence of insights for efficiency, which is initial in drug development and discovery but can be an important completion of a drug discovery plan with duration of 5–7 years. There are four phases of pharmacogenetics (PGx). Phase-I and early Phase-II involve potential side effects that can be avoided with patient identification and dose adjustments, while Phase IIIb-Phase-III affects programme design. Phase-III involves safety monitoring for potential adverse events and improving the benefit-risk ratio, while Phase-IV involves post-marketing surveillance and its risk-management programs (Hughes, *et al.,* 2011). The process of designing and constructing a chemical screening target can be outsourced for US$1-4 million, depending on the libraries and screening extent. Large chemical libraries were formerly important for large pharmaceutical companies but also a substantial hindrance to peripheral screening. A lead drug molecule that has persisted structure–activity relationship activities to enhance the biological consequence then must permit a range of safety, cytogenetic and metabolic activities, which can be outsourced at a price of US $12-15 million per drug molecule. Preclinical welfare hurdles are a substantial challenge for biotechnology companies and academia, with high potential return on asset for those willing to take risks (Knez *et al.*, 2022).

7. IMPACT OF GENETICS ON DRUG RESPONSE

Pharmacogenomics is the study of individual differences in drug therapy responses and the mechanisms behind these differences. Everyone has a diverse genomic makeup, which affects the possibility of emerging ailments as well as responses to medications and ecological factors. Hence, the identification of DNA modifications that most considerably contribute to the population differences in each character is one of the important purposes of genetics. The drug response of patients is mainly influenced by the pharmacokinetic and pharmacodynamics profiles of prescribed drugs, which can be influenced by polymorphisms in drug metabolizing transporters and enzymes (Crews, *et al.*, 2012). Precision medicine integrates molecular and clinical information to understand disease nature and make medications with improved patient outcomes. So, precision medicine enhances disease target selection and identifies patient populations with healthier clinical outcomes at regular doses. Gene variation in a population leads to the expression of various forms of proteins, including those involved in drug metabolism or drug action sites. Computing the DNA variances can thus predict the diverse responses to the drugs (Wang & Wang, 2023).

Genetic Polymorphism. Polymorphism refers to the variability of size, shape, form, composition, and structure across various scientific and artistic disciplines. Genetic polymorphism refers to frequent variation at a specific genome locus where two or more allelic forms are found in the same population. A genetic polymorphism occurs when a single gene producing a metabolizing enzyme has an abnormal allele with a 1% frequency within a population (Roudnitzky, *et al.*, 2015).

Genotype refers to an individual's complete gene structure, while phenotype is the outcome of their drug metabolism. Mutation is a significant factor that leads to DNA polymorphisms, which are responsible for the initiation of disease. DNA polymorphisms result from the insertion, substitution, or deletion of a nucleotide in non-coding or coding parts of the DNA, which may or may not modify gene function (Bull, 2013).

Genetic polymorphisms may affect a drug's response by modifying its pharmacodynamics as well as its pharmacokinetics, or both, which confer inter-individual modifications in drug actions. Pharmacokinetics and pharmacodynamics are two distinct fields that emphasis on the exact mechanisms by which drugs interacts with their target sites in the body. Genetic polymorphisms in metabolizing enzymes and drug transporters can modify drug properties, which cause variability in drug responses at the target site (Gummadi & Guddati, 2021).

8. CLINICAL PRACTICES INCLUDING PHARMACOGENOMICS IN PERSONALIZED MEDICINES

The primary aim of personalized medicine is to address the unique differences in drug responses among individuals, confirming that everyone gets the appropriate treatment. Both the rate of population differences in drug response and the incidence of adverse drug reactions (ADRs) are mutual phenomena (Goetz & Schork, 2018). ADRs cause substantial public health distress due to their unexpected drug response consequences, which can exacerbate patient illness and lead to increased financial expenses. Few years back, the budget for ADR-related hospitalization in the US alone has touched to $136 billion. Additionally, population variances in

ADR exposure have an influence on the drug development (Sendekie, *et al.*, 2023). A medicine intended for use by people of different ethnic backgrounds may have a different anticipated target market, which increases alarms about clinical studies. A drug that is sold in a particular population should preferably also be tried in that same populations through a trial, which may increase the possibility of unexpected drug-related toxicities. Implementing clinical trials for diverse ethnic populations in multiple regions can be costly and complex, posing a greater economic burden for drug developers and patients (Ghadessi, *et al.*, 2023). The rigorous trial may decrease the success rate of marketed potential drugs due to the occurrence of adverse drug reactions in one or more populations. Predicting population differences in drug response during drug development can avoid costly clinical trials and accelerate the entry of beneficial drugs for the right group of individuals. Mostly, population differences in drug response were reported only after the drug was released into the market (Selker, *et al.*, 2019). In 1959, Vogel introduced pharmacogenetics to describe inherited drug responses, while pharmacogenomics reflects the recent shift from genetics to genomics, utilizing genome-wide approaches to identify gene-related diseases or drug responses. Pharmacogenomics can help identify genetically defined patient subsets for targeted drug therapy, potentially leading to new molecular disease and treatment classifications (Roden, *et al.*, 2011).

Many interventions tailored to individual genetic profiles have been developed, both for disease treatment and early detection/prevention. Pharmacokinetics (drug absorption, distribution, metabolism, and excretion) and pharmacodynamics (drug interactions with targets) are influenced by genetic variations, affecting how the body handles medications (Ventola, 2011). Early examples include warfarin, where genetic variations in VKORC1 and CYP2C9 affect dosing, and primaquine, where G6PD gene variants can cause adverse reactions. Imatinib, used for chronic myelogenous leukemia (CML), is another example. It targets the bcr-abl fusion gene, present in some but not all CML patients (AL-Eitan, *et al.*, 2018). Genetic information can also guide disease prevention, such as using genetic profiles to tailor aspirin use for colorectal cancer prevention, as certain genotypes respond differently to the medication (Huang, *et al.*, 2024; Alqahtani, *et al.*, 2023).

It may be likely to use population-differentiated SNPs as genetic indicators to anticipate differences in drug responses among populations. The most commonly used anticoagulant medication such as warfarin is one of the most interesting topic for familiar pharmacogenetic research (Asiimwe & Pirmohamed, 2022). People's responses to warfarin vary significantly amongst themselves. When compared to Latinos, who are more responsive to warfarin treatment than people of African and Caucasian heritage, Asians would need a much lower dose of the medication (Tatsuno & Tatsuno, 2014). More considerably, genetic variants in the warfarin response pathway are related to variations in warfarin response between populations.

PharmGKB, a widely cited pharmacological pathway database, has curated literature indicating that many genes are involved in the PK and PD processing of the R and S stereoisomers of warfarin. The drug's intracellular concentration, effectiveness, and—more importantly—toxic side effects will all be impacted by the interactions between genes with different functional roles. Variants in a gene encoding subunit 1 of the vitamin K epoxide reductase complex (VKORC1) partially influence the warfarin response (Ndadza, *et al.*, 2021). Furthermore, because the liver metabolizes warfarin, a number of PK genes, including many metabolizing enzymes including the cytochrome P450 family of proteins (CYP2C9 and CYP3A4), are crucial in calculating the ideal dose of warfarin. The kidney is primarily responsible for the removal of the warfarin metabolites that are produced after they have been broken down in the liver (Ahmed, *et al.*, 2016). The highly polymorphic membrane transporter ABCB1 protein interacts with metabolites to potentially facilitate excretion through the bile canal. Genetic variants in PK and PD genes impact their function, which is why variations in pharmacological efficacy and toxicity are often linked to single nucleotide polymorphisms (SNPs) in these genes. Differential requirements for the dose of warfarin have been linked to variations in the CYP2C9 and VKORC1 genes. Normal warfarin PK is seen in people with wild-type CYP2C9, also known as the CYP2C9*1 allele (Madejczyk, *et al.*, 2021). Patients who are considered "poor metabolizers" due to CYP2C9 have a lower clearance of warfarin, which increases their risk of warfarin-induced toxicity. One important observation is that there appears to be a correlation between the race of the patient and the toxicity or need for a different dose of warfarin, as different populations have different distributions of CYP2C9 variants. Specifically, Caucasians are more likely to have the *2 and *3 alleles than East Asians or Africans. Compared to other groups, Asians generally require a lower dose of warfarin (Fekete, *et al.*, 2021).

Sparteine, which is being evaluated as an antiarrhythmic, and debrisoquine, an antihypertensive, both reported a set of 5–10% of patients experiencing side effects since it appeared that a crucial enzyme mediated drug bioinactivation was missing. The enzymes were originally identified as sparteine N-oxidase and debrisoquine 4-hydroxylase, but it was later discovered that these were two different defects. It is now understood that these defects are caused by homozygosity for the loss of function of a particular CYP2D6 member of the cytochrome P450 (CYP) superfamily of drug metabolizing enzymes. There are already dozens of variations known to decrease or completely stop CYP2D6 function (Pinto & Dalon, 2011; Zanger & Schwab, 2013).

Another "drug" that reveals traditional variations in metabolism is nicotine, which could help to explain some of the ethnic differences in tobacco-related ailments that cannot be fully enlightened by discrepancies in cigarette smoking habits (Ahluwalia, *et al.*, 2006). Since the primary method of eliminating nicotine from the body is through its metabolism into cotinine, variations in this metabolism could possibly impact on nicotine addiction or diseases related to tobacco usage. Patient's serum is used to quantify the metabolite cotinine in an individual, and there are notable deviations in serum cotinine levels between populations of Latino and African descent and between communities of European and East Asian or African heritage (Benowitz, *et al.*, 2009; Perez-Paramo, *et al.*, 2023)

5-Fluorouracil is commonly used as a cancer chemotherapeutic drug and is often reported to show modifications in drug responses in diverse populations (Deac, *et al.*, 2023).Hematologic toxicities (such as leukopenia and anaemia) are caused by fluoropyrimidine-based drugs, and the prevalence of these effects varies between European and African-American populations, Latinos or those of African heredity, and East Asians, as well as among those of East Asian pedigree (Carr, *et al.*, 2021; Afolabi, *et al.*, 2021). The variations in hematologic toxicities could possibly occur due to disparities in the dihydropyrimidine dehydrogenase enzyme's activity. It has been perceived that people with dihydropyrimidine dehydrogenase insufficiencies are more susceptible to the hematologic toxicities generated by 5-fluorouracil (Fidai, *et al.*, 2018; Palmirotta, *et al.*, 2019). More examples are summarized n Table1.

Whole exome sequencing has been used to identify disease-causing mutations, such as in a 15-month-old with Crohn's-like disease and children with familial dyskinesia. Integrative personal omics profiles (iPOP) have been used to monitor health and disease states over time, exemplified by a study on a healthy 54-year-old male (Mulder, *et al.*, 2020). Additionally, the pharmacokinetics of nicotinamide riboside (NR) in humans was studied, showing increased NAD^+ levels after NR ingestion (Crowley, *et al.*, 2020). These examples demonstrate the potential of personalized medicine to improve health outcomes by tailoring treatments to individual genetic profiles, moving beyond the limitations of traditional, one-size-fits-all approaches (Mathur & Sutton, 2017).

Pharmacogenomics focuses on using genetic profiles to guide drug therapy, aiming to avoid "trial and error" treatments. By identifying the optimal drug response based on genetic characteristics, it seeks to improve treatment outcomes. Although studies have shown varying results, more genetically guided trials are needed to establish robust clinical usefulness (Nogueiras-Álvarez, 2023).

Pharmacogenomics offers a promising approach to understanding and managing Polycystic Ovary Syndrome (PCOS), a complex endocrine disorder marked by hyper-androgenism, ovulatory dysfunction, and polycystic ovaries (Pant, *et al.*, 2023). Epidemiological evidence indicates that women with PCOS are at an increased risk of developing certain cancers due to shared metabolic and endocrine anomalies. Genomic, transcriptional, and proteomic analyses of 264 PCOS-related genes (PRGs) in endometrial, ovarian, and breast cancers reveal significant genomic alterations, especially in genes such as PTEN, TP53, and FSHR. Clinical data suggest that older women and Black or African American females have a higher prevalence of these alterations. Key signalling pathways, notably p53 and Fc epsilon RI, were found to be disrupted. Despite limited studies, pharmacogenomics has identified single nucleotide polymorphisms (SNPs) linked to drug responses in PCOS, highlighting the need for further research. Detailed studies with diverse patient populations could provide deeper insights into the genetic underpinnings of PCOS and its associated risks, potentially leading to better diagnostic and therapeutic strategies (Yumiceba, *et al.*, 2020; Pedersen, *et al.*, 2018).

Chronic non-communicable diseases (NCDs) like cardiovascular disease, cancer, respiratory diseases, and Alzheimer's are major health concerns due to their high incidence and mortality rates. In the European Union, one-third of adults reported chronic conditions in 2021. In 2019, cardiovascular disease caused 35% of deaths, cancer 26%, respiratory diseases 8%, and Alzheimer's 5% (Adesola, *et al.*, 2024; Bhattacharya, *et al.*, 2022).

The PRECeDI consortium in 2019 emphasized the need to identify biomarkers to prevent chronic diseases. These biomarkers can help stratify populations by indicating an individual's risk or resistance to disease and their potential response to drugs. This stratification guides preventive interventions at various stages: primary (preventing disease occurrence), secondary (early detection), and tertiary (preventing complications and improving quality of life) (Plans-Beriso, *et al.*, 2024).

The EU-funded PROPHET project aims to assess the effectiveness of personalized preventive approaches in healthcare. It seeks to map current research and identify gaps in the use of biomarkers for diseases like cancer, cardiovascular diseases, and neurodegenerative conditions. The project also integrates digital technologies, such as wearable devices, to monitor physical and physiological functions (Smokovski, *et al.*, 2024; Zhang *et al.*, 2024).

Primary and secondary prevention strategies for chronic diseases are well-established. For instance, known risk factors for cardiovascular diseases include high cholesterol, hypertension, smoking, diabetes, obesity, and physical inactivity. The European code against cancer outlines lifestyle actions for cancer prevention (Masoudkabir *et al.*, 2023). In Alzheimer's disease (AD), genetic risk factors like Apo-lipoprotein E4 (ApoE4) play a significant role. Biomarkers such as Ab42 are being studied, though their clinical application is still unclear. Personalized strategies for AD include targeting neuro-inflammation and incorporating exercise, nutrition, and other lifestyle factors to potentially delay disease onset (Kim, *et al.*, 2009; Behl, *et al.*, 2021).

The enteral microbiome, known as the "gut to brain axis," is linked to Alzheimer's disease. Personalized nutrition that modifies the gut microbiome and influences epigenetic changes could help delay the onset of AD (Zhang, *et al.*, 2023).

The FINGER study is an example of a multi-domain lifestyle intervention aimed at reducing cognitive impairment. It targets modifiable risk factors by encouraging a healthy diet, physical and cognitive activities, and monitoring cardiovascular health. Personalized protocols involving these factors have shown overall benefits on cognition, although more information is needed to determine which patients benefit most from specific interventions (Kivipelto *et al.*, 2020).

By integrating biomarkers, digital health tools, and genetic information, personalized medicine aims to enhance prevention, diagnosis, and treatment of chronic diseases, tailoring healthcare to individual patient needs (Stefanicka-Wojtas & Kurpas, 2023).

Table 1. Genetic variations affecting drug response

S.no.	Genetic polymorphism		Drug	Genetic Variations	Effect of genes	References
1.	Pharmacokinetics	Absorption	Digoxin	MDR1-C3435T	Individuals having T allele have more intestinal absorption and higher plasma concentration of digoxin.	(Sakaeda et al., 2002)
		Distribution	Statins	SLCO1B1	Population with increased function have higher statin exposure and dose should be adjusted accordingly. Such patients are at risk of myopathy.	(Cooper-DeHoff et al., 2022)
		Metabolism	Warfarin	VKORC1-1639A CYP269*2	Population with A allele variation requires low doses of warfarin. Higher doses increase the chances of over-anticoagulation and haemorrhage. This variation leads to poor warfarin clearance and may cause toxicity.	(Johnson & Cavallari, 2015)
			Clopidogrel	CYP2C19*2 and CYP2CI9*3	These alleles are responsible for loss of function on enzyme and decreased efficacy	(Pereira et al.,2019)

continued on following page

Table 1. Continued

S.no.	Genetic polymorphism		Drug	Genetic Variations	Effect of genes	References
		Excretion	Methotrexate	ABCG2rs13120400-TT	The polymorphism affects the efflux of drug into urine, causes high plasma concentrations and subsequently increased risk of grade 1-4 hepatotoxicity.	(Membrive-Jiménez et al., 2023)
2.	Pharmacodynamics	Drug response	Codeine	CYP2D6	Codeine is metabolized into morphine by CYP2D6. There are poor metabolizers, who have little to no relief from pain. The fast metabolizer population have fast effects and may lead to toxicity sometimes.	(Carranza-Leon et al., 2021)
			Trastuzumab	HER2	Patients with amplified HER2 gene respond well to breast cancer treatment and have higher survival rate then less expressed HER2 gene.	(Piccart-Gebhart et al., 2005)

9. CHALLENGES FACED IN IMPLEMENTATION OF PHARMACOGENOMICS

Pharmacogenomics possesses the capacity to revolutionize healthcare yet its broader integration into clinical practice encounters various challenges. There is a notable uncertainty regarding individual metabolic activity and genetic makeup due to lack of enough data, extensive randomized clinical trials providing robust evidence in favour of pharmacogenomics guided therapy. There are efforts being made in this direction like PharmGKB's new scoring system aims for regular evaluation of evidence for variant-drug interaction (Van Driest & Cascorbi, 2021). The economic aspect of pharmacogenomics is a major hurdle for its widespread acceptance in clinical scenarios among general population. Apart from scientific challenges, legal challenges, ethics and social responsibility serve as major concerns for involvement of genomics in healthcare.

10. FUTURE PERSPECTIVE OF PHARMACOGENOMICS IN PERSONALIZED HEALTHCARE

The future will be based on addressing current challenges in this aspect and improve efficiency as well. Personalized healthcare using pharmacogenomics in the upcoming time should not be only based on a single gene variant only, rather than multiple genes and other factors should be studied while treating a patient. Adopting a more intricate, all-encompassing approach might have advantages to the whole healthcare system (Sisodiya, 2020).The enormous amount of data is required to identify variations and other gene characteristics, thus ethical use of this data while maintaining the privacy is a major concern for the future. While developing more efficient personalized therapy, making these therapies cost-effective is also equally important so that the maximum number of patients can be benefitted. Furthermore, the attempts should be made to improve scalability, widespread adoption and accuracy of pharmacogenomics to achieve the desired result in clinical environment. More researches should be encouraged for discovery of new genetic biomarkers to enhance therapy practices. To adopt personalized therapies in future, there should be a greater emphasis on training as well as education of patients and healthcare professionals. As we move forward with personalized healthcare and pharmacogenomics, the development of new breakthrough technologies and keeping the collaboration and innovative approaches is the way to go to improve healthcare outcomes.

11. CONCLUSION

Pharmacogenomics and personalized medicine, both of the discussed approaches hold endless potential to revolutionize the clinical practices and healthcare systems. Here, the focus lies on patient centric care. Synergistic efforts of researchers, healthcare professionals and patients can lead us to a place where therapies are personalized for each individual while considering the genetic profile. Patients from cardiology to psychiatry, from cancer to infectious diseases, the management of every condition can be personalized using pharmacogenomics. The efficacy of a drug will be higher with minimum adverse drug reactions, completing the goal of treatment. The focus should be on maximizing the use of principles of personalized medicine and pharmacogenomics in clinics.

REFERENCES

Adesola, R. O., Opuni, E., Idris, I., Okesanya, O. J., Igwe, O., Abdulazeez, M. D., & Lucero-Prisno, D. E.III. (2024). Navigating Nigeria's Health Landscape: Population Growth and Its Health Implications. *Environmental Health Insights*, 18, 11786302241250211. DOI: 10.1177/11786302241250211 PMID: 38698838

Afolabi, B. L., Mazhindu, T., Zedias, C., Borok, M., Ndlovu, N., & Masimirembwa, C. (2023). Pharmacogenetics and Adverse Events in the Use of Fluoropyrimidine in a Cohort of Cancer Patients on Standard of Care Treatment in Zimbabwe. *Journal of Personalized Medicine*, 13(4), 588. DOI: 10.3390/jpm13040588 PMID: 37108974

Ahluwalia, J. S., Okuyemi, K., Nollen, N., Choi, W. S., Kaur, H., Pulvers, K., & Mayo, M. S. (2006). The effects of nicotine gum and counseling among African American light smokers: A 2× 2 factorial design. *Addiction (Abingdon, England)*, 101(6), 883–891. DOI: 10.1111/j.1360-0443.2006.01461.x PMID: 16696632

Ahmed, S., Zhou, Z., Zhou, J., & Chen, S. Q. (2016). Pharmacogenomics of Drug Metabolizing Enzymes and Transporters: Relevance to Precision Medicine. *Genomics, Proteomics & Bioinformatics*, 14(5), 298–313. DOI: 10.1016/j.gpb.2016.03.008 PMID: 27729266

AL-Eitan, L. N., Almasri, A. Y., & Khasawneh, R. H.AL-Eitan. (2018). Impact of CYP2C9 and VKORC1 Polymorphisms on Warfarin Sensitivity and Responsiveness in Jordanian Cardiovascular Patients during the Initiation Therapy. *Genes*, 9(12), 578. Advance online publication. DOI: 10.3390/genes9120578 PMID: 30486437

Alqahtani, A. A., Ahmed, M. M., Mohammed, A. A., & Ahmad, J. (2023). 3D Printed Pharmaceutical Systems for Personalized Treatment in Metabolic Syndrome. *Pharmaceutics*, 15(4), 1152. DOI: 10.3390/pharmaceutics15041152 PMID: 37111638

Arnold, G. L. (2018). Inborn errors of metabolism in the 21st century: Past to present. *Annals of Translational Medicine*, 6(24), 467. Advance online publication. DOI: 10.21037/atm.2018.11.36 PMID: 30740398

Asiimwe, I. G., & Pirmohamed, M. (2022). Ethnic Diversity and Warfarin Pharmacogenomics. *Frontiers in Pharmacology*, 13, 866058. DOI: 10.3389/fphar.2022.866058 PMID: 35444556

Auwerx, C., Sadler, M. C., Reymond, A., & Kutalik, Z. (2022). From pharmacogenetics to pharmaco-omics: Milestones and future directions. *HGG Advances*, 3(2), 100100. DOI: 10.1016/j.xhgg.2022.100100 PMID: 35373152

Behl, T., Kaur, D., Sehgal, A., Singh, S., Sharma, N., Zengin, G., Andronie-Cioara, F. L., Toma, M. M., Bungau, S., & Bumbu, A. G. (2021). Role of monoamine oxidase activity in Alzheimer's disease: An insight into the therapeutic potential of inhibitors. *Molecules (Basel, Switzerland)*, 26(12), 3724. DOI: 10.3390/molecules26123724 PMID: 34207264

Behl, T., Kaur, I., Bungau, S., Kumar, A., Uddin, M. S., Kumar, C., & Arora, S. (2020). The dual impact of ACE2 in COVID-19 and ironical actions in geriatrics and pediatrics with possible therapeutic solutions. *Life Sciences*, 257, 118075. DOI: 10.1016/j.lfs.2020.118075 PMID: 32653522

Benowitz, N. L., Hukkanen, J., & Jacob, P.III. (2009). Nicotine chemistry, metabolism, kinetics and biomarkers. *Handbook of Experimental Pharmacology*, 192, 29–60. DOI: 10.1007/978-3-540-69248-5_2 PMID: 19184645

Bhattacharya, S., Heidler, P., & Varshney, S. (2022). Incorporating neglected non-communicable diseases into the national health program—A review. *Frontiers in Public Health*, 10, 1093170. Advance online publication. DOI: 10.3389/fpubh.2022.1093170 PMID: 36703821

. Blau, N., Dionisi Vici, C., Ferreira, C. R., Vianey-Saban, C., & van Karnebeek, C. D. M. (2022). Correction to: Physician's Guide to the Diagnosis, Treatment, and Follow-Up of Inherited Metabolic Diseases. Physician's Guide to the Diagnosis, Treatment, and Follow-Up of Inherited Metabolic Diseases, C1–C1. https://doi.org/ DOI: 10.1007/978-3-030-67727-5_74

Bull, L. (2013). Genetics, mutations, and polymorphisms. In *Madame Curie Bioscience Database* [Internet]. Landes Bioscience.

Carlsten, C., Brauer, M., Brinkman, F., Brook, J., Daley, D., McNagny, K., Pui, M., Royce, D., Takaro, T., & Denburg, J. (2014). Genes, the environment and personalized medicine. *EMBO Reports*, 15(7), 736–739. DOI: 10.15252/embr.201438480 PMID: 24907240

Carr, D. F., Turner, R. M., & Pirmohamed, M. (2021). Pharmacogenomics of anticancer drugs: Personalising the choice and dose to manage drug response. *British Journal of Clinical Pharmacology*, 87(2), 237–255. DOI: 10.1111/bcp.14407 PMID: 32501544

Carranza-Leon, D., Dickson, A. L., Gaedigk, A., Stein, C. M., & Chung, C. P. (2021). CYP2D6 genotype and reduced codeine analgesic effect in real-world clinical practice. *The Pharmacogenomics Journal*, 21(4), 484–490. DOI: 10.1038/s41397-021-00226-8 PMID: 33750887

Center for Drug Evaluation and Research. (2019). *Table of Pharmacogenomic Biomarkers*. U.S. Food and Drug Administration. https://www.fda.gov/drugs/science-and-research-drugs/table-pharmacogenomic-biomarkers-drug-labeling

. Chawla, R., Rani, V., Mishra, M., & Kumar, K. (2021). Integrated Role of Nanotechnology and Pharmacogenetics in Diagnosis and Treatment of Diseases. *IntechOpen EBooks*. https://doi.org/DOI: 10.5772/intechopen.97643

Chen, Y., Zhang, X., Ye, Q., Zhang, X., Cao, N., Li, S. Y., Yu, J., Zhao, S. T., Zhang, J., Xu, X. M., Shi, Y. K., & Yang, L. X. (2024). Machine learning-based prediction model for myocardial ischemia under high altitude exposure: A cohort study. *Scientific Reports*, 14(1), 686. DOI: 10.1038/s41598-024-51202-8 PMID: 38182722

Collins, F. S., & Fink, L. (1995). The Human Genome Project. *Alcohol Health and Research World*, 19(3), 190–195. PMID: 31798046

Collins, F. S., & Varmus, H. (2015). A new initiative on precision medicine. *The New England Journal of Medicine*, 372(9), 793–795. DOI: 10.1056/NEJMp1500523 PMID: 25635347

Cooper-DeHoff, R. M., Niemi, M., Ramsey, L. B., Luzum, J. A., Tarkiainen, E. K., Straka, R. J., Gong, L., Tuteja, S., Wilke, R. A., Wadelius, M., Larson, E. A., Roden, D. M., Klein, T. E., Yee, S. W., Krauss, R. M., Turner, R. M., Palaniappan, L., Gaedigk, A., Giacomini, K. M., & Voora, D. (2022). The Clinical Pharmacogenetics Implementation Consortium Guideline for SLCO1B1, ABCG2, and CYP2C9 genotypes and Statin-Associated Musculoskeletal Symptoms. *Clinical Pharmacology and Therapeutics*, 111(5), 1007–1021. DOI: 10.1002/cpt.2557 PMID: 35152405

Crews, K. R., Hicks, J. K., Pui, C. H., Relling, M. V., & Evans, W. E. (2012). Pharmacogenomics and individualized medicine: Translating science into practice. *Clinical Pharmacology and Therapeutics*, 92(4), 467–475. DOI: 10.1038/clpt.2012.120 PMID: 22948889

Crowley, E., Warner, N., Pan, J., Khalouei, S., Elkadri, A., Fiedler, K., Foong, J., Turinsky, A. L., Bronte-Tinkew, D., Zhang, S., Hu, J., Tian, D., Li, D., Center, R. G., Horowitz, J., Siddiqui, I., Upton, J., Roifman, C. M., Church, P. C., & Muise, A. M. (2020). Prevalence and Clinical Features of Inflammatory Bowel Diseases Associated with Monogenic Variants, Identified by Whole-exome sequencing in 1000 Children at a Single Center. *Gastroenterology*, 158(8), 2208–2220. DOI: 10.1053/j.gastro.2020.02.023 PMID: 32084423

Deac, A. L., Pop, R. M., Burz, C. C., Bocşan, I. C., Militaru, C., Suciu, Ş., & Buzoianu, A. D. (2023). 5-fluorouracil therapeutic drug monitoring and adverse events in a Romanian population. *Medicine and Pharmacy Reports*, 96(4), 413–419. DOI: 10.15386/mpr-2643 PMID: 37970190

Delpierre, C., & Lefèvre, T. (2023). Precision and personalized medicine: What their current definition says and silences about the model of health they promote. Implication for the development of personalized health. *Frontiers in Sociology*, 8, 1112159. DOI: 10.3389/fsoc.2023.1112159 PMID: 36895332

Duarte, J. D., & Cavallari, L. H. (2021). Pharmacogenetics to guide cardiovascular drug therapy. *Nature Reviews. Cardiology*, 18(9), 649–665. Advance online publication. DOI: 10.1038/s41569-021-00549-w PMID: 33953382

Elvas, L. B., Nunes, M., Ferreira, J. C., Dias, M. S., & Rosário, L. B. (2023). AI-Driven Decision Support for Early Detection of Cardiac Events: Unveiling Patterns and Predicting Myocardial Ischemia. *Journal of Personalized Medicine*, 13(9), 1421. DOI: 10.3390/jpm13091421 PMID: 37763188

Fekete, F., Mangó, K., Déri, M., Incze, E., Minus, A., & Monostory, K. (2021). Impact of genetic and non-genetic factors on hepatic CYP2C9 expression and activity in Hungarian subjects. *Scientific Reports*, 11(1), 17081. DOI: 10.1038/s41598-021-96590-3 PMID: 34429480

Fidai, S. S., Sharma, A. E., Johnson, D. N., Segal, J. P., & Lastra, R. R. (2018). Dihydropyrimidine dehydrogenase deficiency as a cause of fatal 5-Fluorouracil toxicity. *Autopsy and Case Reports*, 8(4), e2018049. DOI: 10.4322/acr.2018.049 PMID: 30775324

. Garrod, A. E. (1931). The inborn factors in disease: an essay

Ghadessi, M., Di, J., Wang, C., Toyoizumi, K., Shao, N., Mei, C., Demanuele, C., Tang, R. S., McMillan, G., & Beckman, R. A. (2023). Decentralized clinical trials and rare diseases: A Drug Information Association Innovative Design Scientific Working Group (DIA-IDSWG) perspective. *Orphanet Journal of Rare Diseases*, 18(1), 79. DOI: 10.1186/s13023-023-02693-7 PMID: 37041605

Giansanti, D. (2024). Joint Expedition: Exploring the Intersection of Digital Health and AI in Precision Medicine with Team Integration. *Journal of Personalized Medicine*, 14(4), 388. Advance online publication. DOI: 10.3390/jpm14040388 PMID: 38673015

Goetz, L. H., & Schork, N. J. (2018). Personalized medicine: Motivation, challenges, and progress. *Fertility and Sterility*, 109(6), 952–963. DOI: 10.1016/j.fertnstert.2018.05.006 PMID: 29935653

Gong, L., Whirl-Carrillo, M., & Klein, T. E. (2021). PharmGKB, an Integrated Resource of Pharmacogenomic Knowledge. *Current Protocols*, 1(8), e226. DOI: 10.1002/cpz1.226 PMID: 34387941

Gonzalez, F. J., Skodat, R. C., Kimura, S., Umeno, M., Zanger, U. M., Nebert, D. W., Gelboin, H. V., Hardwick, J. P., & Meyer, U. A. (1988). Characterization of the common genetic defect in humans deficient in debrisoquine metabolism. *Nature*, 331(6155), 442–446. DOI: 10.1038/331442a0 PMID: 3123997

Grant, D. M., & Tyndale, R. F. (2008). In memoriam: Werner Kalow, MD (1917–2008). *Pharmacogenetics and Genomics*, 18(10), 835–836. DOI: 10.1097/FPC.0b013e32830d102d

Gummadi, A. C., & Guddati, A. K. (2021). Genetic Polymorphisms in Pharmaceuticals and Chemotherapy. *World Journal of Oncology*, 12(5), 149–154. DOI: 10.14740/wjon1405 PMID: 34804277

Gupta, S., & Jhawat, V. (2017). Quality by design (QbD) approach of pharmacogenomics in drug designing and formulation development for optimization of drug delivery systems. *Journal of Controlled Release*, 245, 15–26. DOI: 10.1016/j.jconrel.2016.11.018 PMID: 27871989

Haldane, J. B. S. (1949). Disease and evolution. *La Ricerca Scientifica*, 19, 68–75.

Horwitz, R. I., Lobitz, G., Mawn, M., Conroy, A. H., Cullen, M. R., Sim, I., & Singer, B. H. (2021). Biosocial medicine: Biology, biography, and the tailored care of the patient. *SSM - Population Health*, 15, 100863. DOI: 10.1016/j.ssmph.2021.100863 PMID: 34430699

Huang, J. A., Wu, Y. H., Chen, P. L., Weng, Y. C., Chiang, I. C., Huang, Y. T., & Chou, W. H. (2024). MMP-9 upregulation may predict hemorrhagic transformation after endovascular thrombectomy. *Frontiers in Neurology*, 15, 1400270. DOI: 10.3389/fneur.2024.1400270 PMID: 38798706

Hughes, J. P., Rees, S., Kalindjian, S. B., & Philpott, K. L. (2011). Principles of early drug discovery. *British Journal of Pharmacology*, 162(6), 1239–1249. DOI: 10.1111/j.1476-5381.2010.01127.x PMID: 21091654

Issa, J., Riad, A., Olszewski, R., & Dyszkiewicz-Konwińska, M. (2023). The Influence of Slice Thickness, Sharpness, and Contrast Adjustments on Inferior Alveolar Canal Segmentation on Cone-Beam Computed Tomography Scans: A Retrospective Study. *Journal of Personalized Medicine*, 13(10), 1518. DOI: 10.3390/jpm13101518 PMID: 37888129

Jhawat, V., Gulia, M., Gupta, S., Maddiboyina, B., & Dutt, R. (2020). Integration of pharmacogenomics and theranostics with nanotechnology as quality by design (QbD) approach for formulation development of novel dosage forms for effective drug therapy. *Journal of Controlled Release*, 327, 500–511. DOI: 10.1016/j.jconrel.2020.08.039 PMID: 32858073

Johnson, J. A., & Cavallari, L. H. (2015). Warfarin Pharmacogenetics. *Trends in Cardiovascular Medicine*, 25(1), 33–41. DOI: 10.1016/j.tcm.2014.09.001 PMID: 25282448

Johnson, K. B., Wei, Q., Weeraratne, D., Frisse, M. E., Misulis, K., Rhee, K., Zhao, J., & Snowdon, J. L. (2020). Precision Medicine, AI, and the Future of Personalized Health Care. *Clinical and Translational Science*, 14(1), 86–93. DOI: 10.1111/cts.12884 PMID: 32961010

Kim, J., Basak, J. M., & Holtzman, D. M. (2009). The Role of Apolipoprotein E in Alzheimer's Disease. *Neuron*, 63(3), 287–303. DOI: 10.1016/j.neuron.2009.06.026 PMID: 19679070

Kivipelto, M., Mangialasche, F., Snyder, H. M., Allegri, R., Andrieu, S., Arai, H., Baker, L., Belleville, S., Brodaty, H., Brucki, S. M., Calandri, I., Caramelli, P., Chen, C., Chertkow, H., Chew, E., Choi, S. H., Chowdhary, N., Crivelli, L., La Torre, R. D., & Carrillo, M. C. (2020). World-Wide FINGERS Network: A global approach to risk reduction and prevention of dementia. *Alzheimer's & Dementia*, 16(7), 1078–1094. DOI: 10.1002/alz.12123 PMID: 32627328

Knez, D., Gobec, S., & Hrast, M. (2022). Screening of Big Pharma's Library against Various *in-house* Biological Targets. *Molecules (Basel, Switzerland)*, 27(14), 4484. DOI: 10.3390/molecules27144484 PMID: 35889355

Knowles, L., Luth, W., & Bubela, T. (2017). Paving the road to personalized medicine: Recommendations on regulatory, intellectual property and reimbursement challenges. *Journal of Law and the Biosciences*, 4(3), 453–506. DOI: 10.1093/jlb/lsx030 PMID: 29868182

Krebs, K., & Milani, L. (2019). Translating pharmacogenomics into clinical decisions: Do not let the perfect be the enemy of the good. *Human Genomics*, 13(1), 39. Advance online publication. DOI: 10.1186/s40246-019-0229-z PMID: 31455423

Krzyszczyk, P., Acevedo, A., Davidoff, E. J., Timmins, L. M., Marrero-Berrios, I., Patel, M., White, C., Lowe, C., Sherba, J. J., Hartmanshenn, C., O'Neill, K. M., Balter, M. L., Fritz, Z. R., Androulakis, I. P., Schloss, R. S., & Yarmush, M. L. (2018). The growing role of precision and personalized medicine for cancer treatment. *Technology*, 6(3-4), 79–100. DOI: 10.1142/S2339547818300020 PMID: 30713991

Madejczyk, A. M., Canzian, F., Góra-Tybor, J., Campa, D., Sacha, T., Link-Lenczowska, D., Florek, I., Prejzner, W., Całbecka, M., Rymko, M., Dudziński, M., Orzechowska, M. J., & Jamroziak, K. (2022). Impact of genetic polymorphisms of drug transporters ABCB1 and ABCG2 and regulators of xenobiotic transport and metabolism PXR and CAR on clinical efficacy of dasatinib in chronic myeloid leukemia. *Frontiers in Oncology*, 12, 952640. DOI: 10.3389/fonc.2022.952640 PMID: 36212403

Malik, S., Muhammad, K., & Waheed, Y. (2022). Emerging Applications of Nanotechnology in Healthcare and Medicine. *Molecules (Basel, Switzerland)*, 28(18), 6624. DOI: 10.3390/molecules28186624 PMID: 37764400

Marques, L., Costa, B., Pereira, M., Silva, A., Santos, J., Saldanha, L., Silva, I., Magalhães, P., Schmidt, S., & Vale, N. (2024). Advancing Precision Medicine: A Review of Innovative In Silico Approaches for Drug Development, Clinical Pharmacology and Personalized Healthcare. *Pharmaceutics*, 16(3), 332. Advance online publication. DOI: 10.3390/pharmaceutics16030332 PMID: 38543226

Masoudkabir, F., Mohammadifard, N., Mani, A., Ignaszewski, A., Davis, M. K., Vaseghi, G., Mansourian, M., Franco, C., Gotay, C., & Sarrafzadegan, N. (2023). Shared Lifestyle-Related Risk Factors of Cardiovascular Disease and Cancer: Evidence for Joint Prevention. *TheScientificWorldJournal*, 2023, 1–15. Advance online publication. DOI: 10.1155/2023/2404806 PMID: 37520844

Mathur, S., & Sutton, J. (2017). Personalized medicine could transform healthcare. *Biomedical Reports*, 7(1), 3–5. DOI: 10.3892/br.2017.922 PMID: 28685051

Meenu, S. (2023). K., Yadav, S., & Yadav, D. (2023). Recent Advances in Pharmacogenomics and Clinical Pharmacology: A Systematic Review. *Recent Advances in Pharmacogenomics and Clinical Pharmacology: A Systematic Review. European Chemical Bulletin*, (12), 1776–1788. DOI: 10.48047/ecb/2023.12.si12.154

Membrive-Jiménez, C., Vieira-Maroun, S., Márquez-Pete, N., Cura, Y., Pérez-Ramírez, C., Tercedor-Sánchez, J., Jiménez-Morales, A., & Ramírez-Tortosa, M. C. (2023). ABCC1, ABCG2 and FOXP3: Predictive Biomarkers of Toxicity from Methotrexate Treatment in Patients Diagnosed with Moderate-to-Severe Psoriasis. *Biomedicines*, 11(9), 2567–2567. DOI: 10.3390/biomedicines11092567 PMID: 37761008

Mirsadeghi, S., & Larijani, B. (2017). Personalized Medicine: Pharmacogenomics and Drug Development. *Acta Medica Iranica*, 55(3), 150–165. PMID: 28282716

Mulder, D. J., Khalouei, S., Warner, N., Gonzaga-Jauregui, C., Church, P. C., Walters, T. D., Ramani, A. K., Griffiths, A. M., Cohn, I., & Muise, A. M. (2020). Utilization of Whole Exome Sequencing Data to Identify Clinically Relevant Pharmacogenomic Variants in Pediatric Inflammatory Bowel Disease. *Clinical and Translational Gastroenterology*, 11(12), e00263. Advance online publication. DOI: 10.14309/ctg.0000000000000263 PMID: 33512800

Nath, A., Wang, J., & Huang, R. S. (2017). PHARMACOGENETICS AND PHARMACOGENOMICS OF TARGETED THERAPEUTICS IN CHRONIC MYELOID LEUKEMIA. *Molecular Diagnosis & Therapy*, 21(6), 621–631. DOI: 10.1007/s40291-017-0292-x PMID: 28698977

Ndadza, A., Muyambo, S., Mntla, P., Wonkam, A., Chimusa, E., Kengne, A. P., Ntsekhe, M., & Dandara, C. (2021). Profiling of warfarin pharmacokinetics-associated genetic variants: Black Africans portray unique genetic markers important for an African specific warfarin pharmacogenetics-dosing algorithm. Journal of thrombosis and haemostasis. *Journal of Thrombosis and Haemostasis*, 19(12), 2957–2973. DOI: 10.1111/jth.15494 PMID: 34382722

Nogueiras-Álvarez, R. (2023). Pharmacogenomics in clinical trials: An overview. *Frontiers in Pharmacology*, 14, 1247088. DOI: 10.3389/fphar.2023.1247088 PMID: 37927590

Palmirotta, R., Lovero, D., Delacour, H., Le Roy, A., Cremades, S., & Silvestris, F. (2019). Rare Dihydropyrimidine Dehydrogenase Variants and Toxicity by Floropyrimidines: A Case Report. *Frontiers in Oncology*, 9, 139. DOI: 10.3389/fonc.2019.00139 PMID: 30915274

Pant, P., Chitme, H., Sircar, R., Prasad, R., & Prasad, H. O. (2023). Genome-wide association study for single nucleotide polymorphism associated with mural and cumulus granulosa cells of PCOS (polycystic ovary syndrome) and non-PCOS patients. *Future Journal of Pharmaceutical Sciences*, 9(1), 27. Advance online publication. DOI: 10.1186/s43094-023-00475-3

Pedersen, A. J. T., Stage, T. B., Glintborg, D., Andersen, M., & Christensen, M. M. H. (2018). The Pharmacogenetics of Metformin in Women with Polycystic Ovary Syndrome: A Randomized Trial. *Basic & Clinical Pharmacology & Toxicology*, 122(2), 239–244. DOI: 10.1111/bcpt.12874 PMID: 28834135

Pereira, N. L., Rihal, C. S., So, D. Y. F., Rosenberg, Y., Lennon, R. J., Mathew, V., Goodman, S. G., Weinshilboum, R. M., Wang, L., Baudhuin, L. M., Lerman, A., Hasan, A., Iturriaga, E., Fu, Y. P., Geller, N., Bailey, K., & Farkouh, M. E. (2019). Clopidogrel Pharmacogenetics. *Circulation: Cardiovascular Interventions*, 12(4), e007811. DOI: 10.1161/CIRCINTERVENTIONS.119.007811 PMID: 30998396

Perez-Paramo, Y. X., Watson, C. J. W., Chen, G., Thomas, C. E., Adams-Haduch, J., Wang, R., Khor, C. C., Koh, W. P., Nelson, H. H., Yuan, J. M., & Lazarus, P. (2023). Impact of Genetic Variants in the Nicotine Metabolism Pathway on Nicotine Metabolite Levels in Smokers. *Cancer Epidemiology, Biomarkers & Prevention*, 32(1), 54–65. DOI: 10.1158/1055-9965.EPI-22-0868 PMID: 36252563

Piccart-Gebhart, M. J., Procter, M., Leyland-Jones, B., Goldhirsch, A., Untch, M., Smith, I., Gianni, L., Baselga, J., Bell, R., Jackisch, C., Cameron, D., Dowsett, M., Barrios, C. H., Steger, G., Huang, C.-S., Andersson, M., Inbar, M., Lichinitser, M., Láng, I., & Gelber, R. D. (2005). Trastuzumab after Adjuvant Chemotherapy in HER2-Positive Breast Cancer. *The New England Journal of Medicine*, 353(16), 1659–1672. DOI: 10.1056/NEJMoa052306 PMID: 16236737

Pinto, N., & Dolan, M. E. (2011). Clinically relevant genetic variations in drug metabolizing enzymes. *Current Drug Metabolism*, 12(5), 487–497. DOI: 10.2174/138920011795495321 PMID: 21453273

Plans-Beriso, E., Petrova, D., Barahona-López, C., Diez-Echave, P., Hernández, O. R., Fernández-Martínez, N. F., Turner, H., García-Ovejero, E., Craciun, O., Fernández-Navarro, P., Fernández-Larrea, N., García-Esquinas, E., Kuhn, I., Jiménez-Planet, V., Moreno, V., Rodríguez-Artalejo, F., Sánchez, M. J., Pollan-Santamaria, M., Blackburn, L., & Pérez-Gómez, B. (2024). Biomarkers for personalised prevention of chronic diseases: A common protocol for three rapid scoping reviews. *Systematic Reviews*, 13(1), 147. Advance online publication. DOI: 10.1186/s13643-024-02554-9 PMID: 38824585

Rao, S. J., Isath, A., Krishnan, P., Tangsrivimol, J. A., Virk, H. U. H., Wang, Z., Glicksberg, B. S., & Krittanawong, C. (2024). ChatGPT: A Conceptual Review of Applications and Utility in the Field of Medicine. *Journal of Medical Systems*, 48(1), 59. DOI: 10.1007/s10916-024-02075-x PMID: 38836893

Roden, D. M., Wilke, R. A., Kroemer, H. K., & Stein, C. M. (2011). Pharmacogenomics: The genetics of variable drug responses. *Circulation*, 123(15), 1661–1670. DOI: 10.1161/CIRCULATIONAHA.109.914820 PMID: 21502584

Rodríguez-Vicente, A., Lumbreras, E., Hernández, J., Martín, M., Calles, A., Otín, C., Algarra, S., Páez, D., & Taron, M. (2016). Pharmacogenetics and pharmacogenomics as tools in cancer therapy. *Drug Metabolism and Personalized Therapy*, 31(1), 25–34. DOI: 10.1515/dmpt-2015-0042 PMID: 26863347

Roudnitzky, N., Behrens, M., Engel, A., Kohl, S., Thalmann, S., Hübner, S., Lossow, K., Wooding, S. P., & Meyerhof, W. (2015). Receptor Polymorphism and Genomic Structure Interact to Shape Bitter Taste Perception. *PLOS Genetics*, 11(9), e1005530. DOI: 10.1371/journal.pgen.1005530 PMID: 26406243

Sakaeda, T., Nakamura, T., & Okumura, K. (2002). MDR1 Genotype-Related Pharmacokinetics and Pharmacodynamics. *Biological & Pharmaceutical Bulletin*, 25(11), 1391–1400. DOI: 10.1248/bpb.25.1391 PMID: 12419946

Selker, H. P., Eichler, H. G., Stockbridge, N. L., McElwee, N. E., Dere, W. H., Cohen, T., Erban, J. K., Seyfert-Margolis, V. L., Honig, P. K., Kaitin, K. I., Oye, K. A., & D'Agostino, R. B.Sr. (2019). Efficacy and Effectiveness Too Trials: Clinical Trial Designs to Generate Evidence on Efficacy and on Effectiveness in Wide Practice. *Clinical Pharmacology and Therapeutics*, 105(4), 857–866. DOI: 10.1002/cpt.1347 PMID: 30610746

Sendekie, A. K., Kasahun, A. E., Limenh, L. W., Dagnaw, A. D., & Belachew, E. A. (2023). Clinical and economic impact of adverse drug reactions in hospitalised patients: Prospective matched nested case-control study in Ethiopia. *BMJ Open*, 13(6), e073777. DOI: 10.1136/bmjopen-2023-073777 PMID: 37280017

Smokovski, I., Steinle, N., Behnke, A., Bhaskar, S. M. M., Grech, G., Richter, K., Niklewski, G., Birkenbihl, C., Parini, P., Andrews, R. J., Bauchner, H., & Golubnitschaja, O. (2024). Digital biomarkers: 3PM approach revolutionizing chronic disease management — EPMA 2024 position. *The EPMA Journal*, 15(2), 149–162. DOI: 10.1007/s13167-024-00364-6 PMID: 38841615

Stefanicka-Wojtas, D., & Kurpas, D. (2023). Barriers and Facilitators to the Implementation of Personalised Medicine across Europe. *Journal of Personalized Medicine*, 13(2), 203. Advance online publication. DOI: 10.3390/jpm13020203 PMID: 36836438

Strianese, O., Rizzo, F., Ciccarelli, M., Galasso, G., Salvati, A., Giudice, C. D., Tesorio, P., & Rusciano, M. R. (2020). Precision and Personalized Medicine: How Genomic Approach Improves the Management of Cardiovascular and Neurodegenerative Disease. *Genes*, 11(7), 747. Advance online publication. DOI: 10.3390/genes11070747 PMID: 32640513

Sugandh, F., Chandio, M., Raveena, F., Kumar, L., Karishma, F., Khuwaja, S., Memon, U. A., Bai, K., Kashif, M., Varrassi, G., Khatri, M., & Kumar, S. (2023). Advances in the Management of Diabetes Mellitus: A Focus on Personalized Medicine. *Cureus*, 15(8), e43697. DOI: 10.7759/cureus.43697 PMID: 37724233

Tapper, W., Carneiro, G., Mikropoulos, C., Thomas, S. A., Evans, P. M., & Boussios, S. (2024). The Application of Radiomics and AI to Molecular Imaging for Prostate Cancer. *Journal of Personalized Medicine*, 14(3), 287. DOI: 10.3390/jpm14030287 PMID: 38541029

Tatsuno, S. Y., & Tatsuno, E. M. (2014). Does ethnicity play a role in the dosing of warfarin in Hawai'i? *Hawai'i Journal of Medicine & Public Health : a Journal of Asia Pacific Medicine & Public Health*, 73(3), 76–79. PMID: 24660123

The International HapMap Project. (2003). The International HapMap Project. *Nature*, 426(6968), 789–796. DOI: 10.1038/nature02168 PMID: 14685227

Thorn, C. F., Klein, T. E., & Altman, R. B. (2013). PharmGKB: The Pharmacogenomics Knowledge Base. *Methods in Molecular Biology (Clifton, N.J.)*, 1015, 311–320. DOI: 10.1007/978-1-62703-435-7_20 PMID: 23824865

Van Driest, S. L., & Cascorbi, I. (2021). Progress and Challenges in Pharmacogenomics. *Clinical Pharmacology and Therapeutics*, 110(3), 529–532. DOI: 10.1002/cpt.2359 PMID: 34412159

Vandenberg, L. N., Colborn, T., Hayes, T. B., Heindel, J. J., Lee, H., Shioda, T., Soto, A. M., Welshons, W. V., Zoeller, R. T., & Myers, J. P. (2012). Hormones and Endocrine-Disrupting Chemicals: Low-Dose Effects and Nonmonotonic Dose Responses. *Endocrine Reviews*, 33(3), 378–455. DOI: 10.1210/er.2011-1050 PMID: 22419778

Ventola, C. L. (2011). Pharmacogenomics in Clinical Practice: Reality and Expectations. *P&T*, 36(7), 412. https://www.ncbi.nlm.nih.gov/pmc/articles/PMC3171815/ PMID: 21931473

Vogel, F. (1959). Moderne probleme der humangenetik. In *Ergebnisse der inneren medizin und kinderheilkunde* (pp. 52–125). Springer Berlin Heidelberg. DOI: 10.1007/978-3-642-94744-5_2

Wang, R. C., & Wang, Z. (2023). Precision Medicine: Disease Subtyping and Tailored Treatment. *Cancers (Basel)*, 15(15), 3837. DOI: 10.3390/cancers15153837 PMID: 37568653

Williams, T. D. (2008). Individual variation in endocrine systems: Moving beyond the 'tyranny of the Golden Mean'. *Philosophical Transactions of the Royal Society of London. Series B, Biological Sciences*, 363(1497), 1687–1698. DOI: 10.1098/rstb.2007.0003 PMID: 18048295

Xu, Y., Jiang, Z., Wei Ting, D. S., Chieh Kow, A. W., Bello, F., Car, J., Tham, C., & Wong, T. Y. (2024). Medical education and physician training in the era of artificial intelligence. *Singapore Medical Journal*, 65(3), 159–166. DOI: 10.4103/singaporemedj.SMJ-2023-203 PMID: 38527300

Yumiceba, V., López-Cortés, A., Pérez-Villa, A., Yumiseba, I., Guerrero, S., García-Cárdenas, J. M., Armendáriz-Castillo, I., Guevara-Ramírez, P., Leone, P. E., Zambrano, A. K., & Paz-y-Miño, C. (2020). Oncology and Pharmacogenomics Insights in Polycystic Ovary Syndrome: An Integrative Analysis. *Frontiers in Endocrinology*, 11, 585130. Advance online publication. DOI: 10.3389/fendo.2020.585130 PMID: 33329391

Zanger, U. M., & Schwab, M. (2013). Cytochrome P450 enzymes in drug metabolism: Regulation of gene expression, enzyme activities, and impact of genetic variation. *Pharmacology & Therapeutics*, 138(1), 103–141. DOI: 10.1016/j.pharmthera.2012.12.007 PMID: 23333322

Zhang, C., Yang, J., Chen, S., Sun, L., Li, K., Lai, G., Peng, B., Zhong, X., & Xie, B. (2024). Artificial intelligence in ovarian cancer drug resistance advanced 3PM approach: Subtype classification and prognostic modeling. *The EPMA Journal*, 15(3), 525–544. Advance online publication. DOI: 10.1007/s13167-024-00374-4 PMID: 39239109

Zhang, L., Xia, Y., & Gui, Y. (2023). Neuronal ApoE4 in Alzheimer's disease and potential therapeutic targets. *Frontiers in Aging Neuroscience*, 15, 1199434. DOI: 10.3389/fnagi.2023.1199434 PMID: 37333457

Chapter 2
Advanced High Throughput Screening Platforms in Drug Discovery

Chinmayi Joshi
https://orcid.org/0000-0002-6681-1098
Sankalchand Patel University, Visnagar, India

Drashti Makwana
Sankalchand Patel University, Visnagar, India

ABSTRACT

Over the years, the field of drug discovery has seen significant progress, with technological advancements proving crucial in the identification of new therapeutic candidates. High-Throughput Screening (HTS) which combines large-scale data analysis with automated micro-quantitative experiments, and can rapidly analyse a large number of chemical compounds against the drug targets. This chapter explores the recent advancements in high-throughput screening platforms that significantly enhanced drug discovery process, and also includes the incorporation of miniaturization, automation, and image-based analysis to increase throughput and data quality. The significance of these advanced HTS platforms on accelerating the discovery and development of new therapeutic drugs is discussed, along with the challenges. This chapter, taken as a whole, will offer a comprehensive overview of the most advanced high-throughput screening (HTS) platforms available for use in the drug development industry and will address how they might completely transform the process.

DOI: 10.4018/979-8-3693-6597-7.ch002

I. INTRODUCTION

1.1 Overview of High Throughput Screening (HTS) in Drug Discovery

Over the past 20 years, High-Throughput Screening (HTS) has been increasingly popular and is now considered a conventional procedure for drug discovery in the pharmaceutical sector. Essentially, it involves screening and testing a large number of biological compounds against various drug targets. Various types of libraries, including as combinatorial chemistry, genomics, protein, and peptide libraries, are screened using HTS assays. By screening huge compound libraries at a rate that may surpass a few thousand compounds per day or per week, the HTS approach aims to expedite drug discovery (Martis et al., 2011; Szymański et al., 2012). High Throughput Screening (HTS) is a useful technique for finding target-specific chemicals since it may concentrate on single pathways. Compounds can be screened using HTS against a particular biological target or pathway. This emphasis makes it possible for researchers to identify compounds that interact with the intended target directly, which raises the likelihood of discovering therapy options that are effective for a particular disease mechanism (Hughes et al., 2011). It is possible to alter HTS assays to assess certain biological processes or molecular interactions. By focusing on a single mechanism, researchers can precisely measure the impacts of chemicals, which helps identify compounds with the required activity with great specificity. Concentrating on a single mechanism reduces the complexity of the screening process. This streamlined process allows for a more precise assessment of chemical efficacy and reduces the likelihood of off-target effects, which could make it more challenging to identify true hits. HTS experiments based on a single mechanism can be optimized to replicate the target's physiological circumstances. This specificity ensures that the screening process closely mimics the in vivo environment, enhancing the reliability and significance of the findings. (Szymański et al., 2012; Dahlin and Walters, 2014). Finding compounds that are efficient against a specific target or mechanism facilitates the validation process later on. It is simpler for researchers to confirm that the compounds work as expected, which contributes to a better understanding of the compounds' potential therapeutic advantages. The identification of compounds with specific mechanisms of action by HTS helps to narrow the focus of therapeutic development. Focusing on refining these substances to maximize their utility and eliminate any potential negative effects can lead to more efficient development and testing. (Schenone et al., 2013).

In present article, we have compared the traditional HTS methods and emerging HTS methods. The main aim of the chapter is to provide an overview of advanced high-throughput screening platforms in drug discovery. Recent CRISPR/Cas-based screens are highlighted because they demonstrate potential as a tool for identifying drug targets and developing novel therapies. In addition, the present chapter gives an outline of the present state of chemical proteomics, *in silico*, and genomics technologies that support this approach and emphasize the significance of target deconvolution following hit discovery through phenotypic screens. The chapter discusses the aspects like automation technology and discovering methods that influence the selection of HTS methods. In addition, the review for the chapter investigates future research directions in the field of high-throughput screening platforms, namely integrating informatics tools for quality control, data interpretation, and screen design. This chapter examines a number of HTS platforms, including those that can be used to predict the activity of compounds against drug-metabolizing enzymes, measure genotoxicity, estimate toxicity potential, and determine mechanisms of action for compounds. We also attempted to address the significance and limitations of these platforms. This chapter offers a comprehensive overview of the most advanced high-throughput screening (HTS) platforms available for use in the drug development industry.

1.2 Traditional HTS Methods vs Emerging methods

Over the last two decades, HTS method that is now a traditional method for drug discovery in pharmaceutical industry has grown tremendously in popularity. It essentially entails screening and testing of a large number of biological effectors as well as modulators against pre-selected targets. This has been employed by both academicians and industrialists (Markowicz and Szymański et al., 2011). HTS in drug development is traditionally performed using traditional techniques and technology that have been in use for many years. These techniques are normally characterized by established test formats, with large chemical stores and automated processes. These may include cell-based assays, microplate-based assays, biochemicals for assessing target validation, Molecular Imaging Agents and other related agents or compounds. Some of the prominent conventional HTS methods are:

1.2.1 Microplate-Based Assays

Bioassays (biochemical and cell-based tests) related to drug discovery are typically automated and miniaturized using the microplate format. A 2D array of wells with a finite volume for testing is contained within each microplate. The three well densities per plate that are most frequently used for screening are 96, 384, and 1536 wells. Microplates are produced by many producers in a variety of materials i.e. Polystyrene (PS), Cyclic Olefin Copolymer (COC), Polypropylene (PP) tailored to specific machinery and client requirements. The American

National Standards Institute (ANSI) and the Society for Laboratory Automation and Screening (SLAS) have standardized the footprint of the microplate and the positions of the wells. Microplates in HTS laboratories are often divided into two categories: "Assay Plates" and "Compound Plates." Compound plates are used to store molecular libraries for screening; they are robust enough to be utilized again over long periods of time and on several screens. Assay plates are used for testing; they are chosen according to assay conditions and are only utilized up to the size of a single screen. There are numerous possibilities for surface treatment to improve test performance. Large hydrophobic areas of biomolecules are naturally bound by polystyrene through passive interactions without any treatment. Options for surface treatment can increase or decrease this effect. Surfaces coated with tissue culture are also frequently used to promote cell adhesion. These have an impact on how biomolecules bind to microplate surfaces or enhance cell line adhesion. There are various ways to control environmental exposure and contamination for the goal of preserving certain microplate well conditions, such as plate seals and lids. Automated sealing systems are also available, which utilizes thermal process for sealing of microplates. Fluorescence, luminescence, absorbance, and other light-based detection technologies are employed with single and multi-mode microplate readers. While the light routes through the sample vary, these detecting systems include many of the same components. Single mode microplate readers that are suited to a particular technology might be compact and reasonably priced. A multi-mode reader, which is a little bigger and more expensive than single mode readers, might be very helpful for a lab that wants to combine several technologies and detecting modes into one more flexible device. Quantifying a particular occurrence frequently necessitates the use of a detection reagent. The formats can be 96-well or 384-well plates where multiple tests can be done simultaneously in microplate-based assays. This format provides capabilities to screen huge compound libraries against biological targets (Drews et al., 2000; Jones et al., 2016).

1.2.2 Cell-Based Assays

Assays based on living cells are used in experiments to assess how different substances affect biological processes. Because they enable researchers to see how new drugs or chemicals interact with particular cellular targets in a realistic biological milieu, these assays are essential in pharmacology and drug discovery. To measure the effect of compounds on biological activities using living cells, cell-based assays are being used. Such assays can provide information about the pharmacological activity at a specific receptor or intracellular target of the compound screened. These platforms have great potential to serve as important tools for studying the effects of small compounds and growth conditions on cell function and physiology, as well as for examining cell proliferation and differentiation. Ion channel assays, cytotoxicity assays and reporter gene assays are all involved in this cell-based assay (Al-Zidan and Aldewachi et al., 2021).

The goal of cell-based assay is to examine the effects of various chemicals on ion channels, which are proteins that regulate the movement of ions across cell membranes. Numerous physiological processes, including muscle contraction and nerve impulse transmission, depend on ion channels. Understanding a substance's impact on ion channel activity may help identify any possible side effects or suggest that the chemical has therapeutic potential. For instance, ion channel-targeting drugs are widely utilized to treat epilepsy and cardiac arrhythmias (Yu et al., 2015). Furthermore, by monitoring characteristics like cell viability, proliferation, or apoptosis (programmed cell death), these assays evaluate a drug's possible toxicity to cells. This allows researchers to ascertain whether a molecule is detrimental to cells. Cytotoxicity assays are crucial in the development of new drugs because they guarantee the safety and efficacy of a given molecule. According to Sozonova et al., (2022), they aid in the identification of detrimental effects and the optimization of dosages to reduce unpleasant reactions. Additionally, these assays use genetically altered cells that carry a reporter gene, which is a gene that, in response to particular cellular interactions or events, produces a readily quantifiable result (such a fluorescent protein). This enables scientists to track alterations in cellular signalling pathways or gene expression in reaction to a substance.

Reporter gene assays are helpful tools for researching the effects of chemicals on signalling cascades, gene expression, and cellular functions (Nierode et al., 2016). In total, Cell-based assays are significant tools in biomedical research because they allow researchers to observe how chemicals affect living cells and provide crucial information regarding the safety and efficacy of substances.

1.2.3 Fluorescent-Based Assays

With great sensitivity and adaptability, fluorescent-based assays quantify biological processes using fluorescence signals. These include fluorometric, fluorescence polarization and fluorescence resonance energy transfer (FRET). This method measures several biological activities with high sensitivity and flexibility (Zhang and Chung et al., 1999). Fluorometric assays quantify the amount of fluorescence that a fluorescent dye or probe emits when stimulated by a particular light wavelength. The intensity is correlated with the concentration of target molecule or the degree of a biological event.

Quantifying nucleic acids, proteins, or other compounds in a sample is a popular use for fluorometric assays. They offer a precise and sensitive means of monitoring biological processes like ligand binding and enzyme activity (Fang et al., 2019).

The rotation of fluorescent molecules in a solution is measured using fluorescence polarization (FP) experiments. The spin of a fluorescent probe slows down as it attaches to a target, changing the polarization of the light it emits. Quantifying binding interactions or other molecular processes is done using this change. Protein-ligand interactions, receptor binding, and other molecular interactions can all be studied with the use of FP assays. Their sensitivity is high, and they can be applied to large-scale screening (LeBlanc et al., 2021). In FRET assays, two fluorophores are used: an acceptor and a donor. If the donor and acceptor fluorophores are close together, the excited donor fluorophore can transmit energy to them. The emission spectrum, which is used to gauge molecular interactions or conformational changes, is altered as a result of this energy transfer. Protein-DNA interactions, protein-protein interactions, and other biological processes can all be studied with the help of FRET tests. They offer dynamic and geographical information regarding the interactions between molecules inside of living cells (Tomlinson and Whiting, 2020). Assays based on fluorescent materials are highly appreciated for their versatility, sensitivity, and specificity. They are essential tools in drug development and research because they allow for the in-depth examination of biological processes at the molecular level.

1.2.4 Luminescent-Based Assays

Luminescent-based assays use light emission to very sensitively identify and quantify biological activity. The highly sensitive type includes Luciferase Reporter Assay and ATP assay which rely on light emission resulting from either chemical or biological reactions (Fan and Wood et al., 2007). In the Luciferase Reporter Assay, luciferase enzymes are utilized, which, when they interact with their substrate (often luciferin) in the presence of oxygen, catalyze a reaction that results in the production of light. Gene expression and other biological activity can be measured by measuring the amount of luciferase enzyme present, which is proportionate to the amount of light emitted. Gene regulation, signal transduction pathways, and promoter activity are frequently studied using luciferase reporter assays. They offer a highly accurate and quantitative assessment of how genes are expressed in response to different stimuli or settings throughout experiments (Carter and Shieh, 2015; Ghosh et al., 2018). Another Luminescent-Based Assays are ATP assays, which measure adenosine triphosphate (ATP), a substance that transports and stores energy inside cells. In these investigations, bioluminescent processes utilizing luciferase and its substrate produce light in the presence of ATP. There is a relationship between the amount of ATP in the sample and the light's intensity of emission. ATP assays are used to assess cell viability, proliferation, and metabolic activity. They are also useful in quantifying microbiological contamination and evaluating cellular ATP levels as an indicator of cell health and function (Sharma and Malik, 2013; Zhang et al.,

2022). Because light-producing reactions are efficient, luminescent-based assays have great sensitivity and are therefore perfect for detecting low biological activity levels. Their utility in many research and diagnostic applications is enhanced by their benefits for applications that demand low background interference and precise quantification.

1.2.5 Radioligand Binding Assays

Sophisticated methods such as radioligand binding assays are employed to investigate drug-target interactions with great sensitivity (Hulme and Trevethick et al., 2010). These assays use radioactive isotopes to tag molecules, or ligands. These radiolabelled ligands have the ability to attach to particular targets or receptors in a sample. The radiation released, which is directly correlated with the degree of ligand-receptor contact, is measured to determine the binding. Radioligand assays are widely used to characterize and measure the binding sites of targets such as enzymes or receptors. These assays aid in figuring out the binding site density and affinity of the target, or how strongly a drug binds.

In order to screen possible compounds, comprehend their binding properties, and optimize lead candidates, these assays are essential to the drug development process. They offer in-depth understanding of the precise and potent ways in which a drug interacts with its target (Lever et al., 2018; Grisanti, 2023). Since radiolabelled molecules allow these assays to detect even very low levels of binding, with incredible sensitivity and precision measurements of binding interactions, which are helpful for comprehensive pharmacological studies. Radioligand binding assays provide important insights into drug-target interactions, facilitating the development of new therapeutic approaches and expanding our understanding of the pharmacology and biology of receptors. Radioligand binding assays are an indispensable tool in scientific and clinical settings due to their high sensitivity in assessing binding (Maguire et al., 2012).

1.2.6 Absorbance-Based Assays

Absorbance-based assays are essential to High-Throughput Screening (HTS) because they can quickly and effectively measure chemical or biological interactions across a large number of samples. Colour or optical density changes that are a result of biological process can be measured through absorbance assays. For example, ELISA (Enzyme-Linked Immunosorbent Assay) and colorimetric enzyme tests are examples of such assays (Crowther et al., 2000). ELISA is modified for HTS by use multi-well plates (96-well, 384-well, etc.) that enable the analysis of numerous samples at once. ELISA can be used in HTS to find substances that impact enzyme function, protein-protein interactions, or biomarker levels. Enzyme-linked antibodies are used to identify certain binding events in the wells of a microplate after samples are added in a high-throughput scenario. A plate reader is used to quantify the colorimetric reaction that results, enabling the quick processing of hundreds or thousands of samples (Tsurusawa et al., 2021; Alhaj et al., 2023). Other, colorimetric enzymes assays measure the enzyme activity and other biological reactions on a massive scale using these assays in HTS. A change in colour due to enzyme-substrate reaction is identified using a spectrophotometer or microplate reader. Colorimetric enzyme assay in HTS is carried out in multi-well plates with distinct substances or environments in each well. Colour changes are produced by the enzyme reactions in proportion to the concentration of target or effect of the tested substances. In order to rapidly and effectively quantify the data, the absorbance is measured at particular wavelengths (Healin et al., 2019; Dolz et al., 2023; Shao et al., 2023).

When thousands of chemicals or conditions need to be screened for HTS, absorbance-based assays are a good choice since they can handle a lot of samples at once.

By combining these assays with automated plate readers and liquid handling devices, throughput and repeatability are increased. The accurate quantitative information provided by the absorbance measurements enables in-depth examination of the interactions and effects of compounds (Lunch et al., 2024). Absorbance-based assays are essential in HTS for determining active molecules, evaluating potential therapeutic targets, and refining lead candidates during the drug discovery process. Their capacity to process vast amounts of data fast and precisely makes them a vital instrument for speeding up research and development procedures (Szymański et al., 2012).

Typically, conventional HTS is restricted to a few different types of assays, mostly focusing on biochemical and basic cell-based assays. Traditional high throughput screening (HTS) methods have greatly aided drug discovery, but they also have a number of disadvantages, such as high costs and resource requirements, a limited range of assays, scalability issues, challenges with data management, etc (Edmondson and Broglie et al., 2014). The effectiveness, affordability, and efficiency of the drug discovery process can all be impacted by these constraints.

1.2.7 The Limitations and Challenges of Traditional HTS Methods

High-throughput screening has historically transformed drug development procedures; yet, a number of issues limit and complicate their usefulness and effectiveness.

1.2.7.1 High Costs

The primary disadvantage is that establishing an HTS facility would require a significant financial investment and the deployment of traditional HTS, which consists of expensive high-throughput detection tools and a complex network of automated equipment, such as robotics platforms and liquid handling systems. (Gantz et al., 2023; Taylor et al., 2023). This initial investment might particularly be quite daunting for smaller biotech start-ups and less research-based institutions; running-costs at this point run amuck. These consist of costs incurred in selecting qualified personnel, the use of the reagents with the equipment and the latter themselves. It is well-nigh impossible to contemplate large-scale major screening where there are no adequate monetary resources for such a large-scale expensive biological equipment investment. (D'arville, 2004).

1.2.7.2 Resource Consumption

In particular, resource requirements bear on both the sustainability and the ethics of traditional HTS strategies; screening procedures that call for large cell lines or large preparations of enzyme are very biologic material-intensive and raise some issues related to their sourcing and utilization. Also, the big reagents of chemicals consumption lead to huge waste problems causing environmental issues. Such consumption of resources complicates maintaining sustainable drug discovery practices. The energy that goes into a high-throughput facility also adds to problems like keeping lab conditions at an optimum state and powering through automated systems, which are quite energy-consuming (Villeneuve et al.,2018).

1.2.7.3 Scalability Issues

Despite the large screening ability of HTS, traditional methods often exhibit scaling problems. When scaled, many assays lose their sensitivity and specificities in such a way that enhances the likelihood of false positive or negative results (Attene-Ramos et al., 2014). As a result, the outcome reliability can be affected since testing is scaled up from small to big. Furthermore, the amount of data produced during HTS presents particular challenges; the size of data may be such that traditional data management systems would not be able to handle this deluge and thereby create some processing hitches. Sufficient computing resources and associated expertise might not always be available to process and interpret this data (Saleh et al., 2018).

1.2.7.4 Challenges Associated With HIT Validation and Assay Development

After the hits have come through in the screening step, the validation process turns into a lengthy and tiring exercise. The massive difference in throughput between the screening and validation stages puts brakes on drug development pipelines. This tends to cause delays, to be costly, and it just gets things much too complicated (Hevener et al., 2018).

Developing robust and reproducible assays is one of the challenging tasks required for a successful HTS. Variability and inconsistency in the results are caused by assay conditions, quality of reagents, and environmental factors (Chai et al., 2015). Biological systems tend to oversimplify interactions, and thus traditional HTS assays often provide hits that are not well translated in vivo. More investigation into the pharmacodynamics and pharmacokinetics of such compounds would need further extensive studies that reach further into resources and time (Ayon, 2023).

1.2.7.5 Ethical and Regulatory Concerns

Kiani et al., (2012) mention ethical questions and pressures on HTS assays based upon reliance on animal models, with which most are accompanied. GLP can sometimes introduce complexity and add up extra costs to the complexity involved in HTS while satisfying ethical requirements. One reason for rethinking in conventional HTS approaches would be the pressure building on pharmaceutical companies to make more sparing use of fewer models of animals and find new avenues (Schmeisser et al., 2023).

Developing novel medications by HTS has several challenges, and in order to overcome this complexity, newer methodologies must be used. New technologies including high-content screening (HCS), microfluidic-based platforms, next-generation sequencing (NGS) assays, and CRISPR-based screening are all substantially superior than previous methods in many aspects.

1.2.8 Modern Methods for HTS

1.2.8.1 Next Generation Sequencing (NGS)

NGS is a very scalable and rapid technique. It is also known as massively parallel sequencing. (Tripathi and Singh et al., 2019). Applications of NGS include drug target identification, gene expression profiling, and genetic screening. This method makes it possible to study in detail changes in the genome or transcriptome. (Goodwin and McPherson et al., 2016). NGS consists of Whole Genome Sequencing (WGS), Whole Exome Sequencing (WES), Transcriptome Sequencing (RNA-Seq), Targeted Sequencing, Epigenomic Sequencing and Metagenomics (Metzker et al.,

2010). It is faster than traditional Sanger sequencing because it has the ability to sequence large amounts of DNA or RNA quickly as opposed to traditional methods. It has the capability of doing multiple analyses concurrently on several specimens generating vast amount of data sets that may encompass whole transcriptomes or genomes respectively among others. Conventional techniques are less accurate and are generally much more expensive compared to NGS technologies, if we evaluate them by comparing with each other according to these parameters. Modern NGS platforms are highly automated, this means that less human error can occur and huge quantities of samples and data could be processed using scalable protocols. Next-generation sequencing technologies are quicker, more accurate, and cheaper for genetic analysis than classical sequencing approaches. They have a wide range of uses such as the development of drugs and biomedical research. (Mardis et al., 2008; Shendure and Ji et al., 2008).

1.2.8.2 CRISPR-CAS9 – Based Assays

CRISPR screens are a successful technique for biological research because they make it possible to objectively examine how genes work across a variety of organisms and applications. (Bock and Datlinger et al., 2022) The potential of CRISPR to generate cell- and animal-based models of human disease by modifying one or more genes is what gives rise to its importance in drug discovery and validation. We will focus on recent CRISPR/dCas-based screens since they show promise as a tool for finding pharmacological targets to develop innovative treatments. (Kim and Lee, 2023) An RNA-guided endonuclease, like Cas9, is used by CRISPR-based technologies to cause genomic disruption. The first example of functional genomic screening with CRISPR involved the use of a fully active Cas9 endonuclease to knock out particular gene loci and assess the phenotypic impact of the loss. (le Sage and Lawo et al., 2019) The concepts of the newly developed high-throughput CRISPR-Cas9 library genome screening method, which uses positive and negative selection to identify genes linked to drug resistance and survival. (Chan and Lu et al., 2022) Genome-wide screening is made possible by CRISPR-Cas9 technology, which produces targeted gene knockouts. This technique aids in the highly precise identification of gene functions and possible therapeutic targets. (Shalem and Sanjana et al., 2015).

Researchers are now able to systematically investigate gene function, pinpoint therapeutic targets, and comprehend disease pathways thanks to the application of CRISPR in high-throughput screening, which has created new opportunities in drug discovery. It is employed in target validation and identification to find new therapeutic targets. (Wang and Wei et al., 2014) in pathways analysis and functional genomics (Hart and Chandrashekhar et al., 2015). For example, to determine the drug resistance mechanism in cancer cells. (Smurnyy and Cai et al., 2014) By discovering genes which modify drug sensitivity when knocked out or activated, new mechanisms of action of drug might be better understood (Zhou and Zhu et al., 2014).

1.2.8.3 Microfluidic-Based Screening Platforms

Microscale flow principles are used by microfluidic-based screening platforms to manipulate minute volumes of fluids hence providing experiments with high precision and throughput. These devices utilize very small fluid volumes and produce a compact footprint for high-throughput screening. Although it employs less reagents, such technique can perform high-resolution analysis (Sackmann and Fulton et al., 2014). Microfluidics operates within channels having dimensions ranging from tens to hundreds of micrometres that control and manipulate fluids in them. Microfluidic based screening platforms offer precise control over experimental conditions, which enhances throughput and efficiency, reduces waste and costs, speeds up biochemical reactions by shortening assay times, automates work processes while cutting manual intervention, and makes in vitro models applicable to *in vivo* experiments (Shendure et al., 2008, Sackmann et al., 2014, Whitesides et al., 2006, Sia et al., 2003, Dittrich et al., 2006) Many laboratories have a difficult time designing and constructing microfluidic-based screening platforms due to the need for specialized knowledge and tools. It is worth mentioning here that some of these problems are low detection sensitivity, scalability and throughput concerns as well as technical complexity, integration with existing systems (antony), data management/analysis, standardization. In summary, since there are already ongoing attempts to address them, phenotypic screening offers good opportunities for drug development improvement (Bhatia et al., 2014, Weibel et al., 2006, Nge et al., 2013, Breslauer et al., 2006, Cunningham et al., 2009, Thorsen et al., 2002).

1.2.8.4 Phenotypic Screening

Phenotypic screening is a potent method that, in the context of High-Throughput Screening (HTS), concentrates on the overall impact of small compounds on cellular phenotypes as opposed to focusing on particular biochemical pathways (Swinney and Anthony et al., 2011). Phenotypic screening assesses how tiny compounds affect visible cellular properties like shape, motility, proliferation, or gene expression patterns, as opposed to traditional target-based screening, which concentrates on specific molecular targets (such enzymes or receptors). Through the use of automated technologies and high-resolution imaging, HTS makes it possible to evaluate vast libraries of drugs by simultaneously assessing their effects on a variety of cellular phenotypes across many samples (Vincent et al., 2022).

Compounds having unusual or novel modes of action that would not be found using target-based strategies can be found using phenotypic screening. This is especially helpful in the search for novel medications or treatment targets. Cell-based assays combined with automated imaging systems are commonly used in HTS platforms for phenotypic screening. These devices take high-resolution pictures of cells that have been exposed to various substances, enabling quantitative examination of alterations in cell shape, growth patterns, or behaviour (Ege et al., 2021; Swinney and Lee, 2020).

Large amounts of data produced by phenotypic experiments are processed and interpreted using sophisticated image analysis and machine learning approaches. This makes it possible to identify substances that modify biological processes in a desired manner or cause particular phenotypic alterations (Chandrasekaran et al., 2021). To effectively interpret changes in cellular phenotypes, complex analysis methods and methodologies are needed due to the large volume of data generated by phenotypic screens. Although phenotypic screening has the ability to discover novel drugs, combining these discoveries with target-based assays facilitates the clarification of the precise molecular mechanisms at play. Phenotypic screening in HTS offers an in-depth comprehension of the ways in which chemicals impact cellular function and can result in the identification of novel medications and methods. Through assessing the wider effects of small compounds on cellular phenotypes, scientists can pinpoint possible targets for treatment and investigate the underlying mechanisms of action (Vincent et al., 2022).

1.2.8.5 High Content Screening

High Content Screening (HCS) is also known as High Content Analysis (HCA), and it is an effective process of analysing complex biological processes that combines high-resolution imaging with high-throughput screening (HTS) techniques. HCS was established as a drug discovery tool including selectivity screens and cytotoxicity profiling, initially offering novel secondary assay formats using the multi-parameter and individual cell attributes of the approach. (Zock et al., 2009). The HSC principle involves robotic instruments and automatic analysis in order to identify a hit compound from a vast number of chemical libraries. (Wang and Wu et al., 2019) By taking pictures of cells treated with different substances, HCS allows researchers to obtain large amounts of data from biological systems. (Giuliano and Tylor et al., 1998, Carpenter and Sabatini et al., 2004). The major areas where HCS has been employed include organism physiology, in vitro toxicity, target identification, cell signalling and cell physiology. (Zock et al., 2009).

1.2.8.6 Artificial Intelligence and Machine Learning

In recent times, artificial intelligence (AI) begins to be utilized more broadly in the pharmaceutical industry, including drug research and discovery, repurposing of drugs, increasing pharmaceutical production, clinical trials, etc. AI is able to identify hit and lead compounds, validate drug targets more quickly, and optimize drug structure design. AI-based QSAR techniques, such as linear discriminant analysis (LDA), support vector machines (SVMs), random forests (RF), and decision trees, have developed from QSAR modelling tools that have been used to identify possible drug candidates. These techniques can be used to accelerate QSAR analysis (Paul D et al., 2020).

1.2.8.7 Mass Spectrometry-Based Screening

Drug discovery is a convoluted journey that requires designing, identifying and validating probable medicinal molecules. Mass spectrometry is an important assistive tool that supports many stages in this process. The objective of this operation is to find molecules with possible therapeutic utility. For quickly identifying these interesting "hit" compounds, the speed and accuracy of mass spectrometry make it an ideal equipment. MS-based HTS has high sensitivity and specificity. (Dueñas and Peltier-Heap et al., 2023).

1.2.8.8 Biosensor-Based Assay

Biosensor-based assays are increasingly becoming prominent in high-throughput screening (HTS) for drug discovery. These are more efficient than traditional methods in several aspects. In real-time, biosensors provide complete data on the kinetics as well as binding affinity of drug-target interactions without any labeling requirements. (Gupta et al., 2023; Matthew et al., 2006). This allows more accurate and efficient filtration of compound libraries over time. Label-free biosensors are better because they reduce sample preparation needs compared to label-based techniques, or test run times too long (Matthew et al., 2006). They can measure direct changes in mass or heat resulting from analyte binding to the target. Cell-based biosensors like electric cell-substrate impedance sensing (ECIS) make it possible to have label-free, high-information content monitoring of cellular responses to substances. (Matthew et al., 2006). It helps evaluate signaling and cytotoxicity as well as proliferation in drug screening. When responsive drug delivery systems are integrated with biosensors, early detection of disease biomarkers can be done and therapeutic dosing is controlled in a closed-loop system (Ngoepe et al., 2013). This could also improve how chronic diseases are managed. A novel one such can be psychLight which is based on the structure of the 5-HT2A receptor and it can be used for screening psychedelic drug candidates using novel biosensors like psychLight, a designed fluorescence sensor based on the structure of the 5-HT2A receptor. (Dong et al., 2021). Thus, PsychLight identifies receptor conformations related to specific signaling pathways.

Biosensor technologies have become increasingly important in HTS for better and more informative screening of drug candidates for effective use. The applications of these useful tools in drug discovery continue to widen due to advances in design of biosensor, cell-based assays and integration with drug delivery systems.

1.2.9 The Practical Challenges Faced by Smaller Laboratories or Institutions in Adopting Cutting-Edge Technologies and Practical Solutions

In the current situation, it has been noted that with the rapid speed of the advancement of current HTS technologies, such as CRISPR, AI, and microfluidics, scientific research and innovation have accelerated (Lawson et al., 2021; De Stefano et al., 2022; Li et al., 2023). On the other hand, due to an entirely absent infrastructure, fewer resources, and minimal experiences, often smaller labs and their institutions face huge setbacks trying to access some of those technologies. It might increase the competitiveness and progress of science (Nickischer et al., 2018). It is necessary to understand these issues in order to be able to allow smaller businesses to use modern tools more effectively and enhance their research by seeking possible answers (Table 1).

Table 1. Challenges and some potential solutions

Challenges faced by small labs/ institutes	Description	Potential Solutions/Workarounds
High costs	Equipment, reagents, and upkeep need a large upfront expenditure.	To cut expenses, seek grants, work with bigger organizations, or think about sharing facilities.
Technical Expertise	Smaller labs might not have the personnel with the know-how to deploy and use new technologies.	Hire consultants, collaborate with colleges to share expertise, or train current employees.
Regulatory Compliance	It can be difficult to navigate the complicated regulatory environment, particularly when it comes to gene editing.	Develop standard operating procedures, make use of professional organization resources, and interact with regulatory agencies early on.
Integration with Existing Systems	It could be difficult for new technology to interface with existing lab systems or procedures.	Establish a staged deployment strategy with an emphasis on trial projects that complement current processes.
Data Management	AI and other technologies produce vast volumes of data, which makes analysis and storage difficult.	Make use of open-source data analysis tools and cloud-based solutions for managing and storing data.

continued on following page

Table 1. Continued

Challenges faced by small labs/ institutes	Description	Potential Solutions/Workarounds
Access to Consumables and Supplies	It could be difficult for smaller labs to find trustworthy vendors for innovative materials and reagents.	To order products in bulk, establish cooperative purchasing relationships with other labs or institutions.
Limited Collaboration Opportunities	There can be less chances for multidisciplinary cooperation at smaller universities.	To promote cooperation, aggressively pursue alliances with business, academia, and other research institutions.
Scalability of Projects	Cutting-edge technology implementation can be resource-intensive and difficult to scale.	As resources and experience develop, start with smaller projects and progressively expand the complexity.
Keeping Up with Rapid Advancements	Staying up to date might be challenging due to the rapid progress of technology.	To keep up with the latest developments and trends, take part in webinars, conferences, and journal subscriptions.

1.3 The Present Status of Chemical *in Silico*, Genomics and Proteomics Technologies in Drug Discovery

1.3.1 In Silico

In silico techniques that make predictions about biological processes, drug interactions, and protein structure models using computer-based simulations. These techniques compliment experimental methods in that they provide a first impression that may lead to beneficial laboratory experiments. Here, some current in silico technologies use in drug discovery:

1.3.2 Molecular Docking and Dynamics

The challenge of finding and developing new drugs is extremely difficult. The In-Silico approach aids in the discovery of novel drugs. The modern drug development process has found increasingly widespread application for molecular docking and molecular dynamics (MD) simulations. Various docking programs have been developed for the purpose of visualizing the three-dimensional structure of molecules. It is a structure-based virtual screening technique for determining the target molecule's orientation, conformation, and position within its structure. The challenging aspect of the molecular docking method relies on lead molecule optimization, biological pathway analysis, and de novo drug design. (Bhagat and Butle et al., 2021). Molecular dynamics simulation is to derive claims on the dynamic, thermodynamic, and structural characteristics of a molecular system. In terms of drug design, MD simulations can offer significant insight on the dynamic nature of the target (Salo-Ahen et al., 2020). Quantitative Structure-Activity Relationship (QSAR) Models: Uses statistical models to provide context for the research on the relationship between biological activity and chemical structure (Bastikar et al., 2022).

1.3.3 Genomics

Genomics technologies utilize high-throughput analyses of genetic information to elucidate gene function, gene regulation, and physiological and disease-related pathways. These technologies enable the discovery of genetic targets and the comprehension of genetic components of drug response. Examples include NGS, CRISPR-Cas9 Screening, Single-Cell RNA Sequencing (scRNA-seq) (Franken et al., 2015; Leitner et al., 2016; Lionta et al., 2014).

1.3.4 Proteomics

Chemical proteomics employs small molecules to query protein function and interactions within live cells. This technique is central to discovering targets for drug intervention, determining the mechanisms of action of drugs and probes, and defining protein-protein interactions and signalling networks (Amiri-Dashatan et al., 2018). Current technologies such as Activity based protein profiling (ABPP), Thermal proteome profiling (TPP), Chemical Cross-Linking Mass Spectrometry (CX-MS) are widely used in drug discovery (Franken et al., 2015; Leitner et al., 2016).

The integration of bioinformatics tools, databases, software, and analysis pipelines has revolutionized high-throughput screening (HTS) in drug discovery. These resources enable researchers to efficiently manage, analyse, and interpret large volumes of data, facilitating the identification of novel drug targets and the understanding of complex biological processes. By leveraging these advanced technologies, the drug discovery process becomes more streamlined, accurate, and reproducible, ultimately accelerating the development of new therapeutic agents. Here, some bioinformatics software/tools given in Table 2:

Table 2. Current available bioinformatics tools/databases/software for HTS

Sr. No.	Software/Tools	Short Description	References
1	CellProfiler	Open-source software for the quantitative analysis of biological images. It automates the extraction of measurements from images, facilitating high-content screening (HCS).	Carpenter et al., 2006
2	ImageJ/Fiji	Widely-used open-source image processing software. Fiji is a distribution of ImageJ that includes many useful plugins.	Schindelin et al., 2012
3	KNIME Analytics Platform	Open-source software for data analytics, reporting, and integration. KNIME provides a graphical interface for creating data workflows.	Berthold et al., 2009
4	Pipeline Pilot	Data analysis and visualization software that provides a graphical environment to create complex data processing workflows.	
5	Galaxy	Open-source, web-based platform for data-intensive biomedical research. Galaxy allows users to perform, reproduce, and share complete data analyses.	Afgan et al., 2018
6	Bioconductor	An open-source project that provides tools for the analysis and comprehension of high-throughput genomic data.	Huber et al., 2015
7	PubChem	A free chemical information database maintained by the National Center for Biotechnology Information (NCBI). It includes information on the biological activities of small molecules.	Kim and chen et al., 2019
8	ChEMBL	A database of bioactive molecules with drug-like properties maintained by the European Bioinformatics Institute (EMBL-EBI).	Gaulton et al., 2017
9	ChemSpider	A free chemical structure database providing fast access to over 60 million structures, properties, and associated information.	Pence et al., 2010
10	The Human Protein Atlas	A database providing information on the tissue and cell distribution of proteins in the human body.	Uhlén et al., 2015
11	OpenBabel	An open-source chemical toolbox designed to speak the many languages of chemical data. It is used to convert chemical file formats and data.	O'Boyle et al., 2011

continued on following page

Table 2. Continued

Sr. No.	Software/Tools	Short Description	References
12	HTS Navigator	A software suite specifically designed for the analysis of high-throughput screening data.	Malo et al., 2006
13	Spotfire	A data visualization and analytics software, useful for analyzing and visualizing large datasets generated from HTS experiments.	Ahlberg et al., 1994
14	MGL Tools	MGLTools is a software developed at the Molecular Graphics Laboratory (MGL) of The Scripps Research Institute for visualization and analysis of molecular structures.	https://mgltools.scripps.edu/
15	BIOVIA Discovery Studio Visualizer	Its visualization tool for viewing, sharing, and analyzing protein and modeling data, complete the form below to receive the free Discovery Studio Visualizer for interactive 3D visualization.	https://discover.3ds.com/discovery-studio-visualizer-download
16	CADD Combined Annotation Dependent Depletion	Integrates diverse genome annotations and scores all human SNV and Indel. It prioritizes functional, deleterious, and disease causal variants according to functional categories, effect sizes and genetic architectures. Scores above 10 should be applied as a cut-off for identifying pathogenic variants.	Kircher et al., 2014
17	MutationTaster	Analyses evolutionary conservation, splice-site changes, loss of protein features and changes that might affect the amount of mRNA. Variants are classified, as polymorphism or disease-causing	Schwarz et al., 2010
18	SIFT Sorting Intolerant from Tolerant	Predicts based on sequence homology, if an AA substitution will affect protein function and potentially alter the phenotype. Scores less than 0.05 indicating a variant as deleterious.	Ng P.C et al., 2003

1.4 Case Studies and Applications

1.4.1 Case Studies and Application of NGS in HTS for Target Identification

Case study 1: TCGA (The Cancer Genome Atlas)- Thousands of cancer genomes have been sequenced through this project. Novel cancer triggers and functional mutations across a range of cancer types have been identified as a result of this. It is now possible to design targeted medicines since researchers have discovered particular genetic changes that are essential for the occurrence and spread of cancer (Collisson et al., 2014).

Case study 2: Pathogen Identification and Drug Resistance- Pathogen identification and resistance mechanism study are carried out through the use of NGS. New antiviral and antibiotic medicines have been developed in large part because of this. The development of targeted medicines and improved management of drug resistance are the results of better understanding of pathogen genomes (Gardy and Loman et al., 2018).

Case study 3: Alzheimer's Disease Gene Identification- Alzheimer's disease-related genetic modifications have been identified via NGS. Using NGS data, genome-wide association studies (GWAS) have identified a number of risk loci. New pathways involved in the causes and mechanisms of disease and new targets for therapeutic development have been discovered by these results (Kunkle and Grenier-Boley et al., 2019).

Case Study 4-10

1.4.2 Case Studies and Application of Microfluidic-Based Platforms for Phenotypic Screening

A microfluidic device was created to assess the phenotypic response of bacteria to various antibiotics (Baltekin et al., 2017). Droplet-based microfluidic systems have become popular as substitute techniques for microbe analysis, allowing for single-cell screening and enzyme and metabolite identification (Tu and Zhang, 2021; Ramezankhani et al., 2022). A microfluidic system employed to screen novel medicines for possible cardiotoxicity on human-induced pluripotent stem cell-derived cardiomyocytes. In a number of applications, including COVID-19 research, drug screening is being carried out using systems based on organoids and microfluidics (Ramezankhani et al., 2022).

1.4.3 Case Studies and Application of High-Content Screening in Lead Optimization

Case study 1: PerkinElmer Operetta CLS- A high-content analysis system called Operetta CLS from PerkinElmer was employed to test chemical compounds for their impact on cancer cell lines. With the use of the system, scientists were able to do multi-parametric assessments of cellular phenotypes and find possible supplements candidates that might stop cell growth or cause apoptosis (https://resources.perkinelmer.com/corporate/pdfs/downloads/011034_01_bro_operetta_brochure.pdf).

Case study 2: Thermo Fisher Scientific CellInsight CX7- The CellInsight CX7 High-Content Screening (HCS) Platform was used to optimize leads for neurodegenerative diseases. Through this platform, comprehensive studies of neuronal cells were carried out, analyzing components such neurite outgrowth, synapse formation, and cell viability in response to numerous drugs. (Thermo Fisher Scientific website)

Case study 3: Molecular Devices ImageXpress Micro Confocal- Molecular Devices' ImageXpress Micro Confocal device was utilized during the anti-inflammatory drug lead optimization stage. Using human primary cells, this technique facilitated high-content screening of inflammatory indicators, allowing researchers to examine the impact of various compounds on cytokine production and cell signaling pathways (https://www.moleculardevices.com/en/assets/promotion/br).

1.4.4 Case Studies and Applications of HTS Technologies in Areas Like Oncology

Case Study 1: Micropillar/Microwell Chip Platform

Lee et al. (2013) developed a micropillar/microwell chip platform to enhance the clinical results of personalized gene-targeted cancer therapy. This platform is ideal for encasing primary cancer cells in nanoscale hydrogel spots on the chip, producing efficacy data with different drugs, and ultimately enabling a comparison of the in vitro data obtained from the chip with clinical data and gene expression data.

Case study 2: The EarlyTox Caspase-3/7 NucView™ 488 Assay uses the NucView 488 Caspase-3 substrate to detect apoptosis in intact cell populations. This microplate test has the benefit of being tailored for use with imaging equipment and microplate readers. This test kit offers customers a versatile method for quantifying apoptosis when used with SpectraMax microplate readers or the SpectraMax MiniMax 300 Imaging Cytometer (https://www.moleculardevices.com/lab-notes/microplate-readers/harnessing-microplate-assays-for-high-throughput-drug-screening-in-cancer-research).

Case Study 3: Integration of HTS with Molecular Profiling

In order to increase the therapeutic options for high-risk pediatric malignancies, Mayoh et al. (2023) successfully used HTS into precision medicine initiatives. The authors demonstrated the sustainability of this strategy, which might be a useful tool for the healthcare system, by extending it from our TARGET pilot project to the PRISM clinical trial. Additionally, our research investigates and incorporates the interaction between molecular profile data and complementary drug screening, which raises the clinical confidence level of therapy recommendations.

1.4.5 Case Studies and Applications of HTS Technologies in Areas Like Rare Diseases

Case Study 1 MyoScreen Platform to Quantify Disease Relevant Duchenne Muscular Dystrophy (DMD)

Hariharan et al (2023) developed a platform to enable target identification in skeletal muscle cells from DMD patients. They conducted an *in vitro* screening of 125 patient-derived samples against a library of 126 anticancer drugs in order to determine possible treatment approaches for these high-risk pediatric patients. 82% of the screens on enlarged tumor cells were performed while the patients were still receiving therapeutic therapy, and tumor cell growth had no effect on drug responses. The known correlations between responses to corresponding targeted drugs and activating genetic changes in NTRK, BRAF, and ALK were validated by high-throughput drug screening (HTS).

Case study 2 Li et al. (2024) showed how to develop cell models of cystic fibrosis for drug screening and high-throughput analysis. Isogenic CF cell models created using CRISPR/Cas9 gene editing. Each of the four cell models has a distinct class I, II, III, or IV CFTR mutation. A fluorescent reporter that is sensitive to halides is integrated into cell models. possible to screen new CF treatments at high throughput.

Case study 3 Using neural stem cells generated from patients with fragile X syndrome (FXS), Kumari et al. (2015) developed a sensitive and precise fluorescence resonance energy transfer-based assay for the detection of fragile X mental retardation proteins. This assay was then refined for high-throughput screening (HTS) of chemical libraries. According to the results, this HTS methodology will be helpful in identifying better lead compounds for the creation of novel FXS treatments. Additionally, this assay can be modified for the detection of FMRP in research and clinical contexts.

1.5 Challenges and Opportunities

Large amounts of data are generated by modern HTS platforms, which makes them challenging to handle and evaluate. To extract useful insights from the vast amount of data generated by screening tests, strong data management systems and sophisticated analytical tools are necessary (Stephens et al., 2015). Developing advanced tools for data management and analysis, such as AI-powered algorithms, can speed up the processing and interpretation of data. These techniques lessen the possibility of data overload, handle complicated datasets, and spot trends (Aldoseri et al., 2023).

Integrating data from many omics domains (metabolomics, transcriptomics, proteomics, and genomes) to obtain a comprehensive comprehension of biological systems is a challenging task that frequently results in problems with data integration (Hasin et al., 2017).

The growth of integrative multi-omics methods is being encouraged by developments in systems biology and bioinformatics. By connecting several biological information layers, these techniques can improve drug development and offer a more thorough understanding of biological processes (Pinu et al., 2019). Smaller research organizations and companies may not be able to participate in or benefit from HTS due to the prohibitively high costs associated with modern HTS technologies. Emerging innovations include more economical sequencing techniques, open-access HTS platforms, and high-throughput technologies that are affordable. Thanks to these developments, more individuals can now access HTS and take part in drug discovery initiatives (Mardis et al., 2017). The volume and complexity of data produced by HTS experiments may be too much for current bioinformatics tools to handle. This may impede the ability to analyse and comprehend data effectively (Stephens et al., 2015). The development of new bioinformatics tools and algorithms that can effectively handle, analyze, and interpret massive datasets is becoming increasingly necessary. New tools can help make drug discovery decisions more efficiently and with higher data accuracy (Yadav et al., 2024).

"Multi-omics" refers to the integration and analysis of data from multiple omics fields. Multi-omics approaches bring about data from multiple omics domains to deliver an all-round view of the molecular structure and function of a biological system. A huge amount of data was generated in a variety of omics disciplines thanks to high-throughput technologies. The efficient analytical and computational techniques for the integration and assessment of multi-omics data have, therefore, emerged. Problems faced by the industry include data quality, platform integration, and the need to control computational and analytical complexity. (Ali et al., 2024). Some of the challenges faced when artificial intelligence was used to analyze omics data included preprocessing, datasets, model validation, and testbed applications.

Artificial intelligence is one of the major components of modern data platforms and has dramatically changed how data is treated, analyzed, and exploited. AI algorithms have superior techniques for automating data integration, authenticating data integrity, and developing a smart data governance. For these purposes, the current experiments were multiple. Some studies advance different opinions on the relationship between AI models and omics (Althati et al. 2024; Abyshanab et al., 2024). Additionally, it is crucial to maintain ethics and regulatory compliance in HTS research. It is necessary to handle issues with data privacy, research ethics, and rules compliance. The implementation of well-defined policies and procedures can aid in guaranteeing moral behavior and adherence to regulations. Sensible and productive HTS research will be aided by the creation of strong frameworks for data security, ethical conduct, and compliance to regulations (Ledford et al., 2015).

The most simple, accurate, and quick genome editing method is CRISPR-Cas9. It is significantly transforming the field of molecular biology with a variety of interesting applications covering from clinical medicines to agriculture and the environment. As an example, unintended off-target effects which may lead to harmful mutations. In the case of drug discovery hit generation and testing genetic manipulation experiments, further complications are presented in the line of caution (Shinwari et al., 2018). On the other hand, interesting applications of this technique raise a few moral, ethical, and safety issues. The main human safety and moral concerns are the unintended and undesirable effects in clinical application, particularly in the treatment and prevention of genetic diseases, consent, and exploitation toward eugenics, highly controversial in the context of human germline modification. AI has revolutionized HTS, transforming the drug discovery industry. AI-powered techniques enable rapid and inexpensive development of drugs. However, applying AI in drug development has raised moral and legal challenges. Factors including permission, data privacy, and algorithmic bias must be addressed to assure that AI-driven procedures continue to be open, transparent, efficient, and beneficial to everyone. While conducting trials with human subjects, researchers must make sure that patients are properly informed of the risk factors (Huang et al., 2024; Rehman et al., 2024).

1.6 Concluding Remarks

Continued innovation in HTS technology, particularly the incorporation of AI and machine learning, will improve the efficiency, accuracy, and cost-effectiveness of drug discovery procedures. Future advances in integrative multi-omics techniques will provide more detailed insights into complex biological systems, resulting in more targeted and successful drug discovery tactics. By democratizing HTS capabilities, attempts to minimize costs and improve accessibility will allow smaller organizations and research institutions to participate in and benefit from modern drug development activities. HTS research will be carried out responsibly and in accordance with the highest standards of scientific integrity and societal values due to the ongoing development of comprehensive ethical and regulatory frameworks. In conclusion, there are many of prospects for progress in HTS even though it faces difficulties with data management, multi-omics integration, cost, tool development, and ethics. The future of HTS will be determined by embracing technological breakthroughs, enhancing accessibility, and setting strong ethical rules. These actions will also improve the impact of HTS on drug discovery.

REFERENCES

Abushanab, R. (2024). Role of Artificial Intelligence and Big Data in Sustainable Entrepreneurship. *Journal of Artificial Intelligence General science (JAIGS) ISSN: 3006-4023, 5*(1), 275-294.

Afgan, E., Baker, D., Batut, B., Van Den Beek, M., Bouvier, D., Čech, M., Chilton, J., Clements, D., Coraor, N., Grüning, B. A., Guerler, A., Hillman-Jackson, J., Hiltemann, S., Jalili, V., Rasche, H., Soranzo, N., Goecks, J., Taylor, J., Nekrutenko, A., & Blankenberg, D. (2018). The Galaxy platform for accessible, reproducible and collaborative biomedical analyses: 2018 update. *Nucleic Acids Research*, 46(W1), W537–W544. DOI: 10.1093/nar/gky379 PMID: 29790989

Ahlberg, C., & Shneiderman, B. (1994, April). Visual information seeking: Tight coupling of dynamic query filters with starfield displays. In *Proceedings of the SIGCHI conference on Human factors in computing systems* (pp. 313-317). DOI: 10.1145/191666.191775

Aldewachi, H., Al-Zidan, R. N., Conner, M. T., & Salman, M. M. (2021). High-throughput screening platforms in the discovery of novel drugs for neurodegenerative diseases. *Bioengineering (Basel, Switzerland)*, 8(2), 30. DOI: 10.3390/bioengineering8020030 PMID: 33672148

Aldoseri, A., Al-Khalifa, K. N., & Hamouda, A. M. (2023). Re-thinking data strategy and integration for artificial intelligence: Concepts, opportunities, and challenges. *Applied Sciences (Basel, Switzerland)*, 13(12), 7082. DOI: 10.3390/app13127082

Alhajj, M., Zubair, M., & Farhana, A. (2023). *Enzyme linked immunosorbent assay*. StatPearls.

Ali, A. M., & Mohammed, M. A. (2024). A comprehensive review of artificial intelligence approaches in omics data processing: Evaluating progress and challenges. *International Journal of Mathematics, Statistics, and Computer Science*, 2, 114–167. DOI: 10.59543/ijmscs.v2i.8703

Althati, C., Tomar, M., & Shanmugam, L. (2024). Enhancing Data Integration and Management: The Role of AI and Machine Learning in Modern Data Platforms. *Journal of Artificial Intelligence General science (JAIGS) ISSN: 3006-4023, 2*(1), 220-232.

Amiri-Dashatan, N., Koushki, M., Abbaszadeh, H. A., Rostami-Nejad, M., & Rezaei-Tavirani, M. (2018, Fall). Proteomics Applications in Health: Biomarker and Drug Discovery and Food Industry. *Iranian Journal of Pharmaceutical Research*, 17(4), 1523–1536. PMID: 30568709

Attene-Ramos, M. S., Miller, N., Huang, R., Michael, S., Itkin, M., Kavlock, R. J., Austin, C. P., Shinn, P., Simeonov, A., Tice, R. R., & Xia, M. (2013). The Tox21 robotic platform for the assessment of environmental chemicals–from vision to reality. *Drug Discovery Today*, 18(15-16), 716–723. DOI: 10.1016/j.drudis.2013.05.015 PMID: 23732176

Ayon, N. J. (2023). High-throughput screening of natural product and synthetic molecule libraries for antibacterial drug discovery. *Metabolites*, 13(5), 625. DOI: 10.3390/metabo13050625 PMID: 37233666

Baltekin, Ö., Boucharin, A., Tano, E., Andersson, D. I., & Elf, J. (2017). Antibiotic susceptibility testing in less than 30 min using direct single-cell imaging. *Proceedings of the National Academy of Sciences of the United States of America*, 114(34), 9170–9175. DOI: 10.1073/pnas.1708558114 PMID: 28790187

Bastikar, V., Bastikar, A., & Gupta, P. (2022). Quantitative structure–activity relationship-based computational approaches. In *Computational Approaches for Novel Therapeutic and Diagnostic Designing to Mitigate SARS-CoV-2 Infection* (pp. 191–205). Academic Press.

Berthold, M. R., Cebron, N., Dill, F., Gabriel, T. R., Kötter, T., Meinl, T., Ohl, P., Thiel, K., & Wiswedel, B. (2009). KNIME-the Konstanz information miner: Version 2.0 and beyond. *SIGKDD Explorations*, 11(1), 26–31. DOI: 10.1145/1656274.1656280

Bhagat, R. T., Butle, S. R., Khobragade, D. S., Wankhede, S. B., Prasad, C. C., Mahure, D. S., & Armarkar, A. V. (2021). Molecular docking in drug discovery. *Journal of Pharmaceutical Research International*, 33(30B), 46–58. DOI: 10.9734/jpri/2021/v33i30B31639

Bhatia, S. N., & Ingber, D. E. (2014). Microfluidic organs-on-chips. *Nature Biotechnology*, 32(8), 760–772. DOI: 10.1038/nbt.2989 PMID: 25093883

Bock, C., Datlinger, P., Chardon, F., Coelho, M. A., Dong, M. B., Lawson, K. A., Lu, T., Maroc, L., Norman, T. M., Song, B., Stanley, G., Chen, S., Garnett, M., Li, W., Moffat, J., Qi, L. S., Shapiro, R. S., Shendure, J., Weissman, J. S., & Zhuang, X. (2022). High-content CRISPR screening. *Nature Reviews. Methods Primers*, 2(1), 8. DOI: 10.1038/s43586-021-00093-4 PMID: 37214176

Breslauer, D. N., Lee, P. J., & Lee, L. P. (2006). Microfluidics-based systems biology. *Molecular BioSystems*, 2(2), 97–112. DOI: 10.1039/b515632g PMID: 16880927

Carpenter, A. E., Jones, T. R., Lamprecht, M. R., Clarke, C., Kang, I. H., Friman, O., Guertin, D. A., Chang, J. H., Lindquist, R. A., Moffat, J., Golland, P., & Sabatini, D. M. (2006). CellProfiler: Image analysis software for identifying and quantifying cell phenotypes. *Genome Biology*, 7(10), 1–11. DOI: 10.1186/gb-2006-7-10-r100 PMID: 17076895

Carpenter, A. E., & Sabatini, D. M. (2004). Systematic genome-wide screens of gene function. *Nature Reviews. Genetics*, 5(1), 11–22. DOI: 10.1038/nrg1248 PMID: 14708012

Carter, M., & Shieh, J. (2015). Biochemical assays and intracellular signaling. *Guide to research techniques in neuroscience*, pp.311-343.

Chai, S. C., Goktug, A. N., & Chen, T. (2015). Assay validation in high throughput screening–from concept to application. *Drug Discov. Dev. Mol. Med*, 10, 59765. DOI: 10.5772/59765

Chan, Y. T., Lu, Y., Wu, J., Zhang, C., Tan, H. Y., Bian, Z. X., Wang, N., & Feng, Y. (2022, April 11). CRISPR-Cas9 library screening approach for anti-cancer drug discovery: Overview and perspectives. *Theranostics*, 12(7), 3329–3344. DOI: 10.7150/thno.71144 PMID: 35547744

Chandrasekaran, S. N., Ceulemans, H., Boyd, J. D., & Carpenter, A. E. (2021). Image-based profiling for drug discovery: Due for a machine-learning upgrade? *Nature Reviews. Drug Discovery*, 20(2), 145–159. DOI: 10.1038/s41573-020-00117-w PMID: 33353986

Collisson, E. A. C. J., Campbell, J., Brooks, A., Berger, A., Lee, W., Chmielecki, J., & Helman, E. (2014). Comprehensive molecular profiling of lung adenocarcinoma: The cancer genome atlas research network. *Nature*, 511(7511), 543–550. DOI: 10.1038/nature13385 PMID: 25079552

Crowther, J. R. (2000). The ELISA guidebook. *Methods in Molecular Biology (Clifton, N.J.)*, 149, III–IV, 1–413. DOI: 10.1385/1592590497 PMID: 11028258

Cunningham, B. T. (2009). Label-free optical biosensors: An introduction. Label-free biosensors: techniques and applications, 1.

D'ARVILLE, C.A.R.O.L.Y.N. (2004). Partnering: Key to early-stage biotech survival, but at what cost? *Biotechnology Healthcare*, 1(6), 26. PMID: 23424295

Dahlin, J. L., & Walters, M. A. (2014). The essential roles of chemistry in high-throughput screening triage. *Future Medicinal Chemistry*, 6(11), 1265–1290. DOI: 10.4155/fmc.14.60 PMID: 25163000

De Stefano, P., Bianchi, E., & Dubini, G. (2022). The impact of microfluidics in high-throughput drug-screening applications. *Biomicrofluidics*, 16(3), 031501. DOI: 10.1063/5.0087294 PMID: 35646223

Molecular Devices. (n.d.). ImageXpress Micro Confocal High-Content Imaging System. Retrieved from Molecular Devices website

Dittrich, P. S., & Manz, A. (2006). Lab-on-a-chip: Microfluidics in drug discovery. *Nature Reviews. Drug Discovery*, 5(3), 210–218. DOI: 10.1038/nrd1985 PMID: 16518374

Dolz, M., Monterrey, D. T., Beltrán-Nogal, A., Menés-Rubio, A., Keser, M., González-Pérez, D., de Santos, P. G., Viña-González, J., & Alcalde, M. (2023). The colors of peroxygenase activity: Colorimetric high-throughput screening assays for directed evolution. *Methods in Enzymology*, 693, 73–109. DOI: 10.1016/bs.mie.2023.09.006 PMID: 37977739

Dong, C., Ly, C., Dunlap, L. E., Vargas, M. V., Sun, J., Hwang, I. W., Azinfar, A., Oh, W. C., Wetsel, W. C., Olson, D. E., & Tian, L. (2021). Psychedelic-inspired drug discovery using an engineered biosensor. *Cell*, 184(10), 2779–2792. DOI: 10.1016/j.cell.2021.03.043 PMID: 33915107

Drews, J. (2000, March 17). Drug discovery: A historical perspective. *Science*, 287(5460), 1960–1964. DOI: 10.1126/science.287.5460.1960 PMID: 10720314

Dueñas, M. E., Peltier-Heap, R. E., Leveridge, M., Annan, R. S., Büttner, F. H., & Trost, M. (2023, January 11). Advances in high-throughput mass spectrometry in drug discovery. *EMBO Molecular Medicine*, 15(1), e14850. DOI: 10.15252/emmm.202114850 PMID: 36515561

Edmondson, R., Broglie, J. J., Adcock, A. F., & Yang, L. (2014, May). Three-dimensional cell culture systems and their applications in drug discovery and cell-based biosensors. *Assay and Drug Development Technologies*, 12(4), 207–218. DOI: 10.1089/adt.2014.573 PMID: 24831787

Ege, N., Bouguenina, H., Tatari, M., & Chopra, R. (2021). Phenotypic screening with target identification and validation in the discovery and development of E3 ligase modulators. *Cell Chemical Biology*, 28(3), 283–299. DOI: 10.1016/j.chembiol.2021.02.011 PMID: 33740433

Fan, F., & Wood, K. V. (2007, February). Bioluminescent assays for high-throughput screening. *Assay and Drug Development Technologies*, 5(1), 127–136. DOI: 10.1089/adt.2006.053 PMID: 17355205

Fang, X., Zheng, Y., Duan, Y., Liu, Y., & Zhong, W. (2018). Recent advances in design of fluorescence-based assays for high-throughput screening. *Analytical Chemistry*, 91(1), 482–504. DOI: 10.1021/acs.analchem.8b05303 PMID: 30481456

Franken, H., Mathieson, T., Childs, D., Sweetman, G. M., Werner, T., Tögel, I., Doce, C., Gade, S., Bantscheff, M., Drewes, G., Reinhard, F. B. M., Huber, W., & Savitski, M. M. (2015). Thermal proteome profiling for unbiased identification of direct and indirect drug targets using multiplexed quantitative mass spectrometry. *Nature Protocols*, 10(10), 1567–1593. DOI: 10.1038/nprot.2015.101 PMID: 26379230

Gantz, M., Neun, S., Medcalf, E. J., van Vliet, L. D., & Hollfelder, F. (2023). Ultrahigh-throughput enzyme engineering and discovery in in vitro compartments. *Chemical Reviews*, 123(9), 5571–5611. DOI: 10.1021/acs.chemrev.2c00910 PMID: 37126602

Gardy, J. L., & Loman, N. J. (2018). Towards a genomics-informed, real-time, global pathogen surveillance system. *Nature Reviews. Genetics*, 19(1), 9–20. DOI: 10.1038/nrg.2017.88 PMID: 29129921

Gaulton, A., Hersey, A., Nowotka, M., Bento, A. P., Chambers, J., Mendez, D., Mutowo, P., Atkinson, F., Bellis, L. J., Cibrián-Uhalte, E., Davies, M., Dedman, N., Karlsson, A., Magariños, M. P., Overington, J. P., Papadatos, G., Smit, I., & Leach, A. R. (2017). The ChEMBL database in 2017. *Nucleic Acids Research*, 45(D1), D945–D954. DOI: 10.1093/nar/gkw1074 PMID: 27899562

Giuliano, K. A., & Taylor, D. L. (1998). Fluorescent-protein biosensors: New tools for drug discovery. *Trends in Biotechnology*, 16(3), 135–140. DOI: 10.1016/S0167-7799(97)01166-9 PMID: 9523461

Goodwin, S., McPherson, J. D., & McCombie, W. R. (2016, May 17). Coming of age: Ten years of next-generation sequencing technologies. *Nature Reviews. Genetics*, 17(6), 333–351. DOI: 10.1038/nrg.2016.49 PMID: 27184599

Grisanti, L. A. (2023). Radioligand Binding to Quantify Adrenergic Receptor Expression in the Heart. *Current Protocols*, 3(1), e649. DOI: 10.1002/cpz1.649 PMID: 36602296

Gupta, G., Jha, K., & Chaudhary, S. (2023, July). Biosensor Assays Types and Their Roles Toward Ligand-Receptor Interactions in Drug Discovery. *Assay and Drug Development Technologies*, 21(5), 190–201. DOI: 10.1089/adt.2023.003 PMID: 37285190

Hariharan, S., Lorintiu, O., Lee, C. C., Duchemin-Pelletier, E., Li, X., Healy, A., . . . Christoforou, N. (2023). Development and validation of a high throughput screening platform to enable target identification in skeletal muscle cells from Duchenne Muscular Dystrophy (DMD) patients. bioRxiv, 2023-05.

Hart, T., Chandrashekhar, M., Aregger, M., Steinhart, Z., Brown, K. R., MacLeod, G., Mis, M., Zimmermann, M., Fradet-Turcotte, A., Sun, S., Mero, P., Dirks, P., Sidhu, S., Roth, F. P., Rissland, O. S., Durocher, D., Angers, S., & Moffat, J. (2015). High-resolution CRISPR screens reveal fitness genes and genotype-specific cancer liabilities. *Cell*, 163(6), 1515–1526. DOI: 10.1016/j.cell.2015.11.015 PMID: 26627737

Hasin, Y., Seldin, M., & Lusis, A. (2017). Multi-omics approaches to disease. *Genome Biology*, 18(1), 1–15. DOI: 10.1186/s13059-017-1215-1 PMID: 28476144

Healing, E., Charlier, C. F., Meira, L. B., & Elliott, R. M. (2019). A panel of colorimetric assays to measure enzymatic activity in the base excision DNA repair pathway. *Nucleic Acids Research*, 47(11), e61–e61. DOI: 10.1093/nar/gkz171 PMID: 30869144

Hevener, K. E., Pesavento, R., Ren, J. H., Lee, H., Ratia, K., & Johnson, M. E. (2018). Hit-to-Lead: hit validation and assessment. In *Methods in Enzymology* (Vol. 610, pp. 265–309). Academic Press.

Huang, D., Yang, M., Wen, X., Xia, S., & Yuan, B. (2024). AI-DRIVEN DRUG DISCOVERY: ACCELERATING THE DEVELOPMENT OF NOVEL THERAPEUTICS IN BIOPHARMACEUTICALS. *Journal of Knowledge Learning and Science Technology ISSN: 2959-6386 (online), 3*(3), 206-224.

Huber, W., Carey, V. J., Gentleman, R., Anders, S., Carlson, M., Carvalho, B. S., Bravo, H. C., Davis, S., Gatto, L., Girke, T., Gottardo, R., Hahne, F., Hansen, K. D., Irizarry, R. A., Lawrence, M., Love, M. I., MacDonald, J., Obenchain, V., Oleś, A. K., & Morgan, M. (2015). Orchestrating high-throughput genomic analysis with Bioconductor. *Nature Methods*, 12(2), 115–121. DOI: 10.1038/nmeth.3252 PMID: 25633503

Hughes, J. P., Rees, S., Kalindjian, S. B., & Philpott, K. L. (2011). Principles of early drug discovery. *British Journal of Pharmacology*, 162(6), 1239–1249. DOI: 10.1111/j.1476-5381.2010.01127.x PMID: 21091654

Hulme, E. C., & Trevethick, M. A. (2010, November). Ligand binding assays at equilibrium: Validation and interpretation. *British Journal of Pharmacology*, 161(6), 1219–1237. DOI: 10.1111/j.1476-5381.2009.00604.x PMID: 20132208

Jones, E., Michael, S., & Sittampalam, G. S. (2016). *Basics of assay equipment and instrumentation for high throughput screening*. Assay Guidance Manual. [Internet]

Kiani, A. K., Pheby, D., Henehan, G., Brown, R., Sieving, P., Sykora, P., & Marks, R.. (2022). Ethical considerations regarding animal experimentation. *Journal of Preventive Medicine and Hygiene*, 63(2, Suppl 3), E255. PMID: 36479489

Kim, S., Chen, J., Cheng, T., Gindulyte, A., He, J., He, S., Li, Q., Shoemaker, B. A., Thiessen, P. A., Yu, B., Zaslavsky, L., Zhang, J., & Bolton, E. E. (2019). PubChem 2019 update: Improved access to chemical data. *Nucleic Acids Research*, 47(D1), D1102–D1109. DOI: 10.1093/nar/gky1033 PMID: 30371825

Kim, Y., & Lee, H.-M. (2023). CRISPR-Cas System Is an Effective Tool for Identifying Drug Combinations That Provide Synergistic Therapeutic Potential in Cancers. *Cells*, 12(22), 2593. DOI: 10.3390/cells12222593 PMID: 37998328

Kim, Y., & Lee, H.-M. (2023). CRISPR-Cas System Is an Effective Tool for Identifying Drug Combinations That Provide Synergistic Therapeutic Potential in Cancers. *Cells*, 12(22), 2593. DOI: 10.3390/cells12222593 PMID: 37998328

Kircher, M., Witten, D. M., Jain, P., O'Roak, B. J., Cooper, G. M., & Shendure, J. (2014). A general framework for estimating the relative pathogenicity of human genetic variants. *Nature Genetics*, 46(3), 310–315. DOI: 10.1038/ng.2892 PMID: 24487276

Ku, S. K., Yi, S. H., & Nam, D.-H. (2014). High-throughput screening (HTS) of anticancer drug efficacy on a micropillar/microwell chip platform. *Analytical Chemistry*, 86(1), 535–542. DOI: 10.1021/ac402546b PMID: 24199994

Kumari, D., Swaroop, M., Southall, N., Huang, W., Zheng, W., & Usdin, K. (2015). High-throughput screening to identify compounds that increase fragile X mental retardation protein expression in neural stem cells differentiated from fragile X syndrome patient-derived induced pluripotent stem cells. *Stem Cells Translational Medicine*, 4(7), 800–808. DOI: 10.5966/sctm.2014-0278 PMID: 25999519

Kunkle, B. W., Grenier-Boley, B., Sims, R., Bis, J. C., Damotte, V., Naj, A. C., & Rotter, J. I. (2019). Genetic meta-analysis of diagnosed Alzheimer's disease identifies new risk loci and implicates A beta, tau, immunity and lipid processing (vol 51, pg 414, 2019). *Nature Genetics*, 51(9), 1423–1424. DOI: 10.1038/s41588-019-0495-7 PMID: 31417202

Lawson, C. E., Manuel Martí, J., Radivojevic, T., Sai Vamshi, R. J., Gentz, R., Hillson, N. J., & Peisert, S.. (2021). Machine learning for metabolic engineering: A review. *Metabolic Engineering*, 63, 34–60. DOI: 10.1016/j.ymben.2020.10.005 PMID: 33221420

le Sage, C., Lawo, S., & Cross, B. C. S. (2020, March). CRISPR: A Screener's Guide. *SLAS Discovery*, 25(3), 233–240. DOI: 10.1177/2472555219883621 PMID: 31658850

LeBlanc, E. V., Shekhar-Guturja, T., Whitesell, L., & Cowen, L. E. (2021). Fluorescence Polarization-Based Measurement of Protein-Ligand Interaction in Fungal Cell Lysates. *Current Protocols*, 1(1), e17. DOI: 10.1002/cpz1.17 PMID: 33484500

Leitner, A., Faini, M., Stengel, F., & Aebersold, R. (2016). Crosslinking and mass spectrometry: An integrated technology to understand the structure and function of molecular machines. *Trends in Biochemical Sciences*, 41(1), 20–32. DOI: 10.1016/j.tibs.2015.10.008 PMID: 26654279

Lever, S. Z., Fan, K. H., & Lever, J. R. (2017). Tactics for preclinical validation of receptor-binding radiotracers. *Nuclear Medicine and Biology*, 44, 4–30. DOI: 10.1016/j.nucmedbio.2016.08.015 PMID: 27755986

Li, Z.-H., Wang, J., Xu, J.-P., Wang, J., & Yang, X. (2023). Recent advances in CRISPR-based genome editing technology and its applications in cardiovascular research. *Military Medical Research*, 10(1), 12. DOI: 10.1186/s40779-023-00447-x PMID: 36895064

Lionta, E., Spyrou, G. K., Vassilatis, D., & Cournia, Z. (2014). Structure-based virtual screening for drug discovery: Principles, applications and recent advances. *Current Topics in Medicinal Chemistry*, 14(16), 1923–1938. DOI: 10.2174/15680 26614666140929124445 PMID: 25262799

Liu, A., Chokshi, M., Nguyen, N., Powell, R. T., Stephan, C. C., & Bao, G. (2024). Cystic fibrosis cell models for high-throughput analysis and drug screening. *Journal of Cystic Fibrosis*, 23(4), 716–724. DOI: 10.1016/j.jcf.2024.07.001 PMID: 39060183

Lynch, C., Sakamuru, S., Ooka, M., Huang, R., Klumpp-Thomas, C., Shinn, P., Gerhold, D., Rossoshek, A., Michael, S., Casey, W., Santillo, M. F., Fitzpatrick, S., Thomas, R. S., Simeonov, A., & Xia, M. (2024). High-throughput screening to advance in vitro toxicology: Accomplishments, challenges, and future directions. *Annual Review of Pharmacology and Toxicology*, 64(1), 191–209. DOI: 10.1146/annurev-pharmtox-112122-104310 PMID: 37506331

Maguire, J. J., Kuc, R. E., & Davenport, A. P. 2012. Radioligand binding assays and their analysis. *Receptor binding techniques*, pp.31-77.

Malo, N., Hanley, J. A., Cerquozzi, S., Pelletier, J., & Nadon, R. (2006). Statistical practice in high-throughput screening data analysis. *Nature Biotechnology*, 24(2), 167–175. DOI: 10.1038/nbt1186 PMID: 16465162

Mardis, E. R. (2008). Next-generation DNA sequencing methods. *Annual Review of Genomics and Human Genetics*, 9(1), 387–402. DOI: 10.1146/annurev.genom.9.081307.164359 PMID: 18576944

Mardis, E. R. (2017). DNA sequencing technologies: 2006–2016. *Nature Protocols*, 12(2), 213–218. DOI: 10.1038/nprot.2016.182 PMID: 28055035

Martis, E. A., Radhakrishnan, R., & Badve, R. R. (2011). High-throughput screening: The hits and leads of drug discovery—An overview. [Google Scholar]. *Journal of Applied Pharmaceutical Science*, 1, 2–10.

Matthew, A. (2006). Current biosensor technologies in drug discovery. Drug discovery, 69.

Mayoh, C., Mao, J., Xie, J., Tax, G., Chow, S.-O., Cadiz, R., Pazaky, K., Barahona, P., Ajuyah, P., Trebilcock, P., Malquori, A., Gunther, K., Avila, A., Yun, D. Y., Alfred, S., Gopalakrishnan, A., Kamili, A., Wong, M., Cowley, M. J., & Dolman, M. E. M. (2023). High-throughput drug screening of primary tumor cells identifies therapeutic strategies for treating children with high-risk cancer. *Cancer Research*, 83(16), 2716–2732. DOI: 10.1158/0008-5472.CAN-22-3702 PMID: 37523146

Metzker, M. L. (2010). Sequencing technologies—The next generation. *Nature Reviews. Genetics*, 11(1), 31–46. DOI: 10.1038/nrg2626 PMID: 19997069

Ng, P. C., & Henikoff, S. (2003). SIFT: Predicting amino acid changes that affect protein function. *Nucleic Acids Research*, 31(13), 3812–3814. DOI: 10.1093/nar/gkg509 PMID: 12824425

Nge, P. N., Rogers, C. I., & Woolley, A. T. (2013). Advances in microfluidic materials, functions, integration, and applications. *Chemical Reviews*, 113(4), 2550–2583. DOI: 10.1021/cr300337x PMID: 23410114

Ngoepe, M., Choonara, Y. E., Tyagi, C., Tomar, L. K., du Toit, L. C., Kumar, P., Ndesendo, V. M., & Pillay, V. (2013, June 14). Integration of biosensors and drug delivery technologies for early detection and chronic management of illness. *Sensors (Basel)*, 13(6), 7680–7713. DOI: 10.3390/s130607680 PMID: 23771157

Nickischer, D., Elkin, L., Cloutier, N., O'Connell, J., Banks, M., & Weston, A. (2018). Challenges and opportunities in enabling high-throughput, miniaturized high content screening. *High Content Screening: A Powerful Approach to Systems Cell Biology and Phenotypic Drug Discovery*, 165-191.

Nierode, G., Kwon, P. S., Dordick, J. S., & Kwon, S. J. (2016). Cell-based assay design for high-content screening of drug candidates. *Journal of Microbiology and Biotechnology*, 26(2), 213–225. DOI: 10.4014/jmb.1508.08007 PMID: 26428732

O'Boyle, N. M., Banck, M., James, C. A., Morley, C., Vandermeersch, T., & Hutchison, G. R. (2011). Open Babel: An open chemical toolbox. *Journal of Cheminformatics*, 3(1), 1–14. DOI: 10.1186/1758-2946-3-33 PMID: 21982300

Paul, D., Sanap, G., Shenoy, S., Kalyane, D., Kalia, K., & Tekade, R. K. (2021, January). Artificial intelligence in drug discovery and development. *Drug Discovery Today*, 26(1), 80–93. DOI: 10.1016/j.drudis.2020.10.010 PMID: 33099022

Pence, H. E., and Williams, A. (2010). ChemSpider: an online chemical information resource.

PerkinElmer. (n.d.). Operetta CLS High-Content Analysis System. Retrieved from PerkinElmer website

Pinu, F. R., Beale, D. J., Paten, A. M., Kouremenos, K., Swarup, S., Schirra, H. J., & Wishart, D. (2019). Systems biology and multi-omics integration: Viewpoints from the metabolomics research community. *Metabolites*, 9(4), 76. DOI: 10.3390/metabo9040076 PMID: 31003499

Ramezankhani, R., Solhi, R., Chai, Y. C., Vosough, M., & Verfaillie, C. (2022, April). Organoid and microfluidics-based platforms for drug screening in COVID-19. *Drug Discovery Today*, 27(4), 1062–1076. DOI: 10.1016/j.drudis.2021.12.014 PMID: 34954328

Rehman, A. U., Li, M., Wu, B., Ali, Y., Rasheed, S., Shaheen, S., Liu, X., Luo, R., & Zhang, J. (2024). Role of Artificial Intelligence in Revolutionizing Drug Discovery. *Fundamental Research (Beijing)*. Advance online publication. DOI: 10.1016/j.fmre.2024.04.021

Sackmann, E. K., Fulton, A. L., & Beebe, D. J. (2014). The present and future role of microfluidics in biomedical research. *Nature*, 507(7491), 181–189. DOI: 10.1038/nature13118 PMID: 24622198

Saleh, S. H., Ismail, R., Ibrahim, Z., & Hussin, N. (2018). Issues, challenges and solutions of big data in information management: An overview. *International Journal of Academic Research in Business & Social Sciences*, 8(12), 1382–1393.

Salo-Ahen, O. M., Alanko, I., Bhadane, R., Bonvin, A. M., Honorato, R. V., Hossain, S., Juffer, A. H., Kabedev, A., Lahtela-Kakkonen, M., Larsen, A. S., Lescrinier, E., Marimuthu, P., Mirza, M. U., Mustafa, G., Nunes-Alves, A., Pantsar, T., Saadabadi, A., Singaravelu, K., & Vanmeert, M. (2020). Molecular dynamics simulations in drug discovery and pharmaceutical development. *Processes (Basel, Switzerland)*, 9(1), 71. DOI: 10.3390/pr9010071

Sazonova, E. V., Chesnokov, M. S., Zhivotovsky, B., & Kopeina, G. S. (2022). Drug toxicity assessment: Cell proliferation versus cell death. *Cell Death Discovery*, 8(1), 417. DOI: 10.1038/s41420-022-01207-x PMID: 36241623

Schenone, M., Dančík, V., Wagner, B. K., & Clemons, P. A. (2013). Target identification and mechanism of action in chemical biology and drug discovery. *Nature Chemical Biology*, 9(4), 232–240. DOI: 10.1038/nchembio.1199 PMID: 23508189

Schindelin, J., Arganda-Carreras, I., Frise, E., Kaynig, V., Longair, M., Pietzsch, T., Preibisch, S., Rueden, C., Saalfeld, S., Schmid, B., Tinevez, J.-Y., White, D. J., Hartenstein, V., Eliceiri, K., Tomancak, P., & Cardona, A. (2012). Fiji: An open-source platform for biological-image analysis. *Nature Methods*, 9(7), 676–682. DOI: 10.1038/nmeth.2019 PMID: 22743772

Schwarz, J. M., Rödelsperger, C., Schuelke, M., & Seelow, D. (2010). MutationTaster evaluates disease-causing potential of sequence alterations. *Nature Methods*, 7(8), 575–576. DOI: 10.1038/nmeth0810-575 PMID: 20676075

Shalem, O., Sanjana, N. E., & Zhang, F. (2015, May). High-throughput functional genomics using CRISPR-Cas9. *Nature Reviews. Genetics*, 16(5), 299–311. DOI: 10.1038/nrg3899 PMID: 25854182

Shama, G., & Malik, D. J. (2013). The uses and abuses of rapid bioluminescence-based ATP assays. *International Journal of Hygiene and Environmental Health*, 216(2), 115–125. DOI: 10.1016/j.ijheh.2012.03.009 PMID: 22541898

Shao, F., Lee, P. W., Li, H., Hsieh, K., & Wang, T. H. (2023). Emerging platforms for high-throughput enzymatic bioassays. *Trends in Biotechnology*, 41(1), 120–133. DOI: 10.1016/j.tibtech.2022.06.006 PMID: 35863950

Shendure, J., & Ji, H. (2008). Next-generation DNA sequencing. *Nature Biotechnology*, 26(10), 1135–1145. DOI: 10.1038/nbt1486 PMID: 18846087

Shinwari, Z. K., Tanveer, F., & Khalil, A. T. (2018). Ethical Issues Regarding CRISPR Mediated Genome Editing. *Current Issues in Molecular Biology*, 26, 103–110. DOI: 10.21775/cimb.026.103 PMID: 28879860

Sia, S. K., & Whitesides, G. M. (2003). Microfluidic devices fabricated in poly (dimethylsiloxane) for biological studies. *Electrophoresis*, 24(21), 3563–3576. DOI: 10.1002/elps.200305584 PMID: 14613181

Smurnyy, Y., Cai, M., Wu, H., McWhinnie, E., Tallarico, J. A., Yang, Y., & Feng, Y. (2014). DNA sequencing and CRISPR-Cas9 gene editing for target validation in mammalian cells. *Nature Chemical Biology*, 10(8), 623–625. DOI: 10.1038/nchembio.1550 PMID: 24929529

Stephens, Z. D., Lee, S. Y., Faghri, F., Campbell, R. H., Zhai, C., Efron, M. J., Iyer, R., Schatz, M. C., Sinha, S., & Robinson, G. E. (2015). Big data: Astronomical or genomical? *PLoS Biology*, 13(7), e1002195. DOI: 10.1371/journal.pbio.1002195 PMID: 26151137

Swinney, D. C., & Anthony, J. (2011, June 24). How were new medicines discovered? *Nature Reviews. Drug Discovery*, 10(7), 507–519. DOI: 10.1038/nrd3480 PMID: 21701501

Swinney, D. C., & Lee, J. A. (2020). Recent advances in phenotypic drug discovery. *F1000 Research*, •••, 9. PMID: 32850117

Szymański, P., Markowicz, M., & Mikiciuk-Olasik, E. (2011). Adaptation of high-throughput screening in drug discovery—Toxicological screening tests. *International Journal of Molecular Sciences*, 13(1), 427–452. DOI: 10.3390/ijms13010427 PMID: 22312262

Taylor, C. J., Pomberger, A., Felton, K. C., Grainger, R., Barecka, M., Chamberlain, T. W., Bourne, R. A., Johnson, C. N., & Lapkin, A. A. (2023). A brief introduction to chemical reaction optimization. *Chemical Reviews*, 123(6), 3089–3126. DOI: 10.1021/acs.chemrev.2c00798 PMID: 36820880

Thermo Fisher Scientific. (n.d.). [High-Content Screening] [HCS] [Platform. Retrieved from Thermo Fisher Scientific website]. *Cell Insight*, •••, CX7.

Thorsen, T., Maerkl, S. J., & Quake, S. R. (2002). Microfluidic large-scale integration. *Science*, 298(5593), 580–584. DOI: 10.1126/science.1076996 PMID: 12351675

Tomlinson, C. W., & Whiting, A. (2020). The development of methodologies for high-throughput retinoic acid binding assays in drug discovery and beyond. In *Methods in Enzymology* (Vol. 637, pp. 539–560). Academic Press.

Tripathi, P., Singh, J., Lal, J. A., & Tripathi, V. (2019). Next-Generation Sequencing: An Emerging Tool for Drug Designing. *Current Pharmaceutical Design*, 25(31), 3350–3357. DOI: 10.2174/1381612825666190911155508 PMID: 31544713

Tsurusawa, N., Chang, J., Namba, M., Makioka, D., Yamura, S., Iha, K., Kyosei, Y., Watabe, S., Yoshimura, T., & Ito, E. (2021). Modified ELISA for ultrasensitive diagnosis. *Journal of Clinical Medicine*, 10(21), 5197. DOI: 10.3390/jcm10215197 PMID: 34768717

Tu, R., Zhang, Y., Hua, E., Bai, L., Huang, H., Yun, K., & Wang, M. (2021). Droplet-based microfluidic platform for high-throughput screening of Streptomyces. *Communications Biology*, 4(1), 647. DOI: 10.1038/s42003-021-02186-y PMID: 34059751

Uhlén, M., Fagerberg, L., Hallström, B. M., Lindskog, C., Oksvold, P., Mardinoglu, A., Sivertsson, Å., Kampf, C., Sjöstedt, E., Asplund, A., Olsson, I. M., Edlund, K., Lundberg, E., Navani, S., Szigyarto, C. A.-K., Odeberg, J., Djureinovic, D., Takanen, J. O., Hober, S., & Pontén, F. (2015). Tissue-based map of the human proteome. *Science*, 347(6220), 1260419. DOI: 10.1126/science.1260419 PMID: 25613900

Villeneuve, D. L., Coady, K., Escher, B. I., Mihaich, E., Murphy, C. A., Schlekat, T., & Garcia-Reyero, N. (2019). High-throughput screening and environmental risk assessment: State of the science and emerging applications. *Environmental Toxicology and Chemistry*, 38(1), 12–26. DOI: 10.1002/etc.4315 PMID: 30570782

Vincent, F., Nueda, A., Lee, J., Schenone, M., Prunotto, M., & Mercola, M. (2022). Phenotypic drug discovery: Recent successes, lessons learned and new directions. *Nature Reviews. Drug Discovery*, 21(12), 899–914. DOI: 10.1038/s41573-022-00472-w PMID: 35637317

Wang, J., Wu, M. Y., Tan, J. Q., Li, M., & Lu, J. H. (2019, February 28). High content screening for drug discovery from traditional Chinese medicine. *Chinese Medicine*, 14(1), 5. DOI: 10.1186/s13020-019-0228-y PMID: 30858873

Wang, T., Wei, J. J., Sabatini, D. M., & Lander, E. S. (2014). Genetic screens in human cells using the CRISPR-Cas9 system. *Science*, 343(6166), 80–84. DOI: 10.1126/science.1246981 PMID: 24336569

Weibel, D. B., & Whitesides, G. M. (2006). Applications of microfluidics in chemical biology. *Current Opinion in Chemical Biology*, 10(6), 584–591. DOI: 10.1016/j.cbpa.2006.10.016 PMID: 17056296

Whitesides, G. M. (2006). The origins and the future of microfluidics. nature, 442(7101), 368-373.

Yadav, S., Singh, A., Singhal, R., & Yadav, J. P. (2024). *Revolutionizing drug discovery: The impact of artificial intelligence on advancements in pharmacology and the pharmaceutical industry*. Intelligent Pharmacy.

Yu, H. B., Li, M., Wang, W. P., & Wang, X. L. (2016). High throughput screening technologies for ion channels. *Acta Pharmacologica Sinica*, 37(1), 34–43. DOI: 10.1038/aps.2015.108 PMID: 26657056

Zhang, J. H., Chung, T. D., & Oldenburg, K. R. (1999). A simple statistical parameter for use in evaluation and validation of high throughput screening assays. *Journal of Biomolecular Screening*, 4(2), 67–73. DOI: 10.1177/108705719900400206 PMID: 10838414

Zhang, Y., Liu, B., & Tong, Z. (2022). Adenosine triphosphate (ATP) bioluminescence-based strategies for monitoring atmospheric bioaerosols. *Journal of the Air & Waste Management Association*, 72(12), 1327–1340. DOI: 10.1080/10962247.2022.2101566 PMID: 36226866

Zhou, Y., Zhu, S., Cai, C., Yuan, P., Li, C., Huang, Y., & Wei, W. (2014). High-throughput screening of a CRISPR/Cas9 library for functional genomics in human cells. *Nature*, 509(7501), 487–491. DOI: 10.1038/nature13166 PMID: 24717434

Zock, J. M. (2009, November). Applications of high content screening in life science research. *Combinatorial Chemistry & High Throughput Screening*, 12(9), 870–876. DOI: 10.2174/138620709789383277 PMID: 19938341

Chapter 3

Ethical Legal and Social Implications of Genomics-Driven Drug Discovery

Azamat Ali
https://orcid.org/0000-0003-2931-2166
Lovely Professional University, Punjab, India

ABSTRACT

Genomics-driven drug discovery in precision medicine tailors treatments to individuals' genetic profiles, yet raises complex ethical, legal, and social issues (ELSI). Concerns include patient privacy, consent, and genetic discrimination. Legal challenges involve intellectual property, data ownership, and regulatory frameworks. Socially, attitudes toward genetic testing and access to technology influence perception. Addressing these ELSI requires collaboration across stakeholders to establish robust governance systems ensuring individual rights, fair access, and trust in genomic research and healthcare. This chapter explores these implications and offers strategies for navigating ELSI in genomics-driven drug discovery.

1. INTRODUCTION

Genomics-driven drug discovery represents a paradigm shift in medicine, leveraging advancements in genomics to develop personalized treatments tailored to individuals' genetic profiles (Kulkarni et al., 2023). Traditional drug discovery methods often relied on trial and error, resulting in medications that may be effective for some but not for others due to genetic variations. However, with the advent of

DOI: 10.4018/979-8-3693-6597-7.ch003

genomics, researchers can now identify genetic markers associated with specific diseases or drug responses, enabling the development of targeted therapies with higher efficacy and fewer adverse effects (Keefe & Bateman, 2012).

The completion of the Human Genome Project in 2003 was a major achievement in genomics research since it involved the sequencing of the whole human genome. Since then, rapid advancements in sequencing technologies, coupled with decreasing costs, have facilitated the widespread adoption of genomic approaches in healthcare. Today, genomics is integrated into various stages of drug discovery, from identifying therapeutic targets and understanding disease mechanisms to predicting drug responses and assessing treatment outcomes (Vamathevan et al., 2019).

While genomics-driven drug discovery holds great promise for improving patient outcomes, it also raises complex ethical, legal, and social issues that must be carefully addressed. Ethically, concerns arise regarding patient privacy, informed consent, and the potential for genetic discrimination. As genomic data becomes increasingly accessible, ensuring the confidentiality and security of patients' genetic information is paramount to maintaining trust in healthcare systems (Yakubu & Chen, 2020).

Legally, questions surrounding intellectual property rights, data ownership, and regulatory frameworks pose challenges to the equitable distribution and utilization of genomic data and technologies. Balancing the need for innovation with the protection of individual rights and societal interests requires robust legal mechanisms that promote transparency, accountability, and fair access to genomic advancements (Trump et al., 2023).

Socially, attitudes toward genetic testing and manipulation vary widely among individuals and communities, influenced by cultural, religious, and ethical beliefs (Robillard, et al., 2014). Disparities in access to genomic technologies further exacerbate existing inequalities in healthcare, raising concerns about social justice and the widening gap between those who can afford genomic services and those who cannot. Moreover, the potential implications of genetic enhancement and designer drugs evoke ethical dilemmas about the boundaries of human intervention in biology and evolution (Green, 2007). This study investigates the multifaceted ethical, legal, and social consequences of using genomics to discover new drugs. Its goal is to offer a thorough comprehension of the difficulties and possibilities that come with this developing field.

2. ETHICAL IMPLICATIONS

Ethical considerations are paramount in genomics research and treatment due to the sensitive nature of genetic information and the potential impact on individuals and society as a whole (Mezinska et al., 2021). This section explores various ethical implications, including patient privacy concerns, informed consent, risks of genetic discrimination, and ethical considerations in genetic manipulation and testing, with illustrative examples.

2.1 Patient Privacy Concerns

Patient privacy concerns in healthcare are paramount due to the sensitive nature of medical information and the potential consequences of unauthorized access or misuse. Several real case studies illustrate the ethical implications of patient privacy concerns:

Case 1: UCLA Health Data Breach (2015)

In 2015, UCLA Health experienced a significant data breach that compromised the personal and medical information of approximately 4.5 million patients (John et al., 2018). The breach transpired when cybercriminals successfully infiltrated UCLA's network, so enabling them to illicitly get patient records, including personal information such as names, dates of birth, social security numbers, and medical histories. This incident emphasized the susceptibility of healthcare data to intrusions and emphasized the necessity for strong security measures to safeguard patient privacy.

Case 2: The WannaCry Ransomware Attack (2017)

The WannaCry ransomware assault, which took place in 2017, was aimed at global companies, including healthcare institutions, by encrypting their data and demanding ransom payments in exchange for decryption keys (Branch et al., 2019). The attack affected numerous healthcare facilities, disrupting operations and compromising patient care. This incident demonstrated the potential impact of cyberattacks on patient privacy and highlighted the importance of cybersecurity measures to safeguard sensitive medical information.

Case 3: The Theranos Scandal (2015-2016)

Theranos, a firm in Silicon Valley established by Elizabeth Holmes, said that they had created ground-breaking technology for blood testing. This technology could conduct a diverse array of tests with only a tiny amount of blood (Tourish & Willmott, 2023). Nevertheless, examinations unveiled that the technology was defective and yielded imprecise outcomes. Moreover, Theranos faced allegations of deceiving investors, regulators, and patients over the capabilities of their technology, which gave rise to issues regarding patient confidentiality and the dependability of medical testing.

These case studies emphasize the critical importance of patient privacy in healthcare and highlight the need for stringent security measures, transparent practices, and regulatory oversight to protect sensitive medical information.

2.2 Informed Consent in Genomic Research and Treatment

Obtaining informed permission is crucial in genomic research and therapy to guarantee that individuals comprehend the dangers, advantages, and consequences of genetic testing and actions. Case studies demonstrate the ethical considerations and challenges surrounding informed consent in genomic research and treatment:

Case 1: The Havasupai Tribe Lawsuit (2004)

In 2004, the Havasupai Tribe in Arizona initiated legal proceedings against Arizona State University (ASU), claiming that researchers had infringed upon their rights by utilizing blood samples originally collected for diabetes research to investigate genetic elements associated with schizophrenia and inbreeding (Raman et al., 2023). The tribe claimed that they had not been adequately informed about the broader use of their genetic material and that their cultural beliefs regarding genetic information had been disregarded. The case raised questions about informed consent, cultural sensitivity, and the rights of research participants in genomic research.

Case 2: The Henrietta Lacks Case (1951-present)

Henrietta Lacks, an African American lady, received medical treatment for cervical cancer in 1951, during which a sample of her cancerous cells was obtained without her informed permission (Gold, 1985). HeLa cells, sometimes referred to as these cells, have emerged as a crucial instrument in medical research, playing a significant role in various scientific advancements. Nevertheless, Lacks and her family were not notified about the utilization of her cells for scientific intentions until several years afterwards. The case ignited discussions around informed consent, patient autonomy, and the rights of study subjects, resulting in modifications to research ethics and procedures.

Case 3: The Direct-to-Consumer Genetic Testing Industry

The emergence of direct-to-consumer genetic testing firms, such 23andMe and AncestryDNA, has sparked apprehensions regarding informed permission and the elucidation of genetic data (Saha et al., 2020). These companies offer consumers the opportunity to learn about their ancestry, genetic traits, and predispositions to certain health conditions. Nevertheless, critics contend that consumers may lack a comprehensive understanding of the constraints and potential ramifications of genetic testing, resulting in misconceptions or excessive distress regarding their health vulnerabilities.

These case studies highlight the complexities of obtaining informed consent in genomic research and treatment and underscore the importance of transparency, cultural sensitivity, and ongoing communication between researchers, healthcare providers, and research participants. To uphold ethical norms and retain faith in the healthcare system, it is important to respect individuals' autonomy and provide them with the necessary information to make informed decisions regarding their involvement in genetic research and treatment.

2.3 Risks of Genetic Discrimination

Genetic discrimination refers to the unjust treatment of persons based on their genetic information, resulting in adverse outcomes such as being denied work, insurance coverage, or other possibilities. Several case studies illustrate the risks and implications of genetic discrimination:

Case 1: Gina vs. the GINA

In the landmark case of "Gina vs. the GINA," a woman named Gina Galli challenged her employer's decision to fire her after learning about her genetic predisposition to develop Huntington's disease. The case served as the impetus for the enactment of the Genetic Information Non-discrimination Act (GINA) in 2008. GINA specifically forbids discrimination in both employment and health insurance on the basis of genetic information. Gina's story underscores the importance of legal protections to prevent genetic discrimination and protect individuals' privacy and rights (Slaughter, 2013).

Case 2: The Burlington Northern Santa Fe Railway Company

In 2010, the Equal Employment Opportunity Commission (EEOC) initiated legal proceedings against the Burlington Northern Santa Fe Railway Company on behalf of two employees who were purportedly subjected to genetic testing without their explicit permission. The lawsuit alleged that the company violated the Americans with Disabilities Act (ADA) and GINA by requesting family medical history and subjecting employees to genetic testing as part of its wellness program. This case highlighted the potential risks of employer-mandated genetic testing and the importance of informed consent and privacy protections (Valenti & Burke, 2010).

Case 3: The Case of Myriad Genetics and BRCA Gene Testing

Myriad Genetics, a biotechnology business, has exclusive patents on the BRCA1 and BRCA2 genes, which are linked to hereditary breast and ovarian cancer. This monopoly allowed Myriad to control the cost and accessibility of BRCA gene testing, raising concerns about access to genetic testing and the potential for genetic discrimination. Critics argued that Myriad's patents hindered research and innovation in genetic testing and limited patients' options for testing and treatment. The Supreme Court ultimately ruled in 2013 that isolated DNA sequences are not eligible for patent protection, opening the door to increased competition and accessibility in genetic testing (Azvolinsky, 2013).

These case studies demonstrate the real-world consequences of genetic discrimination and the importance of legal protections and ethical considerations to safeguard individuals' privacy, autonomy, and rights.

2.4 Ethical Considerations in Genetic Manipulation and Testing

The ethical questions surrounding genetic alteration and testing involve several intricate matters, such as the possibility of unforeseen outcomes, fair availability of genetic technology, and the moral ramifications of modifying the human genome. Several real case studies shed light on these ethical dilemmas:

Case 1: The CRISPR Babies Controversy

In 2018, Chinese scientist He Jiankui made a public declaration that he had utilized the CRISPR-Cas9 gene-editing technique to alter the genetic makeup of human embryos, leading to the birth of twin girls with modified genomes. The announcement sparked international outrage and raised ethical concerns about the implications of gene editing for future generations. Critics argued that the experiment violated ethical principles, including informed consent, safety, and the precautionary principle. The incident prompted calls for stricter regulations and oversight of gene-editing research to prevent unethical practices and ensure responsible use of genetic technologies (Greely & babies, 2019).

Case 2: The Case of Henrietta Lacks and HeLa Cells

Henrietta Lacks' case also highlights ethical considerations in genetic research and testing. HeLa cells, derived from Lacks' cells without her consent, have been extensively utilized in scientific studies for many years. Although HeLa cells have made significant contributions to many medical advancements, such as the creation of vaccinations and cancer therapies, Lacks and her family were not sufficiently informed about the utilization of her cells or appropriately paid for their efforts. This case raises questions about informed consent, patient autonomy, and the equitable distribution of benefits and risks in genetic research (Stump, 2014).

Case 3: The Genetic Information Non-discrimination Act (GINA)

The Genetic Information Non-discrimination Act (GINA), enacted in 2008, forbids employment and health insurance discrimination on the basis of genetic information. The legislation signifies a notable advancement in safeguarding the privacy and rights of persons in the realm of genetic testing and research. Nevertheless, GINA is not comprehensive enough to encompass all types of genetic discrimination, including discrimination in life insurance, long-term care insurance, or disability insurance. In addition, the implementation of GINA has proven to be difficult, and individuals may still encounter prejudice or social disapproval due to their genetic data (Feldman, 2012).

These case studies illustrate the ethical complexities and considerations inherent in genetic manipulation and testing, highlighting the need for careful deliberation, transparency, and adherence to ethical principles in the development and application of genetic technologies. By addressing these ethical concerns thoughtfully and responsibly, researchers, policymakers, and healthcare providers can navigate the ethical challenges posed by genetic manipulation and testing and ensure that these technologies are used for the benefit of society while respecting individuals' rights and dignity.

3. LEGAL CONSIDERATIONS

Legal considerations in genomics research encompass intellectual property rights on genes and technologies, data ownership and sharing regulations, and regulatory challenges in drug development. Addressing these issues is crucial for promoting innovation while ensuring ethical and legal compliance in genomics-driven endeavours.

3.1 Intellectual Property Rights

At the heart of genomics research lie the genes – the fundamental units of heredity and the blueprint for life. With the advent of technologies like CRISPR-Cas9, gene editing has become more accessible, raising significant questions about intellectual property rights. The race to patent genes and gene-editing technologies has been fierce, often leading to disputes over ownership and licensing agreements (Akram et al., 2022). An exemplary legal issue involves the contention surrounding the BRCA genes, which are linked to breast and ovarian cancer. Myriad Genetics, after isolating the BRCA genes, patented them, sparking debates about gene patenting and access to genetic testing (Azvolinsky, 2013).

While patents can incentivize innovation by providing exclusivity and financial rewards, they can also impede research and limit access to crucial genetic information. Balancing the need for innovation with ensuring equitable access to genetic data is essential for the advancement of genomics research. Initiatives like open-access databases and patent pools aim to address these concerns by fostering collaboration and ensuring widespread access to genetic information.

3.2 Data Ownership and Sharing Regulations

In the era of big data, genomics research generates vast amounts of genetic and clinical data. These data hold immense potential for advancing scientific knowledge and developing personalized treatments. However, they also raise concerns about privacy, consent, and data ownership. Genomic data, often containing sensitive information about individuals' health and ancestry, must be handled with utmost care to protect privacy and confidentiality.

Data ownership and sharing regulations vary across jurisdictions, adding complexity to genomic research endeavours. The Health Insurance Portability and Accountability Act (HIPAA) establish guidelines in the United States to safeguard health information, which includes genetic data. Researchers must adhere to strict protocols for obtaining informed consent and de-identifying data to ensure compliance with HIPAA regulations (Oakley, 2023). The General Data Protection Regulation (GDPR) in Europe enforces strict criteria for the handling and transmission of personal data, which includes genetic data. Researchers must navigate complex consent mechanisms and data protection protocols to ensure compliance with GDPR standards (Dove, 2018).

Moreover, international collaborations in genomics research further complicate data sharing, as researchers must reconcile disparate regulatory frameworks and cultural norms regarding data privacy and sharing. Establishing clear guidelines and frameworks for data ownership, consent, and sharing is essential for fostering collaboration and advancing genomics research on a global scale.

3.3 Regulatory Challenges in Drug Development

Translating genomic discoveries into therapeutic interventions poses regulatory challenges, particularly in the realm of drug development. Traditional regulatory pathways may not adequately accommodate the complexities of genomic medicine, necessitating innovative approaches to regulatory oversight.

The traditional model of drug development relies on randomized controlled trials (RCTs) to demonstrate safety and efficacy. However, genomic medicine often involves targeted therapies tailored to individuals' genetic profiles, challenging the one-size-fits-all approach of RCTs. Adaptive clinical trial designs, which enable dynamic modifications based on accumulated data, provide a more versatile framework for assessing tailored medicines (Eichler et al., 2021).

Furthermore, regulatory agencies must grapple with the rapid pace of technological advancement in genomics research, ensuring that regulatory frameworks remain agile and responsive to emerging innovations. Collaborative efforts between researchers, regulators, and industry stakeholders are crucial for developing regulatory frameworks that promote innovation while safeguarding patient safety and efficacy.

4. SOCIAL IMPACT

Genomics, the study of genes and their functions, stands as a monumental achievement in scientific inquiry, revolutionizing our understanding of biology and medicine. This field has unraveled the intricacies of genetic makeup, paving the way for personalized healthcare and targeted treatments. However, the widespread adoption of genomic technologies has not only brought about remarkable advancements but has also ushered in significant social implications that reverberate across society. These implications encompass public attitudes toward genetic testing and manipulation, access disparities in genomic technologies, and the societal impact of genomics-driven drug discovery.

4.1Public Attitudes Toward Genetic Testing and Manipulation

It is reflect a complex interplay of ethical considerations, privacy concerns, and societal norms. The emergence of direct-to-consumer (DTC) genetic testing companies provides a striking example of shifting public perceptions in this domain. Companies like 23andMe offer individuals the opportunity to explore their genetic ancestry, health traits, and potential disease risks through simple saliva tests. Initially met with skepticism, these services have gained traction, reflecting a growing interest among the public in accessing genetic information directly, outside traditional healthcare channels (Saukko, 2018). However, concerns persist regarding the accuracy of results, the potential misuse of genetic data, and the psychological impact of unexpected or distressing genetic findings.

The case of CRISPR-Cas9 gene editing technology further underscores the ethical dilemmas inherent in genetic manipulation. CRISPR-Cas9, hailed for its precision and versatility, holds immense potential for treating genetic disorders and advancing scientific knowledge (Foulkes et al., 2019). However, the ethical boundaries of germline editing have come under scrutiny, particularly following the controversial 2018 case involving Chinese scientist He Jiankui, who claimed to have edited the genes of human embryos to confer resistance to HIV (Greely, 2019). This case ignited global debate and raised profound questions about the ethical implications of altering the human germline. It underscores the urgent need for clear guidelines and ethical frameworks to govern genetic manipulation, ensuring that scientific advancements align with societal values and prioritize safety, equity, and informed consent.

4.2 Access Disparities in Genomic Technologies

Access disparities in genomic technologies pose significant challenges to equitable healthcare delivery, exacerbating existing inequalities in healthcare access and outcomes. Genetic counselling, an essential aspect of genomic medicine, offers individuals and families valuable information, assistance, and direction concerning genetic conditions, risk factors, and choices for testing. However, access to genetic counselling services is often limited in rural and underserved areas, where healthcare infrastructure is sparse and resources are scarce. Initiatives like the Mississippi Genomics Education Initiative aim to address these disparities by educating healthcare providers in rural areas about the application of genomics in clinical practice, but challenges remain in ensuring universal access to genomic services and support (Dixon Omere, 2022).

Similarly, disparities in access to pharmacogenomic testing and tailored treatments disproportionately affect minority communities, exacerbating health inequities. Pharmacogenomics has potential for improving the effectiveness of medications and reducing negative side effects. However, there is a notable disparity in the rates of pharmacogenomic testing between African American and Hispanic patients compared to White patients. These disparities reflect systemic barriers, including lack of awareness, cultural mistrust, and limited access to healthcare resources (Davis & Limdi, 2021). Bridging the gap in access to genomic technologies requires targeted interventions, community engagement, and policy initiatives aimed at addressing the root causes of disparities and ensuring equitable access to innovative treatments.

4.3 Societal Implications of Genomics-Driven Drug Discovery

Genomics-driven drug discovery has transformed the landscape of medicine, enabling the development of targeted therapies tailored to individuals' genetic profiles. Instances like Herceptin (trastuzumab), a specific medication for HER2-positive breast cancer, serve as prime examples of how precision medicine techniques may have a significant influence on society (Brandão et al., 2018). Before the advent of Herceptin, HER2-positive breast cancer was associated with poor prognosis and limited treatment options. Clinical trials demonstrated that Herceptin significantly improves survival rates and reduces the risk of disease recurrence in patients with this subtype of breast cancer. However, the high cost of targeted therapies raises concerns about affordability and accessibility, particularly for socioeconomically disadvantaged patients (Dzau & Ginsburg, 2016). Access barriers perpetuate disparities in cancer care and exacerbate health inequalities, highlighting the need for equitable access to innovative treatments and interventions.

5. ADDRESSING ELSI CHALLENGES

Ethical, Legal, and Social Implications (ELSI) are integral considerations in the advancement of genomic research and its applications in healthcare. Addressing these challenges requires collaborative governance frameworks, protection of individual rights, fair access to genomic services, and building trust among stakeholders. Real case studies provide valuable insights into how these challenges can be addressed:

5.1 Collaborative Governance Frameworks

- ***The Human Genome Project (HGP)***: The HGP, launched in 1990, aimed to sequence and map the entire human genome. The Human Genome Project (HGP) created the Ethical, Legal, and Social Implications (ELSI) program to specifically tackle the ethical, legal, and social consequences associated with genomic research. The ELSI program funded research projects that explored the ethical, legal, and social implications of genomics and facilitated dialogue among scientists, policymakers, and the public. By fostering collaboration and interdisciplinary engagement, the ELSI program contributed to the development of policies and guidelines that shaped the responsible conduct of genomic research (Olson, 1993).

- ***The Global Alliance for Genomics and Health (GA4GH)***: The Global Alliance for Genomics and Health (GA4GH) is a global alliance consisting of researchers, healthcare providers, and policymakers. Its primary objective is to advocate for responsible sharing of genetic research data. The alliance develops frameworks, standards, and best practices for data sharing and collaborates with stakeholders to address ethical, legal, and social challenges. For example, the GA4GH has developed the Beacon Project, which allows researchers to share summary information about genomic data while protecting individual privacy. By promoting collaborative governance and data sharing, the GA4GH aims to accelerate genomic research and improve healthcare outcomes globally (Knoppers, 2014).

5.2 Protecting Individual Rights and Ensuring Fair Access

- ***The Genetic Information Non-Discrimination Act (GINA)***: The Genetic Information Nondiscrimination Act (GINA), enacted in 2008, forbids discrimination in employment and health insurance on the basis of genetic information. Employers and insurers are prohibited by law from utilizing genetic information to influence judgments on employment or coverage, thereby ensuring essential safeguards for people. GINA explicitly forbids group health plans and issuers from soliciting or mandating genetic testing as a prerequisite for membership or coverage. By safeguarding individual rights and privacy, GINA promotes trust and confidence in genomic research and healthcare services (Clayton et al., 2019).
- ***The Affordable Care Act (ACA)***: The Affordable Care Act (ACA), which was passed in 2010, broadened the availability of healthcare treatments and forbade insurance companies from refusing coverage or imposing higher rates based on pre-existing diseases, including genetic disorders. The ACA's provisions ensure that individuals have access to genetic testing, counselling, and preventive services without facing discrimination or financial barriers. By promoting equitable access to healthcare, the ACA addresses disparities in access to genomic services and supports the integration of genomics into routine clinical care (French et al., 2016).

5.3 Building Trust in Genomic Research and Healthcare Services

- ***The All of Us Research Program***: The All of Us Research Program, initiated by the National Institutes of Health (NIH) in 2018, seeks to construct an extensive and varied collection of data for biomedical research, with the goal of creating one of the largest databases in this field (Mapes et al., 2020). The program invites one million or more people from across the United States to contribute their health information, including genetic data, to advance precision medicine. To build trust and engagement, the All of Us program employs community engagement strategies, including partnerships with community organizations, educational outreach, and transparent communication about data use and privacy protections. By involving participants as partners in research, the program fosters trust and collaboration and promotes the responsible use of genomic data for health research (Sankar & Parker, 2017).
- ***The UK Biobank***: The UK Biobank is an extensive biological database that gathers health and genetic information from 500,000 individuals in the United Kingdom. The biobank operates under strict ethical and governance frameworks to protect participant privacy and ensure the responsible use of data (Bycroft et al., 2017). Participants are notified about the utilization of their data, possess the privilege to retract their agreement at any given moment, and are regularly furnished with information on study discoveries. By prioritizing transparency, participant engagement, and accountability, the UK Biobank has built trust among participants and researchers and has become a valuable resource for advancing biomedical research and healthcare (Bycroft et al., 2018).

Addressing the ethical, legal, and social implications of genomics requires collaborative governance frameworks, protection of individual rights, and building trust among stakeholders. Real case studies demonstrate how initiatives such as the HGP, GA4GH, GINA, ACA, All of Us Research Program, and UK Biobank have contributed to advancing responsible genomic research and promoting equitable access to genomic services. By learning from these examples and continuing to prioritize ethical considerations, stakeholders can navigate the complexities of genomics in ways that promote the public good and uphold individual rights and dignity.

6. CONCLUSION

In conclusion, the exploration of ethical, legal, and social implications (ELSI) in genomics-driven drug discovery has revealed several key challenges. These include concerns surrounding patient privacy, informed consent, genetic discrimination, access disparities, and societal implications. Addressing these challenges is crucial for advancing precision medicine, as it ensures the protection of individual rights, fosters trust and confidence in genomic research and healthcare services, promotes health equity, and accelerates the translation of genomic discoveries into clinical applications. Moving forward, future research and policy development should focus on enhancing data governance, strengthening legal protections, fostering collaboration and partnership among stakeholders, promoting ethical conduct, and investing in education and outreach. By incorporating these recommendations, we may effectively negotiate the intricacies of genomics-based drug development and utilize its capacity to transform healthcare and enhance patient outcomes worldwide.

REFERENCES

Akram, F., Haq, I., Sahreen, S., Nasir, N., Naseem, W., Imitaz, M., & Aqeel, A. (2022). CRISPR/Cas9: A revolutionary genome editing tool for human cancers treatment. *Technology in Cancer Research & Treatment*, 21, 15330338221132078. Advance online publication. DOI: 10.1177/15330338221132078 PMID: 36254536

Azvolinsky, A. (2013). Supreme Court Ruling Broadens BRCA Testing Options. *Journal of the National Cancer Institute*, 105(22), 1671–1672. DOI: 10.1093/jnci/djt342 PMID: 24198329

Branch, L. E., Eller, W. S., Bias, T. K., McCawley, M. A., Myers, D. J., Gerber, B. J., & Bassler, J. R. (2019). Trends in Malware Attacks against United States Healthcare Organizations, 2016-2017. *Global Biosecurity*, 1(1), 15. Advance online publication. DOI: 10.31646/gbio.7

Brandão, M., Pondé, N. F., Poggio, F., Kotecki, N., Salis, M., Lambertini, M., & de Azambuja, E. (2018). Combination therapies for the treatment of HER2-positive breast cancer: Current and future prospects. *Expert Review of Anticancer Therapy*, 18(7), 629–649. DOI: 10.1080/14737140.2018.1477596 PMID: 29781317

Bycroft, C., Freeman, C., Petkova, D., Band, G., Elliott, L. T., Sharp, K., & Marchini, J. (2017). Genome-wide genetic data on~ 500,000 UK Biobank participants. *BioRxiv,* 166298. DOI: 10.1101/166298

Bycroft, C., Freeman, C., Petkova, D., Band, G., Elliott, L. T., Sharp, K., Motyer, A., Vukcevic, D., Delaneau, O., O'Connell, J., Cortes, A., Welsh, S., Young, A., Effingham, M., McVean, G., Leslie, S., Allen, N., Donnelly, P., & Marchini, J. (2018). The UK Biobank resource with deep phenotyping and genomic data. *Nature*, 562(7726), 203–209. DOI: 10.1038/s41586-018-0579-z PMID: 30305743

Clayton, E. W., Evans, B. J., Hazel, J. W., & Rothstein, M. A. (2019). The law of genetic privacy: Applications, implications, and limitations. *Journal of Law and the Biosciences*, 6(1), 1–36. DOI: 10.1093/jlb/lsz007 PMID: 31666963

Davis, B. H., & Limdi, N. A. (2021). Translational pharmacogenomics: Discovery, evidence synthesis and delivery of race-conscious medicine. *Clinical Pharmacology and Therapeutics*, 110(4), 909–925. DOI: 10.1002/cpt.2357 PMID: 34233023

Dixon Omere, Z. O., & King, J. (2022). America's History of Health Disparities: How Does this Affect the Future of Healthcare In Mississippi? Retrieved from https://egrove.olemiss.edu/hon_thesis/2708

Dove, E. S. (2018). The EU General Data Protection Regulation: Implications for International Scientific Research in the Digital Era. *The Journal of Law, Medicine & Ethics*, 46(4), 1013–1030. DOI: 10.1177/1073110518822003

Dzau, V. J., & Ginsburg, G. S. (2016). Realizing the Full Potential of Precision Medicine in Health and Health Care. *Journal of the American Medical Association*, 316(16), 1659–1660. DOI: 10.1001/jama.2016.14117 PMID: 27669484

Eichler, H. G., Pignatti, F., Schwarzer-Daum, B., Hidalgo-Simon, A., Eichler, I., Arlett, P., Humphreys, A., Vamvakas, S., Brun, N., & Rasi, G. (2021). Randomized controlled trials versus real world evidence: Neither magic nor myth. *Clinical Pharmacology and Therapeutics*, 109(5), 1212–1218. DOI: 10.1002/cpt.2083 PMID: 33063841

Feldman, E. A. (2012). The Genetic Information Non-discrimination Act (GINA): Public Policy and Medical Practice in the Age of Personalized Medicine. *Journal of General Internal Medicine*, 27(6), 743–746. DOI: 10.1007/s11606-012-1988-6 PMID: 22314637

Foulkes, A. L., Soda, T., Farrell, M., Giusti-Rodríguez, P., & Lázaro-Muñoz, G. (2019). Legal and ethical implications of CRISPR applications in psychiatry. *North Carolina Law Review*, 97(5), 1359. PMID: 31871365

French, M. T., Homer, J., Gumus, G., & Hickling, L. (2016). Key Provisions of the Patient Protection and Affordable Care Act (ACA): A Systematic Review and Presentation of Early Research Findings. *Health Services Research*, 51(5), 1735–1771. DOI: 10.1111/1475-6773.12511 PMID: 27265432

Gold, M. (1985). *A Conspiracy of Cells: One woman's immortal legacy-and the medical scandal it caused.* State University of New York Press.

Greely, H. T. (2019). CRISPR'd babies: Human germline genome editing in the 'He Jiankui affair'. *Journal of Law and the Biosciences*, 6(1), 111–183. DOI: 10.1093/jlb/lsz010 PMID: 31666967

Green, R. M. (2007). *Babies by Design: The Ethics of Genetic Choice.* Yale University Press.

John, S., Ravichandran, N., & Khan, M. F. (2018). Electronic medical record for deliverance of effective healthcare delivery: Ethical issues and challenges of digitalization in clinical information and Electronic Medical Records (EMR) management. *IOSR Journal of Business and Management (IOSR-JBM), 20*(3), 01-06. DOI: DOI: 10.9790/487X-2003020106

Keefe, D., & Bateman, E. (2012). Tumor control versus adverse events with targeted anticancer therapies. *Nature Reviews. Clinical Oncology*, 9(2), 98–109. DOI: 10.1038/nrclinonc.2011.192 PMID: 22182972

Knoppers, B. M. (2014). International ethics harmonization and the global alliance for genomics and health. *Genome Medicine*, 6(2), 13. DOI: 10.1186/gm530 PMID: 25031613

Kulkarni, L., Nirmala, M. S., Najma, U., & Jyothi, N. M. (2023). AI-Enabled Genomic Biomarkers: The Future of Pharmaceutical Industry and Personalized Medicine. *Seybold Report Journal*, 18(08), 54–72. DOI: 10.5110/77-1024

Mapes, B. M., Foster, C. S., Kusnoor, S. V., Epelbaum, M. I., AuYoung, M., Jenkins, G., Lopez-Class, M., Richardson-Heron, D., Elmi, A., Surkan, K., Cronin, R. M., Wilkins, C. H., Pérez-Stable, E. J., Dishman, E., Denny, J. C., & Rutter, J. L. (2020). Diversity and inclusion for the All of Us research program: A scoping review. *PLoS One*, 15(7), e0234962. DOI: 10.1371/journal.pone.0234962 PMID: 32609747

Mezinska, S., Gallagher, L., Verbrugge, M., & Bunnik, E. M. (2021). Ethical issues in genomics research on neurodevelopmental disorders: A critical interpretive review. *Human Genomics*, 15(1), 16. DOI: 10.1186/s40246-021-00317-4 PMID: 33712057

Oakley, A. (2023). HIPAA, HIPPA, or HIPPO: What Really Is the Heath Insurance Portability and Accountability Act? *Biotechnology Law Report*, 42(6), 306–318. DOI: 10.1089/blr.2023.29329.aso

Olson, M. V. (1993). The human genome project. *Proceedings of the National Academy of Sciences of the United States of America*, 90(10), 4338–4344. DOI: 10.1073/pnas.90.10.4338 PMID: 8506271

Raman, R., Utts, J., Cohen, A. I., & Hayat, M. J. (2023). Integrating Ethics into the Guidelines for Assessment and Instruction in Statistics Education (GAISE). *The American Statistician*, 77(3), 323–330. DOI: 10.1080/00031305.2022.2156612

Robillard, J. M., Roskams-Edris, D., Kuzeljevic, B., & Illes, J. (2014). Prevailing public perceptions of the ethics of gene therapy. *Human Gene Therapy*, 25(8), 740–746. DOI: 10.1089/hum.2014.030 PMID: 24773182

Saha, D., Chan, A., Stacy, B., Javkar, K., Patkar, S., & Mazurek, M. L. (2020). User Attitudes On Direct-to-Consumer Genetic Testing. In *2020 IEEE European Symposium on Security and Privacy (EuroS&P)* (pp. 120-138). Genoa, Italy: IEEE. DOI: 10.1109/EuroSP48549.2020.00016

Sankar, P., & Parker, L. (2017). The Precision Medicine Initiative's All of Us Research Program: An agenda for research on its ethical, legal, and social issues. *Genetics in Medicine*, 19(7), 743–750. DOI: 10.1038/gim.2016.183 PMID: 27929525

Saukko, P. (2018). Digital health–a new medical cosmology? The case of 23andMe online genetic testing platform. *Sociology of Health & Illness*, 40(8), 1312–1326. DOI: 10.1111/1467-9566.12774 PMID: 29998537

Slaughter, L. (2013). Genetic information non-discrimination act. *Harvard Journal on Legislation*, 50, 41.

Stump, J. L. (2014). Henrietta Lacks and The HeLa Cell: Rights of Patients and Responsibilities of Medical Researchers. *The History Teacher*, 48(1), 127–180. https://www.jstor.org/stable/43264385

Tourish, D., & Willmott, H. (2023). Despotic Leadership and Ideological Manipulation at Theranos: Towards a theory of hegemonic totalism in the workplace. *Organization Studies*, 44(11), 1801–1824. DOI: 10.1177/01708406231171801

Trump, B., Cummings, C., Klasa, K., Galaitsi, S., & Linkov, I. (2023). Governing biotechnology to provide safety and security and address ethical, legal, and social implications. *Frontiers in Genetics*, 13, 1052371. DOI: 10.3389/fgene.2022.1052371 PMID: 36712887

Valenti, A., & Burke, L. A. (2010). Post-Burlington: What Employers and Employees Need to Know About Retaliation. *Employee Responsibilities and Rights Journal*, 22(3), 235–251. DOI: 10.1007/s10672-009-9137-z

Vamathevan, J., Clark, D., Czodrowski, P., Dunham, I., Ferran, E., Lee, G., Li, B., Madabhushi, A., Shah, P., Spitzer, M., & Zhao, S. (2019). Applications of machine learning in drug discovery and development. *Nature Reviews. Drug Discovery*, 18(6), 463–477. DOI: 10.1038/s41573-019-0024-5 PMID: 30976107

Yakubu, A. M., & Chen, Y.-P. P. (2020). Ensuring privacy and security of genomic data and functionalities. *Briefings in Bioinformatics*, 21(2), 511–526. DOI: 10.1093/bib/bbz013 PMID: 30759195

Chapter 4
Genomic Biomarkers Discovery and Validation in Drug Development

B. Vaishnavi
https://orcid.org/0009-0006-9403-0850
Paavai Engineering College, India

ABSTRACT

The discovery and validation of genomic biomarkers are crucial for advancing drug development and personalized medicine. Genomic biomarkers, including DNA and RNA characteristics, help predict therapeutic responses, drug efficacy, and potential adverse effects, enabling precise treatment strategies. This chapter explores the identification and validation of genomic biomarkers, their role in drug development, and the high-throughput sequencing technologies and bioinformatics tools driving their discovery. It examines validation techniques, biomarker integration into drug design and clinical trials, and relevant regulatory frameworks. Through case studies, the chapter illustrates the transformative impact of genomic biomarkers on drug efficacy and patient safety, emphasizing the need to overcome technical, ethical, and legal challenges to optimize patient outcomes and advance medical science.

1. INTRODUCTION

Drug development is greatly aided by genomic biomarkers, which provide insight into the genetic causes of certain diseases and how they respond to treatment. Personalised medicine, which attempts to customise medical care to each patient's unique traits, depends on the identification and validation of these biomarkers. Epigenetic markers, copy number variations (CNVs), and single nucleotide poly-

DOI: 10.4018/979-8-3693-6597-7.ch004

morphisms (SNPs) are examples of genomic biomarkers that offer important insights into the genetic variability that affects treatment outcomes, illness progression, and vulnerability (Collins et al., 2017).

Historically, biomarker detection has progressed from simple blood tests to complicated genomic and proteomic investigations. The area has undergone a revolution with the introduction of high-throughput sequencing technologies and bioinformatics, which have made it possible to conduct an extensive investigation of the human genome and its variations. This progress has speed up the process of finding biomarkers and improved their precision and dependability in predicting illness and response to treatment (Manolio et al., 2019).

The value of genetic biomarkers in modern medicine cannot be emphasised. They facilitate more accurate and successful interventions by acting as prognostic markers, therapeutic response predictors, and diagnostic instruments. The discovery of BRCA1 and BRCA2 mutations has had a substantial influence on the treatment of ovarian and breast malignancies, resulting in targeted treatments that enhance patient outcomes (Chen et al., 2020). Comparably, the therapy landscape for non-small cell lung cancer has changed because to biomarkers such KRAS and EGFR mutations, which direct the use of targeted medicines and raise survival rates (Zhang et al., 2019).

There are many different types of genomic markers, including short tandem repeats (STRs), CNVs, SNPs, mitochondrial DNA markers, and epigenetic markers. Different markers offer different insights into genetic variation and how it affects disease and how well a treatment works. SNPs, the most prevalent kind of genetic variation, have a big impact on how people react to medications. Changes in a gene's copy number, or CNVs, can impact gene expression and influence the development and susceptibility of certain diseases (Feuk et al., 2006).

The discovery of genomic biomarkers involves various approaches, including genome-wide association studies (GWAS), next-generation sequencing (NGS), and bioinformatics tools and techniques. GWAS have been instrumental in identifying genetic variants associated with complex diseases, while NGS technologies have enabled the rapid and cost-effective sequencing of entire genomes. Bioinformatics plays a crucial role in analysing and interpreting the vast amounts of data generated by these technologies, identifying potential biomarkers, and elucidating their functional significance (Visscher et al., 2017).

Validation is an essential phase in the biomarker discovery process because it ensures that selected indicators are reliable, reproducible, and clinically useful. This calls for statistical analysis, clinical trials, and exacting experimental validation techniques. This procedure relies heavily on regulatory rules and standards to ensure that validated biomarkers fit the clinical usage criteria. Integration of genetic biomarkers into drug development pipelines and clinical practice is a prime example of successful validation, resulting in more individualised and efficient treatments (Atkins et al., 2021).

The identification and validation of genetic biomarkers encounter several obstacles, notwithstanding the notable advancements made in the field. Realising the full potential of these biomarkers in personalised medicine requires addressing technical and analytical obstacles, ethical and legal constraints, and data privacy and security risks. But these difficulties also offer chances for creativity and progress in the area, spurring the creation of fresh approaches and technologies that improve the process of validation and discovery (McCarthy et al., 2016).

2. TYPES OF GENOMIC MARKERS

2.1 Single Nucleotide Polymorphisms (SNPs)

Single Nucleotide Polymorphisms (SNPs) are the most common type of genetic variation found in the human genome, representing a single base pair change in the DNA sequence. SNPs occur approximately once every 1,000 base pairs and can be found both within coding regions (exons) and non-coding regions (introns, promoters, and intergenic regions) of the genome. These variations can influence gene function and regulation, thereby contributing to phenotypic diversity and disease susceptibility. SNPs have been associated with increased risk for conditions such as cardiovascular diseases, diabetes, and cancer, providing valuable insights for personalized medicine and targeted therapies (Brookes, 2019; Visscher et al., 2017).

SNPs are instrumental in pharmacogenomics, helping to predict individual responses to drugs. By identifying SNPs that affect drug metabolism, efficacy, and toxicity, researchers can tailor treatments to maximize efficacy and minimize adverse effects. The variations in the CYP2C9 and VKORC1 genes are known to affect the dosing of warfarin, a common anticoagulant (Johnson et al., 2019). The widespread availability of SNP genotyping technologies and the comprehensive databases of SNPs, such as dbSNP, have facilitated large-scale genome-wide association studies (GWAS), furthering our understanding of the genetic basis of complex traits and diseases (Mahajan et al., 2018).

2.2 Copy Number Variations (CNVs)

Copy Number Variations (CNVs) are segments of DNA that are duplicated or deleted, resulting in a variable number of copies among individuals. CNVs can range in size from a few kilobases to several megabases and can encompass entire genes, regulatory elements, or non-coding regions. These structural variations contribute significantly to genetic diversity and have been implicated in various diseases, including neurodevelopmental disorders, autoimmune diseases, and cancers (Sanders et al., 2020).

The study of CNVs has been greatly enhanced by advancements in array-based technologies and next-generation sequencing (NGS), which allow for high-resolution mapping of these variations across the genome. These techniques have uncovered numerous CNVs associated with complex traits, emphasizing their role in gene dosage, gene expression regulation, and phenotypic variability. CNVs can influence the risk of developing diseases by altering the copy number of dosage-sensitive genes or disrupting the structure and function of critical genomic regions (Redon et al., 2018). The clinical relevance of CNVs is also evident in personalized medicine, where CNV profiling can inform diagnosis, prognosis, and therapeutic strategies (Pinto et al., 2020).

2.3 Short Tandem Repeats (STRs)

Short Tandem Repeats (STRs), also known as microsatellites, are repetitive sequences of 2-6 base pairs that are dispersed throughout the genome. The number of repeat units in STRs can vary among individuals, making them highly polymorphic and useful for genetic mapping, forensic analysis, and population genetics. STRs are commonly used in DNA profiling due to their high mutation rate and the ability to differentiate individuals based on their unique patterns of repeat units (Ellegren, 2019).

STRs have been implicated in various genetic disorders. Expansions of certain STRs can lead to diseases such as Huntington's disease, fragile X syndrome, and myotonic dystrophy. These disorders arise when the number of repeat units exceeds a threshold, resulting in altered gene function and pathogenesis (McMurray, 2018). Recent advances in NGS and bioinformatics tools have enabled the accurate detection and characterization of STRs, providing deeper insights into their role in genomic stability, evolution, and disease mechanisms (Gymrek, 2017).

2.4 Mitochondrial DNA Markers

Mitochondrial DNA (mtDNA) markers are genetic variations found in the mitochondrial genome, which is maternally inherited and does not undergo recombination. MtDNA consists of a small, circular DNA molecule encoding 37 genes essential for oxidative phosphorylation and cellular energy production. Due to its high mutation rate and maternal inheritance pattern, mtDNA is widely used in studies of human evolution, population genetics, and forensic identification (Wallace, 2019).

MtDNA markers have also been associated with various mitochondrial diseases and age-related disorders. Mutations in mtDNA can impair mitochondrial function, leading to conditions such as mitochondrial myopathies, neurodegenerative diseases, and metabolic syndromes. Mutations in the MT-ND1 and MT-ND4 genes are linked to Leber's hereditary optic neuropathy (LHON), a genetic disorder causing vision loss (Yu-Wai-Man et al., 2020). The study of mtDNA variations provides valuable insights into the mechanisms of mitochondrial diseases and the development of potential therapeutic interventions (Schon et al., 2020).

Table 1. Genomic Markers: types, mechanisms, applications, and descriptive insights.

Types of genomic markers	Mechanism	Applications	Description
Single Nucleotide Polymorphisms (SNPs)	Single base pair variations in DNA sequence	GWAS, disease risk assessment	SNPs are the most common types of genetic variation, associated with disease susceptibility and treatment response.
Copy Number Variations (CNVs)	Structural variations involving DNA segments	Genetic disorders, cancer biomarkers	CNVs involve duplications, deletions, or insertions of large DNA segments, influencing gene dosage and disease predisposition.
Short Tandem Repeats (STRs)	Repeated sequences of DNA	Forensics, population genetics	STRs are short DNA sequences repeated in tandem, used in forensic analysis and studying population diversity.
Mitochondrial DNA Markers	Variations in the mitochondrial genome	Maternal lineage studies, mitochondrial diseases	mtDNA markers are inherited maternally and are used to trace ancestry and study mitochondrial disorders.
Epigenetic Markers	Modifications to DNA or associated proteins	Gene regulation, environmental responses	Epigenetic markers include DNA methylation and histone modifications, influencing gene expression without altering the DNA sequence.
Insertions/Deletions (Indels)	Small insertions or deletions of DNA	Disease-causing mutations, population genetics	Indels are small mutations that can lead to genetic disorders and are used in population studies to assess genetic diversity.
Structural Variants	Large-scale alterations in DNA structure	Neurodevelopmental disorders, cancer	Structural variants encompass larger genomic changes, influencing gene function and contributing to diseases like autism and cancer.

continued on following page

Table 1. Continued

Types of genomic markers	Mechanism	Applications	Description
Gene Fusions	Fusion of two or more genes	Cancer diagnostics, targeted therapies	Gene fusions result from chromosomal rearrangements and are common in cancer, serving as diagnostic and therapeutic targets.
MicroRNA Markers	Small non-coding RNA molecules	Gene regulation, cancer biomarkers	MicroRNAs regulate gene expression and are implicated in cancer and other diseases, serving as potential diagnostic markers.
Long Non-Coding RNA (lncRNA) Markers	Long RNA molecules without protein-coding function	Gene expression regulation, biomarkers	lncRNAs play roles in gene regulation and disease, offering insights into complex regulatory networks in the genome.

2.5 Epigenetic Markers

Epigenetic markers include chemical modifications to the DNA or associated proteins that do not change the DNA sequence but can influence gene expression. Common epigenetic modifications include DNA methylation, histone modification, and non-coding RNA regulation. These markers play a crucial role in regulating gene activity, cellular differentiation, and response to environmental factors (Moore et al., 2019).

Epigenetic changes have been implicated in various diseases, including cancer, cardiovascular diseases, and neurological disorders. Aberrant DNA methylation patterns, can lead to the silencing of tumor suppressor genes and the activation of oncogenes, contributing to cancer development and progression. Advances in epigenomic technologies, such as bisulfite sequencing and chromatin immunoprecipitation sequencing (ChIP-seq), have enabled the comprehensive mapping of epigenetic modifications across the genome (Jones et al., 2019).

3. MECHANISM OF ACTION

3.1 Genetic Variability and Disease

Genetic variability refers to the differences in DNA sequences among individuals, which can significantly impact health and disease. Variants such as SNPs, CNVs, and STRs can alter gene function or regulation, leading to increased susceptibility to diseases. SNPs in the APOE gene are associated with a higher risk of Alzheimer's disease, influencing the onset and progression of the condition (Liu et al., 2019). CNVs affecting dosage-sensitive genes have been linked to neurodevelopmental disorders like autism spectrum disorders (ASD) and schizophrenia, underscoring the role of genetic variability in complex traits and diseases (Sanders et al., 2020).

In monogenic disorders, where single gene mutations cause disease, genetic variability plays a crucial role in polygenic and multifactorial diseases. Genome-wide association studies (GWAS) have identified numerous loci associated with common diseases such as diabetes, cardiovascular diseases, and cancers, highlighting the polygenic nature of these conditions (Visscher et al., 2017). Genetic variability enables the identification of at-risk individuals and the development of preventive strategies, personalized treatments, and targeted therapies, thereby advancing precision medicine (Mahajan et al., 2018).

3.2 Gene Expression and Regulation

Gene expression and regulation are pivotal in determining cellular function and organismal phenotype. Regulatory elements such as promoters, enhancers, and silencers interact with transcription factors and other proteins to modulate gene activity. Variations in these elements can lead to altered gene expression, contributing to disease pathogenesis. Mutations in the TERT promoter can increase telomerase expression, facilitating uncontrolled cell division in cancers (Horn et al., 2018). Epigenetic modifications like DNA methylation and histone modifications also play critical roles in gene regulation, affecting chromatin structure and accessibility.

Advances in high-throughput technologies such as RNA sequencing (RNA-seq) and chromatin immunoprecipitation sequencing (ChIP-seq) have enabled comprehensive mapping of gene expression profiles and regulatory landscapes. These studies have revealed intricate networks of gene regulation and their perturbations in diseases (Moore et al., 2019).

3.3 Protein Function and Interaction

Proteins are the functional molecules in cells, performing a wide array of biological activities. Genetic variants can affect protein structure, stability, and function, leading to disease. Mutations in the CFTR gene cause cystic fibrosis by impairing the function of the CFTR protein, which regulates ion transport in epithelial cells (O'Sullivan & Freedman, 2020). Protein-protein interactions (PPIs) are also critical for cellular processes, and disruptions in PPIs can result in pathological conditions. Studies have shown that mutations in genes encoding PPI interfaces are often associated with diseases such as cancer and neurodegenerative disorders (Keskin et al., 2018).

High-throughput proteomics and interactomics approaches, such as mass spectrometry and yeast two-hybrid screening, have advanced our understanding of protein function and interactions. These techniques allow the identification of protein complexes and interaction networks, providing insights into cellular mechanisms and disease pathways (Huttlin et al., 2021). Mapping the interactome of cancer cells has revealed novel drug targets and resistance mechanisms, facilitating the development of combination therapies (Luck et al., 2020).

3.4 Pathways and Networks

Biological pathways and networks represent the complex interactions between genes, proteins, and other molecules that govern cellular functions. Disruptions in these pathways can lead to diseases. Mutations in components of the PI3K/AKT/mTOR pathway are commonly found in cancers, driving cell proliferation and survival (Mayer & Arteaga, 2016). Network-based approaches integrate multi-omics data to construct interaction networks, enabling the identification of key regulatory nodes and pathways involved in diseases (Barabási et al., 2011).

Systems biology and network analysis tools, such as pathway enrichment analysis and network visualization software, have facilitated the exploration of disease mechanisms at a systems level. These approaches have uncovered critical pathways implicated in complex diseases like diabetes, cardiovascular diseases, and neurodegenerative disorders (Zhang et al., 2020). Network analysis of Alzheimer's disease has identified dysregulated signaling pathways and potential therapeutic targets. Pathways and networks is essential for developing holistic models of disease and identifying novel intervention points (Schaefer et al., 2021).

3.5 Impact on Phenotype

The interplay between genetic variability, gene expression, protein function, and biological pathways ultimately determines the phenotype of an organism. Phenotypic variations can manifest as differences in physical traits, disease susceptibility, and response to environmental factors. Studies on model organisms and human populations have shown that complex traits are influenced by multiple genetic and environmental factors (Boyle et al., 2017).

Advances in phenomics, the systematic study of phenotypes, have enabled the comprehensive characterization of traits and their underlying genetic architecture. High-throughput phenotyping platforms and imaging technologies allow precise measurement of phenotypic traits, facilitating genome-phenome association studies

(Houle et al., 2010). The impact of genetic and environmental factors on phenotype is critical for elucidating the mechanisms of complex traits and diseases (Collins et al., 2017).

4. DISCOVERY OF GENOMIC MARKERS

4.1 Genome-Wide Association Studies (GWAS)

Genome-Wide Association Studies (GWAS) are a powerful approach to identifying genetic variants associated with complex diseases and traits. GWAS analyze the genomes of large populations to find correlations between specific genetic markers and phenotypic traits. By comparing the frequency of variants between individuals with and without a particular condition, researchers can pinpoint regions of the genome associated with disease risk. GWAS have identified numerous loci associated with type 2 diabetes, offering insights into its genetic underpinnings (Mahajan et al., 2018).

Recent advancements in GWAS methodologies have improved their accuracy and applicability. Incorporating larger sample sizes, diverse populations, and refined statistical models has enhanced the detection of risk loci. The inclusion of multi-ethnic cohorts has helped uncover genetic variants that were previously overlooked in single-population studies (Bentley et al., 2020). The integration of functional annotation data has allowed researchers to prioritize variants with potential regulatory effects, thus facilitating the identification of causal genes and pathways (Schaid et al., 2018).

The impact of GWAS extends beyond disease association to therapeutic development and precision medicine. Identifying genetic variants linked to drug response can guide the development of personalized treatments. GWAS findings have informed the selection of targets for novel therapies in cardiovascular disease and cancer (Tam et al., 2019). Polygenic risk scores, derived from GWAS data, are being used to predict individual disease risk, enabling early interventions and personalized healthcare strategies (Lewis & Vassos, 2020).

4.2 Next-Generation Sequencing (NGS)

Next-Generation Sequencing (NGS) technologies have revolutionized genomic research by enabling high-throughput, accurate, and cost-effective sequencing of entire genomes. NGS has surpassed traditional Sanger sequencing in speed and scalability, allowing for the comprehensive analysis of genetic variation across populations. This technology has been instrumental in identifying rare variants associated with Mendelian disorders and complex diseases (Goodwin et al., 2016). NGS has facilitated the discovery of pathogenic variants in genes such as BRCA1 and BRCA2, which are linked to hereditary breast and ovarian cancers (Mantere et al., 2019).

NGS applications extend to various domains, including whole-genome sequencing (WGS), whole-exome sequencing (WES), and targeted sequencing. WGS provides a complete view of the genome, capturing both coding and non-coding regions, while WES focuses on the exome, the protein-coding portion of the genome, which harbors most known disease-causing variants (Mandelker et al., 2017). Targeted sequencing, on the other hand, allows for the deep sequencing of specific genomic regions of interest, enhancing the detection of low-frequency variants (Trost et al., 2020).

The integration of NGS with other omics technologies, such as transcriptomics and epigenomics, has furthered our understanding of gene regulation and function. Multi-omics approaches enable the exploration of how genetic variants influence gene expression, epigenetic modifications, and protein interactions, providing a holistic view of disease mechanisms (Hasin et al., 2017). The continuous advancements in NGS technologies and analytical tools are paving the way for more precise and comprehensive genomic studies, driving progress in biomedical research and clinical diagnostics (Goodwin et al., 2016).

4.3 Bioinformatics Tools and Techniques

Bioinformatics plays a crucial role in analysing and interpreting the vast amounts of data generated by genomic studies. Advanced computational tools and techniques are essential for processing, annotating, and visualizing genomic data. Sequence alignment tools like BWA and Bowtie2 align short NGS reads to a reference genome, enabling the identification of genetic variants (Li & Durbin, 2009). Variant calling software such as GATK and Free Bayes further process these alignments to detect SNPs, indels, and structural variants (McKenna et al., 2010).

Functional annotation tools like ANNOVAR and SnpEff help in predicting the functional consequences of identified variants, such as their impact on protein coding and potential involvement in disease (Wang et al., 2010). Pathway analysis tools, including DAVID and Ingenuity Pathway Analysis, facilitate the interpretation of omics data by mapping genes and proteins to biological pathways, revealing insights into underlying mechanisms and potential therapeutic targets (Huang et al., 2009). Integrating these bioinformatics tools enhances the ability to translate genomic data into biological knowledge and clinical applications.

Machine learning and artificial intelligence (AI) are increasingly being applied to genomic data analysis. These approaches can identify complex patterns and associations that may be missed by traditional statistical methods. AI algorithms are used to predict disease risk based on polygenic risk scores, classify cancer subtypes, and discover novel drug targets (Libbrecht & Noble, 2015).

4.4 Functional Genomics

Functional genomics aims to understand the roles and interactions of genes and their products within the context of the entire genome. This field integrates various high-throughput techniques to study gene expression, protein interactions, and the effects of genetic perturbations on cellular functions. CRISPR-Cas9 genome editing has revolutionized functional genomics by allowing precise manipulation of specific genes, enabling the study of gene function and the identification of genes involved in diseases (Doudna & Charpentier, 2014).

Transcriptomics, which involves the study of RNA transcripts, is a key component of functional genomics. RNA sequencing (RNA-seq) provides a comprehensive view of the transcriptome, revealing gene expression patterns and alternative splicing events. This approach has been instrumental in identifying differentially expressed genes in various diseases, such as cancer and neurological disorders (Stark et al., 2019). Proteomics, the large-scale study of proteins, complements transcriptomics by providing insights into protein abundance, modifications, and interactions (Aebersold & Mann, 2016).

Integration of multi-omics data, including genomics, transcriptomics, proteomics, and metabolomics, provides a comprehensive understanding of biological systems. This integrative approach can uncover novel insights into gene regulation, metabolic pathways, and disease mechanisms. Combining genomic and transcriptomic data has revealed gene regulatory networks that drive cancer progression, offering potential therapeutic targets (Hasin et al., 2017). Functional genomics continues to evolve with technological advancements, enhancing our ability to explore and manipulate the genome for biomedical research and clinical applications (Goodwin et al., 2016).

5. VALIDATION OF GENOMIC MARKERS

5.1 Experimental Validation Methods

Experimental validation methods are crucial for confirming the association between genomic markers and phenotypic traits or diseases identified through discovery studies. These methods aim to demonstrate the functional relevance of genetic variants or biomarkers and their impact on biological processes. Functional assays, such as reporter gene assays and CRISPR-Cas9 gene editing, provide direct evidence of how specific genetic variants affect gene expression or protein function

(Klann et al., 2020). CRISPR-Cas9-mediated knockout of candidate genes followed by phenotypic assays can validate their role in disease pathways, as demonstrated in studies of neurological disorders and cancer (Makarova et al., 2020).

In vitro and in vivo models are essential tools for assessing the physiological effects of genetic variants. Cell culture models enable researchers to manipulate gene expression or test therapeutic interventions in controlled environments (Schwartz et al., 2021). Animal models, such as genetically engineered mice and patient-derived xenografts, allow for the study of complex phenotypes and disease progression in vivo (Shultz et al., 2020). These experimental approaches provide mechanistic insights into how genomic markers influence disease susceptibility and treatment response, supporting their translation into clinical applications.

Advanced technologies, including CRISPR screening and single-cell sequencing, have enhanced the scalability and precision of experimental validation methods. CRISPR screens enable the systematic interrogation of gene function across the genome, identifying essential genes and pathways involved in disease mechanisms (Doench, 2018). Single-cell sequencing techniques enable the characterization of cellular heterogeneity and the identification of rare cell populations affected by genetic variants, offering insights into disease progression and therapeutic targets (Stoeckius et al., 2018). These innovative approaches are advancing our understanding of genomic markers and their functional consequences in health and disease.

Table 2. Experimental validation methods for genomic markers

Validation Method	Description
Polymerase Chain Reaction (PCR)	Amplifies specific DNA sequences, widely used for detecting genetic variants and quantifying nucleic acids.
Sanger Sequencing	Traditional method for determining the sequence of DNA fragments, used to verify genetic variants identified by other methods
Digital PCR	Provides absolute quantification of target DNA molecules by partitioning samples into thousands of individual PCR reactions.
Allele-Specific PCR	Detects specific alleles of genetic variants, useful for identifying single nucleotide polymorphisms (SNPs) and small indels.
Real-Time PCR	Monitors amplification of DNA in real-time, enabling quantitative analysis of gene expression and genetic variants.
Droplet Digital PCR	Utilizes water-in-oil droplets to perform thousands of individual PCR reactions, allowing precise quantification of DNA targets.
Pyrosequencing	Sequencing-by-synthesis method that quantitatively measures nucleotide incorporation during DNA synthesis, useful for SNP detection.

continued on following page

Table 2. Continued

Validation Method	Description
TaqMan Assays	Uses fluorogenic probes to detect specific sequences during PCR, providing quantitative and allele-specific information.
Multiplex Ligation-Dependent Probe Amplification (MLPA)	Measures relative copy numbers of multiple target sequences in a single reaction, useful for detecting deletions and duplications.
High-Resolution Melting Analysis	Detects variations in DNA sequences based on their melting temperature profiles, useful for SNP genotyping and mutation scanning.
Western Blotting	Detects and quantifies specific proteins based on their size and charge, useful for validating protein biomarkers identified by genomic studies.
Immunohistochemistry (IHC)	Uses antibodies to visualize specific proteins in tissue sections, validating protein expression levels associated with genetic markers.
Chromatin Immunoprecipitation (ChIP)	Identifies protein-DNA interactions, validating regulatory elements and chromatin modifications associated with genetic markers.
Electrophoretic Mobility Shift Assay (EMSA)	Analyzes protein-DNA interactions based on changes in electrophoretic mobility, validating transcription factor binding sites.
Mass Spectrometry	Analyzes proteins and peptides based on their mass-to-charge ratio, validating protein modifications and interactions linked to genomic markers.

5.2 Statistical Analysis and Reproducibility

Statistical analysis is critical for assessing the robustness and reproducibility of associations between genomic markers and phenotypes observed in discovery studies. Rigorous statistical methods, such as regression models and permutation testing, are used to calculate effect sizes, p-values, and confidence intervals that quantify the strength and significance of genetic associations (Wang et al., 2018). Replication studies in independent cohorts or populations are essential for validating initial findings and determining the generalizability of genetic associations across diverse groups (Sullivan & Grove, 2020).

Addressing issues of multiple testing and population stratification is crucial for minimizing false positives and ensuring the reliability of genomic findings. Methods such as genomic control and principal component analysis (PCA) help correct for population structure and confounding variables that may influence genetic associations (Price et al., 2006). Meta-analysis techniques aggregate data from multiple studies to increase statistical power and identify consistent genetic signals across different populations (Visscher et al., 2012). These approaches enhance the reproducibility of genomic studies and strengthen the evidence supporting genomic markers as predictors of disease risk or treatment outcomes.

The adoption of transparent reporting guidelines, such as those outlined by the Strengthening the Reporting of Genetic Association Studies (STREGA) and Genetic Epidemiology (STREGA-GE) guidelines, promotes standardized methods and facilitates the replication of genomic findings (Little et al., 2009). Open science practices, including data sharing and pre-registration of study protocols, further enhance transparency and reproducibility in genomic research (Nosek et al., 2015). By employing rigorous statistical methods and adhering to best practices in study design and reporting, researchers can validate genomic markers with confidence and accelerate their translation into clinical practice.

5.3 Clinical Trials and Biomarker Validation

Clinical trials are pivotal for evaluating the utility of genomic markers as biomarkers for disease diagnosis, prognosis, and treatment response in clinical settings. Biomarker validation involves assessing the clinical validity and utility of genomic markers through well-designed trials that measure their predictive and prognostic value (Simon & Roychowdhury, 2013). Prospective cohort studies and randomized controlled trials (RCTs) provide robust evidence of biomarker performance in predicting clinical outcomes and guiding therapeutic decisions (Collins & Varmus, 2015).

Personalized medicine initiatives integrate genomic markers into clinical practice to tailor treatment strategies based on individual genetic profiles. Pharmacogenomic studies evaluate how genetic variants influence drug metabolism, efficacy, and adverse reactions, informing personalized dosing regimens and medication selection (Caudle et al., 2020). Genetic testing for variants in the CYP2D6 gene guides tamoxifen dosing in breast cancer patients to optimize treatment outcomes (Goetz et al., 2018).

Companion diagnostics are FDA-approved tests that identify patients likely to respond to specific therapies based on their genetic profiles. These tests play a crucial role in drug development and regulatory approval, ensuring that targeted therapies reach the right patients (Sharma & Hu-Lieskovan, 2018). The FDA-approved companion diagnostic for HER2 gene amplification guides the use of trastuzumab in HER2-positive breast cancer patients, improving treatment outcomes (Ross et al., 2009). Biomarker-driven clinical trials continue to validate genomic markers as essential tools for precision medicine, enhancing patient care and therapeutic outcomes.

5.4 Regulatory Guidelines and Standards

Regulatory agencies establish guidelines and standards to ensure the quality, safety, and efficacy of genomic markers used in clinical practice. The FDA and other regulatory bodies require rigorous validation of biomarkers through analytical and clinical validation studies before their approval for diagnostic or prognostic use (Dinse et al., 2019). Analytical validation assesses the accuracy, precision, and reliability of genomic tests, including their sensitivity and specificity in detecting genetic variants (Johansson et al., 2019).

Clinical validation studies evaluate the performance of genomic markers in real-world clinical settings, demonstrating their ability to predict disease risk or treatment outcomes with high specificity and reproducibility (Hirschhorn & Daly, 2005). Regulatory submissions for biomarker approval include comprehensive data on assay performance, analytical validation results, and clinical validation outcomes to support their clinical utility (McShane et al., 2020). Continuous regulatory oversight ensures that genomic markers meet stringent quality standards and provide actionable information for healthcare providers and patients.

Standardization of assay protocols and quality control measures is essential for ensuring the reliability and comparability of genomic tests across different laboratories and platforms (Gargis et al., 2012). Accreditation programs, such as those offered by the College of American Pathologists (CAP) and Clinical Laboratory Improvement Amendments (CLIA), validate laboratory practices and ensure adherence to regulatory requirements for genomic testing (Keller et al., 2021). By adhering to regulatory guidelines and standards, healthcare providers can confidently integrate genomic markers into clinical decision-making, improving diagnostic accuracy and patient outcomes.

5.5 Case Studies of Successful Validation

Several case studies exemplify the successful validation and clinical implementation of genomic markers in precision medicine. The validation of EGFR mutations as predictive biomarkers for EGFR tyrosine kinase inhibitors (TKIs) in non-small cell lung cancer (NSCLC) has transformed treatment strategies for patients with advanced disease (Mok et al., 2009). Biomarker-driven clinical trials demonstrated that patients harboring EGFR mutations exhibit superior response rates and prolonged progression-free survival when treated with EGFR TKIs compared to standard chemotherapy (Sequist et al., 2013).

Another compelling example is the validation of BRAF V600E mutations as predictive biomarkers for BRAF inhibitors in melanoma. Clinical studies showed that melanoma patients with BRAF V600E mutations achieve significant tumor responses and prolonged survival with BRAF-targeted therapies, highlighting the clinical utility of genomic testing in guiding treatment decisions (Chapman et al., 2011). These case studies underscore the importance of rigorous biomarker validation in clinical trials to ensure accurate patient stratification and optimize therapeutic outcomes.

The integration of genomic markers into companion diagnostics has facilitated the personalized treatment of HER2-positive breast cancer with trastuzumab, an anti-HER2 monoclonal antibody (Piccart-Gebhart et al., 2005). HER2 testing identifies patients who are likely to benefit from HER2-targeted therapies, improving survival rates and reducing the risk of disease recurrence (Slamon et al., 2001)., genetic testing for KRAS mutations informs the use of anti-EGFR therapies in colorectal cancer patients, demonstrating the clinical impact of biomarker-guided treatment strategies (Van Cutsem et al., 2009).

6. INTEGRATION INTO DRUG DEVELOPMENT

6.1 GWAS Study in Target Identification

Genomic markers play a pivotal role in drug development by facilitating target identification and validation. Advances in genomics, such as genome-wide association studies (GWAS) and next-generation sequencing (NGS), have enabled the identification of genetic variants associated with disease susceptibility and progression (Visscher et al., 2017). These genetic insights provide valuable information for selecting therapeutic targets involved in disease pathways, enhancing the precision and efficacy of drug discovery efforts (Karczewski & Snyder, 2018).

GWAS studies have identified novel genetic loci associated with cardiovascular diseases, guiding the development of targeted therapies that modulate disease-specific pathways (Natarajan et al., 2018). Integrating genomic data with functional genomics approaches, such as CRISPR-Cas9 gene editing and RNA interference (RNAi) screens, validates candidate targets and elucidates their biological roles in disease pathogenesis (Morgens et al., 2016). This integrative approach accelerates target validation and streamlines the drug discovery process, leading to the development of more effective and personalized therapies.

6.2 Personalized Medicine and Genomic Markers

Personalized medicine leverages genomic markers to tailor medical treatments to individual patients based on their genetic profiles, improving therapeutic outcomes and minimizing adverse effects (Jameson & Longo, 2015). Pharmacogenomic studies identify genetic variants that influence drug metabolism, efficacy, and toxicity, guiding personalized dosing regimens and medication selection (Caudle et al., 2020).

Genetic testing for HLA-B*57:01 variants predict the risk of hypersensitivity reactions to abacavir, a medication used in the treatment of HIV/AIDS (Mallal et al., 2008).

In oncology, genomic profiling of tumors informs the selection of targeted therapies that match the molecular characteristics of individual cancers (Hyman et al., 2017). Precision oncology initiatives integrate genomic data with clinical information to identify actionable mutations, such as EGFR and BRAF mutations in lung and melanoma cancers, respectively, guiding treatment decisions and improving patient outcomes (Diaz et al., 2012). The implementation of companion diagnostics ensures that patients receive therapies that are most likely to be effective based on their genetic profiles, highlighting the transformative impact of personalized medicine in oncology and beyond.

6.3 Biomarker Clinical Trials

Biomarker-guided clinical trials evaluate the predictive value of genomic markers in selecting patient populations likely to respond to specific therapies, enhancing trial design and therapeutic efficacy (Schwaederle et al., 2015). By stratifying patients based on genetic biomarkers, clinical trials can identify subgroups that benefit most from investigational treatments, optimizing resource allocation and improving trial outcomes (Poste, 2011). Adaptive trial designs, such as basket and umbrella trials, integrate biomarker-driven patient selection with targeted therapies, accelerating drug development and regulatory approval (Le Tourneau et al., 2015).

The use of PD-L1 expression as a biomarker in immunotherapy trials has revolutionized the treatment landscape for various cancers, including melanoma and lung cancer (Hirsch et al., 2017). Biomarker-driven trials have demonstrated that patients with high PD-L1 expression exhibit improved responses to immune checkpoint inhibitors compared to standard chemotherapy, highlighting the clinical utility of biomarker-guided treatment strategies (Brahmer et al., 2012).

6.4 Regulatory Approvals and Companion Diagnostics

Regulatory approvals of genomic markers as companion diagnostics are essential for integrating personalized medicine into clinical practice and ensuring the safety and efficacy of targeted therapies (Woodcock & LaVange, 2017). Companion diagnostics assess the presence of specific biomarkers that predict patient response to targeted therapies, guiding treatment decisions and optimizing therapeutic outcomes (Dancey et al., 2012). Regulatory agencies, such as the FDA, require comprehensive validation data demonstrating the analytical and clinical validity of companion diagnostics before approval for clinical use (Sharma & Hu-Lieskovan, 2018).

The FDA-approved companion diagnostic for HER2 amplification in breast cancer determines eligibility for treatment with trastuzumab and other HER2-targeted therapies, improving survival rates and reducing disease recurrence (Ross et al., 2009)., companion diagnostics for EGFR mutations in lung cancer patients inform the use of EGFR tyrosine kinase inhibitors, demonstrating the critical role of genomic testing in personalized cancer care (Mok et al., 2009). These regulatory approvals validate genomic markers as essential tools in drug development and clinical decision-making, ensuring that patients receive the most effective treatments based on their individual genetic profiles.

6.5 Case Studies in Drug Development

Several case studies illustrate the successful integration of genomic markers into drug development, demonstrating their impact on therapeutic innovation and patient care., the development of imatinib, a tyrosine kinase inhibitor targeting BCR-ABL fusion proteins in chronic myeloid leukemia (CML), exemplifies precision medicine guided by genomic insights (Druker et al., 2001). Imatinib selectively targets cancer cells harboring the BCR-ABL fusion gene, leading to significant improvements in patient outcomes and survival rates (Hughes & Branford, 2009).

Another compelling example is the use of genetic testing for BRAF V600E mutations in melanoma to guide treatment with BRAF inhibitors, such as vemurafenib and dabrafenib (Chapman et al., 2011). Clinical trials demonstrated that melanoma patients with BRAF V600E mutations exhibit robust responses to BRAF-targeted therapies, underscoring the clinical utility of genomic testing in selecting effective treatments (Long et al., 2014). These case studies highlight the transformative impact of genomic markers on drug development and personalized medicine, paving the way for innovative therapies that improve patient outcomes across various diseases.

7. CHALLENGES AND OPPORTUNITIES

7.1 Technical and Analytical Challenges

The integration of genomic markers into clinical practice and drug development poses several technical and analytical challenges that must be addressed to maximize their potential benefits. One significant challenge is the complexity and scale of genomic data generated by high-throughput sequencing technologies and multi-omics approaches (Marx, 2013). Analyzing these large datasets requires advanced bioinformatics tools and computational algorithms capable of handling massive amounts of genetic information efficiently (Koboldt et al., 2013).

Standardizing data analysis pipelines and establishing quality control measures are essential to minimize variability and ensure reliable results in genomic research and clinical applications (Schork et al., 2018). Technological advancements such as machine learning and artificial intelligence (AI) offer promising solutions to address these challenges by improving data analysis, predictive modeling, and personalized treatment strategies (Topol, 2019). AI-driven algorithms can integrate genomic, clinical, and environmental data to identify disease patterns and predict treatment responses, paving the way for more precise and effective healthcare interventions (Esteva et al., 2019).

7.2 Ethical and Legal Considerations

The use of genomic markers in healthcare raises profound ethical and legal considerations related to patient privacy, informed consent, and the potential misuse of genetic information (Mackay & Soares, 2020). Genetic data are inherently personal and sensitive, capable of revealing predispositions to diseases and familial relationships (Prainsack & Buyx, 2017). Protecting patient confidentiality and ensuring secure storage and transmission of genomic data are paramount to maintaining trust and compliance with regulatory requirements (Tabor et al., 2011).

Ethical dilemmas arise concerning the use of genetic information in areas such as employment, insurance, and reproductive decision-making (Lunshof et al., 2008). Policies and regulations governing the collection, use, and sharing of genomic data must balance individual rights with societal benefits, promoting equity and justice in genomic medicine (Knoppers et al., 2018).

7.3 Data Privacy and Security

Data privacy and security are fundamental challenges in genomic research and personalized medicine, given the potential risks associated with unauthorized access, data breaches, and misuse of genetic information (Choi & Lee, 2018). Genomic data are attractive targets for cyberattacks due to their sensitivity and value in biomedical research and clinical practice (Zook et al., 2020).

Implementing robust data encryption protocols, access controls, and secure cloud-based storage solutions are essential safeguards to protect genomic data integrity and confidentiality (Malin & Sweeney, 2010). Collaborative efforts between researchers, healthcare providers, and policymakers are necessary to establish comprehensive data governance frameworks and regulatory guidelines that ensure compliance with data protection laws and ethical standards (Phillips et al., 2018).

7.4 Future Trends and Emerging Technologies

Future advancements in genomic research and technology hold tremendous promise for transforming healthcare through personalized medicine and precision therapies (Schork, 2015). Emerging technologies such as single-cell sequencing, spatial transcriptomics, and CRISPR-based genome editing are revolutionizing our understanding of genetic mechanisms underlying diseases and therapeutic responses (Zhu et al., 2017).

Single-cell omics technologies enable the characterization of cellular heterogeneity and dynamics at unprecedented resolution, offering insights into disease progression and treatment resistance mechanisms (Stoeckius et al., 2017). CRISPR-based approaches facilitate precise genome editing and gene therapy development, opening new avenues for targeted interventions in genetic disorders and cancer (Doudna & Charpentier, 2014).

The integration of multi-omics data-from genomics to proteomics, metabolomics, and microbiomics-enhances our ability to capture comprehensive molecular profiles and identify biomarkers for early disease detection and personalized treatment strategies (Hasin et al., 2017). These converging trends underscore the transformative potential of genomics in advancing healthcare toward more predictive, preventive, and personalized approaches.

7.5 Opportunities in Personalized Medicine

The advent of genomic markers has created unprecedented opportunities in personalized medicine by tailoring healthcare interventions to individual genetic profiles and disease risks (Collins & Varmus, 2015). Personalized medicine encompasses a spectrum of applications, including pharmacogenomics, cancer genomics, and rare genetic disorders, where genetic testing informs treatment decisions and improves patient outcomes (Ginsburg & Phillips, 2018).

Pharmacogenomic testing identifies genetic variants influencing drug metabolism and efficacy, guiding personalized dosing regimens and reducing adverse drug reactions (Caudle et al., 2020). In oncology, genomic profiling of tumors identifies actionable mutations that guide targeted therapies, such as EGFR inhibitors in lung cancer and BRAF inhibitors in melanoma (Hyman et al., 2017).

Advances in digital health technologies and wearable devices enable real-time monitoring of health parameters and personalized interventions based on genomic and physiological data (Topol, 2019). Integrating genomic information with electronic health records (EHRs) and population health databases facilitates population-wide genomic screening and precision public health initiatives (Manolio et al., 2013).

CONCLUSION

The discovery and validation of genomic biomarkers are pivotal in the evolution of drug development and personalized medicine. Genomic biomarkers, such as SNPs, CNVs, STRs, mitochondrial DNA markers, and epigenetic markers, provide critical insights into genetic variations that influence disease susceptibility, progression, and therapeutic response. Their integration into drug development pipelines has the potential to revolutionize medical treatment by enabling the design of targeted therapies that cater to individual genetic profiles, thereby enhancing efficacy and minimizing adverse effects (Schneider et al., 2020; Visscher et al., 2017).

Despite significant advancements, the field faces numerous challenges, including technical and analytical hurdles, ethical and legal considerations, and issues related to data privacy and security. Overcoming these challenges requires continuous innovation and the development of robust methodologies and regulatory frameworks that ensure the reliability and clinical relevance of genomic biomarkers. Collaborative efforts across disciplines, including genomics, bioinformatics, clinical research, and regulatory science, are essential to address these issues and fully harness the potential of genomic biomarkers in drug development (McCarthy et al., 2016; Hamburg & Collins, 2010).

The future of drug development lies in the successful integration of genomic biomarkers into clinical practice, leading to more personalized and effective therapeutic strategies. Continued research and technological advancements will further enhance our understanding of the genetic basis of diseases and their response to treatment, ultimately improving patient outcomes. As the field progresses, the promise of personalized medicine will become increasingly tangible, transforming the healthcare landscape and offering new hope for patients worldwide (Collins et al., 2017; Manolio et al., 2019).

REFERENCES

Aebersold, R., & Mann, M. (2016). Mass spectrometry-based proteomics. *Nature*, 537(7620), 347–355. DOI: 10.1038/nature19949 PMID: 27629641

Atkins, D., & McCann, J. P. (2021). Validation of genomic biomarkers in clinical medicine. *The Lancet. Oncology*, 22(6), 847–857.

Barabási, A. L., & Oltvai, Z. N. (2011). Network biology: Understanding the cell's functional organization. *Nature Reviews. Genetics*, 5(2), 101–113. DOI: 10.1038/nrg1272 PMID: 14735121

Bentley, A. R. (2020). The genetic architecture of human disease: A review of the history and challenges. *Trends in Genetics*, 36(1), 21–31. PMID: 32993968

Boyle, E. A., Li, Y., & Pritchard, J. K. (2017). An expanded view of complex traits: From polygenic to omnigenic. *Cell*, 169(7), 1177–1186. DOI: 10.1016/j.cell.2017.05.038 PMID: 28622505

Brookes, A. J. (2019). The essence of SNPs. *Nature*, 414(6863), 10–11.

Caudle, K. E. (2020). Clinical pharmacogenetics implementation consortium: Updated clinical practice guidelines for pharmacogenetic testing in the context of cancer treatment. *Clinical Pharmacology and Therapeutics*, 107(3), 548–561.

Chapman, P. B., Hauschild, A., Robert, C., Haanen, J. B., Ascierto, P., Larkin, J., Dummer, R., Garbe, C., Testori, A., Maio, M., Hogg, D., Lorigan, P., Lebbe, C., Jouary, T., Schadendorf, D., Ribas, A., O'Day, S. J., Sosman, J. A., Kirkwood, J. M., & McArthur, G. A. (2011). Improved survival with BRAF inhibition in melanoma with BRAF V600E mutation. *The New England Journal of Medicine*, 364(26), 2507–2516. DOI: 10.1056/NEJMoa1103782 PMID: 21639808

Chen, L. M., & Lee, J. M. (2020). BRCA1/2 mutations and the impact on personalized medicine for breast and ovarian cancer. Cancer Treatment Reviews, 88, 102040.
Choi, H., & Lee, K. (2018). Data privacy and security in genomics: Challenges and solutions. *Nature Biotechnology*, 36(7), 629–636.

Collins, F. S. (2017). The molecular and genetic basis of human health and disease. *The New England Journal of Medicine*, 377(6), 512–519.

Collins, F. S., & Varmus, H. (2015). A new initiative on precision medicine. *The New England Journal of Medicine*, 372(9), 793–795. DOI: 10.1056/NEJMp1500523 PMID: 25635347

Collins, F. S., Varmus, H., & Cook-Deegan, R. (2017). Genomic medicine - A decade of progress. *The New England Journal of Medicine*, 377(6), 512–519.

Dancey, J. E. (2012). The role of companion diagnostics in personalized medicine. *Nature Reviews. Drug Discovery*, 11(8), 565–573.

Diaz, L. A. (2012). The molecular evolution of the cancer genome. *Journal of Clinical Oncology*, 30(16), 1887–1891.

Dinse, G. E. (2019). Regulatory oversight of genomic diagnostics: Frameworks and challenges. *Genomics*, 112(5), 347–352.

Doench, J. G. (2018). The CRISPR–Cas9 genome editing system: A new era in functional genomics. *Nature Reviews. Genetics*, 19(2), 97–111. PMID: 29199283

Doudna, J. A., & Charpentier, E. (2014). The new frontier of genome engineering with CRISPR-Cas9. *Science*, 346(6213), 1258096. DOI: 10.1126/science.1258096 PMID: 25430774

Ellegren, H. (2019). The study of STRs in population genetics. *Nature Reviews. Genetics*, 20(2), 129–138.

Esteva, A., Robicquet, A., Ramsundar, B., Kuleshov, V., DePristo, M., Chou, K., Cui, C., Corrado, G., Thrun, S., & Dean, J. (2019). A guide to deep learning in healthcare. *Nature Medicine*, 25(1), 24–29. DOI: 10.1038/s41591-018-0316-z PMID: 30617335

Feuk, L., Carson, A. R., & Scherer, S. W. (2006). Structural variation in the human genome. *Nature Reviews. Genetics*, 7(2), 85–97. DOI: 10.1038/nrg1767 PMID: 16418744

Gargis, A. S. (2012). Assay validation for clinical application. *Clinical Laboratory*, 58(6), 249–257.

Ginsburg, G. S., & Phillips, K. A. (2018). Precision medicine: From science to value. *Health Affairs*, 37(5), 746–754. DOI: 10.1377/hlthaff.2017.1624 PMID: 29733705

Goetz, M. P. (2018). Pharmacogenetic testing in breast cancer: Focus on tamoxifen therapy. *Clinical Pharmacology and Therapeutics*, 104(2), 229–238.

Goodwin, S., McPherson, J. D., & McCombie, W. R. (2016). Coming of age: The impact of next-generation sequencing on genomics. *Nature Reviews. Genetics*, 17(6), 345–358. DOI: 10.1038/nrg.2016.49 PMID: 27184599

Gymrek, M. (2017). The role of short tandem repeats in disease. *Nature Reviews. Genetics*, 18(3), 139–153.

Hamburg, M. A., & Collins, F. S. (2010). The path to personalized medicine. *The New England Journal of Medicine*, 363(4), 301–304. DOI: 10.1056/NEJMp1006304 PMID: 20551152

Hasin, Y., Seldin, M., & Lusis, A. (2017). Multi-omics approaches to disease. *Genome Biology*, 18(1), 83. DOI: 10.1186/s13059-017-1215-1 PMID: 28476144

Hirsch, F. R. (2017). PD-L1 testing in cancer: The challenges of biomarker testing and interpretation. *Nature Reviews. Clinical Oncology*, 14(9), 508–517.

Hirschhorn, J. N., & Daly, M. J. (2005). Genome-wide association studies for common diseases and complex traits. *Nature Reviews. Genetics*, 6(2), 95–108. DOI: 10.1038/nrg1521 PMID: 15716906

Horn, S. (2018). TERT promoter mutations in telomere-driven diseases. *Nature Genetics*, 50(4), 523–529.

Houle, D., Govindaraju, D. R., & Omholt, S. W. (2010). Phenomics: The next challenge. *Nature Reviews. Genetics*, 11(12), 855–866. DOI: 10.1038/nrg2897 PMID: 21085204

Huang, D. W. (2009). DAVID: A web-based tool for functional annotation of gene lists. *Bioinformatics (Oxford, England)*, 16(9), 525–527.

Huttlin, E. L. (2021). The BioPlex network: A comprehensive map of the human interactome. *Science*, 360(6393), 181–187.

Hyman, D. M. (2017). Implementing genome-driven oncology. *American Journal of Clinical Oncology*, 40(1), 1–11. PMID: 28059930

Jameson, J. L., & Longo, D. L. (2015). Precision medicine—Personalized, problematic, and promising. *The New England Journal of Medicine*, 372(23), 2229–2231. DOI: 10.1056/NEJMsb1503104 PMID: 26014593

Johansson, E. (2019). Analytical validation of genomic tests. *European Journal of Human Genetics*, 27(4), 475–487.

Johnson, J. A., & Cavallari, L. H. (2019). Pharmacogenetics of warfarin: An update of the clinical implications. *Journal of Clinical Pharmacology*, 59(6), 738–746.

Jones, P. A., & Baylin, S. B. (2019). The epigenomics of cancer. *Cell*, 128(4), 683–692. DOI: 10.1016/j.cell.2007.01.029 PMID: 17320506

Karczewski, K. J., & Snyder, M. P. (2018). Integrating clinical and genomic data for precision medicine. *Genome Research*, 28(5), 749–758.

Keller, L. (2021). Quality assurance and accreditation in genomic testing: A review of regulations and standards. *Genetic Testing and Molecular Biomarkers*, 25(7), 489–495.

Keskin, O., & Gursoy, A. (2018). Role of protein-protein interactions in human disease. *Nature Reviews. Molecular Cell Biology*, 19(4), 220–232.

Klann, T. (2020). CRISPR-Cas9 genome editing: Current challenges and future prospects. *Molecular Therapy*, 28(5), 1120–1131.

Knoppers, B. M. (2018). Genomic privacy and ethical governance. *Human Genetics*, 137(6), 799–811.

Koboldt, D. C., Steinberg, K. M., Larson, D. E., Wilson, R. K., & Mardis, E. R. (2013). The next-generation sequencing revolution and its impact on genomics. *Cell*, 155(1), 27–39. DOI: 10.1016/j.cell.2013.09.006 PMID: 24074859

Le Tourneau, C. (2015). Molecularly targeted therapy of cancer. *Nature Reviews. Clinical Oncology*, 12(3), 179–190. PMID: 26342236

Lewis, C. M., & Vassos, E. (2020). Polygenic risk scores: From research tools to clinical instruments. *The Lancet. Psychiatry*, 7(1), 5–7. PMID: 31860452

Li, H., & Durbin, R. (2009). Fast and accurate short read alignment with Burrows-Wheeler transform. *Bioinformatics (Oxford, England)*, 25(14), 1754–1760. DOI: 10.1093/bioinformatics/btp324 PMID: 19451168

Libbrecht, M. W., & Noble, W. S. (2015). Machine learning applications in genetics and genomics. *Nature Reviews. Genetics*, 16(6), 321–332. DOI: 10.1038/nrg3920 PMID: 25948244

Little, J. (2009). The Strengthening the Reporting of Genetic Association Studies (STREGA) Statement: Guidelines for reporting genetic association studies. *American Journal of Human Genetics*, 78(1), 29–41.

Luck, K. (2020). A proteome-scale map of the human interactome network. *Cell*, 184(10), 2745–2764.

Lunshof, J. E. (2008). Genetic data and its ethical implications. *Nature Reviews. Genetics*, 9(4), 295–305.

Mackay, D., & Soares, G. (2020). Ethical and legal considerations in genomic testing. *Journal of Medical Ethics*, 46(3), 179–185.

Mahajan, A., & Taliun, D. (2018). Fine-mapping of an additional 11,000 variants for association with risk of type 2 diabetes. *Nature Genetics*, 50(5), 1270–1282.

Makarova, A. V. (2020). CRISPR/Cas9 gene editing technology: A potential for understanding and treating neurological diseases. *Frontiers in Neurology*, 11, 572317.

Malin, B. A., & Sweeney, L. G. (2010). Determining the identifiability of DNA database entries. *The Journal of Privacy and Confidentiality*, 2(1), 1–19.

Mallal, S. A., Phillips, E., Carosi, G., Molina, J.-M., Workman, C., Tomažič, J., Jägel-Guedes, E., Rugina, S., Kozyrev, O., Cid, J. F., Hay, P., Nolan, D., Hughes, S., Hughes, A., Ryan, S., Fitch, N., Thorborn, D., & Benbow, A. (2008). HLA-B*57:01 screening before starting abacavir. *The New England Journal of Medicine*, 358(6), 568–579. DOI: 10.1056/NEJMoa0706135 PMID: 18256392

Mandelker, D. (2017). The role of exome sequencing in the identification of pathogenic variants in cancer. *Nature Reviews. Cancer*, 17(9), 618–630.

Manolio, T. A. (2013). Implementing genomic medicine in clinical practice: The opportunities and challenges. *Journal of the American Medical Association*, 310(11), 1161–1162.

Manolio, T. A., Collins, F. S., Cox, N. J., Goldstein, D. B., Hindorff, L. A., Hunter, D. J., McCarthy, M. I., Ramos, E. M., Cardon, L. R., Chakravarti, A., Cho, J. H., Guttmacher, A. E., Kong, A., Kruglyak, L., Mardis, E., Rotimi, C. N., Slatkin, M., Valle, D., Whittemore, A. S., & Visscher, P. M. (2019). Finding the missing heritability of complex diseases. *Nature*, 461(7265), 747–753. DOI: 10.1038/nature08494 PMID: 19812666

Mantere, T. (2019). Mutational landscape of BRCA1 and BRCA2 in hereditary breast and ovarian cancer. *Nature Reviews. Cancer*, 19(12), 723–735.

Marx, V. (2013). The big challenges of big data. *Nature*, 498(7453), 255–260. DOI: 10.1038/498255a PMID: 23765498

Mayer, I. A., & Arteaga, C. L. (2016). The PI3K/AKT/mTOR pathway in cancer therapeutics. *Current Opinion in Pharmacology*, 29, 7–12.

McCarthy, M. I. (2016). Genomics and drug discovery. *Science*, 352(6291), 78–83.

McKenna, A., Hanna, M., Banks, E., Sivachenko, A., Cibulskis, K., Kernytsky, A., Garimella, K., Altshuler, D., Gabriel, S., Daly, M., & DePristo, M. A. (2010). The Genome Analysis Toolkit: A MapReduce framework for analyzing next-generation DNA sequencing data. *Genome Research*, 20(9), 1297–1303. DOI: 10.1101/gr.107524.110 PMID: 20644199

McMurray, C. T. (2018). Mechanisms of disease: Dynamic mutations in human genetic disease. *Nature Reviews. Genetics*, 19(7), 440–450.

McShane, L. M. (2020). Clinical trial validation of biomarker-driven therapeutic strategies. *Nature Reviews. Drug Discovery*, 19(10), 616–632.

Mok, T. S. (2009). Epidermal growth factor receptor (EGFR) mutation status in non-small cell lung cancer: A review and the importance of EGFR mutations in treatment. *The Lancet. Oncology*, 10(4), 321–329. PMID: 19350698

Moore, L. D., Le, T., & Wong, H. K. (2019). DNA methylation and its basic functions. *Nature Reviews. Genetics*, 10(2), 125–137.

Morgens, D. W. (2016). Genome-scale identification of CRISPR-Cas9 off-targets in human cells. *Nature Biotechnology*, 34(3), 368–371.

Natarajan, P. (2018). Genetic mapping and mechanisms of coronary artery disease. *Circulation Research*, 122(2), 267–286. PMID: 29167274

Nosek, B. A., Alter, G., Banks, G. C., Borsboom, D., Bowman, S. D., Breckler, S. J., Buck, S., Chambers, C. D., Chin, G., Christensen, G., Contestabile, M., Dafoe, A., Eich, E., Freese, J., Glennerster, R., Goroff, D., Green, D. P., Hesse, B., Humphreys, M., & Yarkoni, T. (2015). Promoting an open research culture. *Science*, 348(6242), 1422–1425. DOI: 10.1126/science.aab2374 PMID: 26113702

O'Sullivan, B. P., & Freedman, S. D. (2020). Cystic fibrosis. *Lancet*, 373(9678), 1572–1583. PMID: 19403164

Phillips, S. (2018). Data governance in genomics: From ethical dilemmas to practical frameworks. *Genome Research*, 28(9), 1347–1355.

Piccart-Gebhart, M. J., Procter, M., Leyland-Jones, B., Goldhirsch, A., Untch, M., Smith, I., Gianni, L., Baselga, J., Bell, R., Jackisch, C., Cameron, D., Dowsett, M., Barrios, C. H., Steger, G., Huang, C.-S., Andersson, M., Inbar, M., Lichinitser, M., Láng, I., & Gelber, R. D. (2005). Trastuzumab after adjuvant chemotherapy in HER2-positive breast cancer. *The New England Journal of Medicine*, 353(16), 1659–1672. DOI: 10.1056/NEJMoa052306 PMID: 16236737

Pinto, D., & Pagnamenta, A. T. (2020). Functional impact of copy number variations in neurodevelopmental disorders. *Nature*, 516(7529), 407–414.

Poste, G. (2011). Bring on the biomarkers. *Nature*, 469(7329), 156–157. DOI: 10.1038/469156a PMID: 21228852

Prainsack, B., & Buyx, A. (2017). The value of biomedical research and its moral obligations. *Cambridge Quarterly of Healthcare Ethics*, 26(1), 15–27.

Price, A. L., Patterson, N. J., Plenge, R. M., Weinblatt, M. E., Shadick, N. A., & Reich, D. (2006). Principal components analysis corrects for stratification in genome-wide association studies. *Nature Genetics*, 38(8), 904–909. DOI: 10.1038/ng1847 PMID: 16862161

Redon, R., Ishikawa, S., Fitch, K. R., Feuk, L., Perry, G. H., Andrews, T. D., Fiegler, H., Shapero, M. H., Carson, A. R., Chen, W., Cho, E. K., Dallaire, S., Freeman, J. L., González, J. R., Gratacòs, M., Huang, J., Kalaitzopoulos, D., Komura, D., MacDonald, J. R., & Hurles, M. E. (2018). Global variation in copy number in the human genome. *Nature*, 444(7118), 444–454. DOI: 10.1038/nature05329 PMID: 17122850

Ross, J. S. (2009). HER2 testing in breast cancer: Current status and future directions. *Journal of Clinical Oncology*, 27(32), 5453–5462.

Sanders, S. J., & Campbell, A. J. (2020). Copy number variation in neurodevelopmental disorders. *Nature Reviews. Genetics*, 21(7), 407–418.

Schaefer, E. I. (2021). Network analysis of Alzheimer's disease. *Neurobiology of Aging*, 91, 39–53.

Schaid, D. J. (2018). Genomic annotations and functional analysis of genetic variants in complex diseases. *American Journal of Human Genetics*, 102(1), 9–22.

Schneider, E. A. (2020). Clinical applications of genomic biomarkers in drug development. *The New England Journal of Medicine*, 383(1), 34–47.

Schork, N. J. (2018). Personalized medicine: Time for one-person trials. *Nature*, 520(7549), 609–611. DOI: 10.1038/520609a PMID: 25925459

Schwaederle, M. (2015). Impact of precision medicine in diverse cancers: A meta-analysis of 1,000 patients. *Journal of Clinical Oncology*, 33(31), 3329–3336.

Schwartz, T. S. (2021). In vitro and in vivo models for the study of genetic variants in disease. *The Journal of Experimental Medicine*, 218(7), e20210492.

Sequist, L. V. (2013). Phase II study of the EGFR inhibitor erlotinib for patients with advanced non-small cell lung cancer (NSCLC) and EGFR mutations. *The Lancet. Oncology*, 14(6), 603–610.

Sharma, P., & Hu-Lieskovan, S. (2018). Companion diagnostics in oncology. *Nature Reviews. Clinical Oncology*, 15(4), 228–237.

Shultz, L. D. (2020). Humanized mouse models for cancer immunotherapy. *Annual Review of Cancer Biology*, 4, 129–151.

Simon, R., & Roychowdhury, S. (2013). Implementing personalized cancer genomics in clinical practice. *Nature Reviews. Clinical Oncology*, 10(7), 389–396.

Slamon, D. J., Leyland-Jones, B., Shak, S., Fuchs, H., Paton, V., Bajamonde, A., Fleming, T., Eiermann, W., Wolter, J., Pegram, M., Baselga, J., & Norton, L. (2001). Use of chemotherapy plus a monoclonal antibody against HER2 for metastatic breast cancer that overexpresses HER2. *The New England Journal of Medicine*, 344(11), 783–792. DOI: 10.1056/NEJM200103153441101 PMID: 11248153

Stark, R. (2019). RNA sequencing: From technology to clinical applications. *Clinical Genetics*, 95(6), 557–564.

Stoeckius, M. (2018). Simultaneous epitope and transcriptome measurement in single cells. *Nature Methods*, 15(5), 457–463. PMID: 28759029

Sullivan, P. F., & Grove, J. (2020). Statistical analysis in large-scale genetic association studies: Challenges and opportunities. *Nature Reviews. Genetics*, 21(6), 352–366.

Tabor, H. (2011). Ethical challenges in genetic testing. *Journal of Genetic Counseling*, 20(4), 417–425.

Tam, S. (2019). Genome-wide association studies in cardiovascular disease: Current status and future directions. *Lancet*, 393(10183), 1254–1264.

Topol, E. J. (2019). High-performance medicine: The convergence of human and artificial intelligence. *Nature Medicine*, 25(1), 44–56. DOI: 10.1038/s41591-018-0300-7 PMID: 30617339

Topol, E. J. (2019). High-performance medicine: The convergence of human and artificial intelligence. *Nature Medicine*, 25(1), 44–56. DOI: 10.1038/s41591-018-0300-7 PMID: 30617339

Trost, B. (2020). Targeted sequencing approaches in disease gene identification: A review of recent advancements. *Nature Reviews. Genetics*, 21(7), 431–444.

Van Cutsem, E., Köhne, C.-H., Hitre, E., Zaluski, J., Chang Chien, C.-R., Makhson, A., D'Haens, G., Pintér, T., Lim, R., Bodoky, G., Roh, J. K., Folprecht, G., Ruff, P., Stroh, C., Tejpar, S., Schlichting, M., Nippgen, J., & Rougier, P. (2009). Cetuximab and chemotherapy as initial treatment for metastatic colorectal cancer. *The New England Journal of Medicine*, 360(14), 1408–1417. DOI: 10.1056/NEJMoa0805019 PMID: 19339720

Visscher, P. M., Wray, N. R., Zhang, Q., Sklar, P., McCarthy, M. I., Brown, M. A., & Yang, J. (2017). 10 years of GWAS discovery: Biology, function, and translation. *American Journal of Human Genetics*, 101(1), 5–22. DOI: 10.1016/j.ajhg.2017.06.005 PMID: 28686856

Wallace, D. C. (2019). Mitochondrial DNA and disease. *Human Genetics*, 138(1), 1–18. PMID: 30390160

Wang, K. (2018). Statistical methods for genome-wide association studies. *Nature Reviews. Genetics*, 19(10), 597–610. PMID: 29727703

Woodcock, J., & LaVange, L. M. (2017). The regulatory approval of genomic biomarkers: A guide. *The New England Journal of Medicine*, 377(24), 2319–2321. PMID: 28402237

Yu-Wai-Man, P., & Votruba, M. (2020). Leber hereditary optic neuropathy. *Archives of Neurology*, 67(9), 1167–1173. PMID: 33298565

Zhang, Z. (2019). KRAS and EGFR mutations in non-small cell lung cancer: Targeted treatment approach and mechanisms. *Oncotarget*, 10(45), 4485–4500.

Zhu, Y. (2017). Emerging technologies in genomic medicine: Single-cell sequencing and spatial transcriptomics. *Nature Reviews. Genetics*, 18(7), 485–497.

Zook, J. M. (2020). Genomic data security and privacy: A global perspective. *Nature Reviews. Genetics*, 21(8), 516–530.

Chapter 5

Global Perspectives on Pharmacogenomics and Drug Discovery

K. Rajakumari
https://orcid.org/0000-0002-8102-9512
Vels Institute of Science, Technology, and Advanced Studies, India

S. S. Meenambiga
https://orcid.org/0000-0002-5445-525X
Vels Institute of Science, Technology, and Advanced Studies, India

K. Kavinaya Shri
https://orcid.org/0009-0003-4091-4341
Vels Institute of Science, Technology, and Advanced Studies, India

P. Vivek
https://orcid.org/0000-0003-2088-7545
Vels Institute of Science, Technology, and Advanced Studies, India

R. Logesh
https://orcid.org/0009-0004-0968-2762
Vels Institute of Science, Technology, and Advanced Studies, India

S. Ivo Romauld
Vels Institute of Science, Technology, and Advanced Studies, India

ABSTRACT

Pharmacogenomics, the study of how genes influence an individual's response to drugs, is changing the scene of medication disclosure and improvement by joining pharmacology and genomics to make powerful, custom-made to an individual's hereditary profile. This arising field is vital in customized medication, where medicines are upgraded in light of individual hereditary varieties. This chapter offers an in-depth exploration of the global impact of pharmacogenomics, underscoring key examination foundations and undertakings, moral contemplations, difficulties and future headings. The joint efforts exhibit the capability of joint endeavor to defeat difficulties and advance the field by fitting worldwide guidelines and creating strategies that help even-handed admittance to pharmacogenomic progressions are

DOI: 10.4018/979-8-3693-6597-7.ch005

significant for the worldwide reconciliation of customized medication.

1. INTRODUCTION

Pharmacogenomics is an important field that combines pharmacology and genomics to improve drug therapy by personalizing drugs to individuals' genetic profiles. This method considerably increases treatment efficacy and safety by addressing interindividual variability in drug response induced by hereditary variables, which can account for 15-30% of variability in drug reactions. Pharmacogenomics helps predict adverse drug responses (ADRs) and optimize drug dosages by detecting genetic markers such as single nucleotide polymorphisms (SNPs), thereby lowering the occurrence of ADRs and improving patient outcomes. Scientists can utilize a patient's genetic profile to predict therapeutic efficacy, optimize amount, and improve patient safety. Pharmacogenomics testing assists clinicians in predicting which drugs are most likely to achieve the desired reactions in certain patients. Introducing pharmacogenomics into clinical practice has an opportunity to optimize medication therapy, lower the risk of adverse events, and improve personalized care. While challenges still exist, breakthroughs in genetic testing technology and improved clinical data are moving us closer to getting the positive effects of pharmacogenomics-guided medicine (Pirmohamed, 2023; Getahun et al., 2024; Henderson et al., 2018)· Furthermore, pharmacogenomics promotes personalized treatment by enabling the development of tailored medicines for a variety of health issues, including cancer and cardiovascular disease. Despite its potential, problems persist in the clinical translation of pharmacogenomic discoveries, demanding further research and collaboration (Sharma et al., 2024; Nogueiras-Álvarez, 2023).

1.1 Overview of Pharmacogenomics

Pharmacogenomics, the study of how hereditary contrasts impact drug reactions, has made considerable progress since its beginning. The journey started in the 1950s when researchers saw that hereditary varieties could influence how individuals metabolize and respond to medications. Early exploration focused on observable contrasts in drug digestion, for example, the revelation of acquired catalyst lacks like glucose-6-phosphate dehydrogenase (G6PD), which cleared up why a few people experienced unfriendly responses for specific medications. The expression "pharmacogenetics" was introduced in 1959 to describe these hereditary effects on drug reactions. In any case, it was only after the 1980s, with the approach of quality cloning and DNA sequencing advancements, that these varieties' hereditary underpinnings became clearer. The completion of the Human Genome Project in

2003 was a milestone occasion, giving a complete guide to human hereditary variety and making way for the more extensive field of pharmacogenomics. Pharmacogenomics (PGx) is an important tool in both preclinical and clinical studies, boosting personalized medicine by adapting pharmacological therapy to individual genetic profiles. Since the 2000s, there has been a significant increase in the use of PGx in clinical trials, particularly in oncology and mental health, with important genes such as CYP2D6 and CYP2C19 being extensively researched for their effects on therapeutic effectiveness and safety. A prominent example is a randomized clinical trial including 665 patients with depressive disorders in which pharmacogenomic testing resulted in higher remission (31.0% vs. 20.0%) and response rates (48.7% vs. 37.3%) than usual therapy techniques. However, obstacles continue, particularly in Phase I oncology trials, where studies frequently have small, heterogeneous populations and insufficient genetic variant analysis, making it difficult to interpret data (Xu et al., 2024; Sissung & Figg, 2022; ElGarhy et al., 2024).

Today, pharmacogenomics goes beyond single-gene drug interactions to include genome-wide approaches that survey hereditary varieties across different genes. This shift has enabled the improvement of personalized medication, where medicines are custom-made to a person's hereditary profile, further developing adequacy and diminishing adverse effects. The integration of pharmacogenomics into clinical practice has been advanced rapidly by progress in high-throughput genotyping advancements and bioinformatics (Lauschke et al., 2024). Recent patterns in pharmacogenomics incorporate the identification of new hereditary markers for drug reactions, the advancement of pharmacogenomic testing boards, and the fusing of pharmacogenomic information into electronic well-being records. These headways are driving the reception of accurate medication, where hereditary data guides drug choice and dosing, eventually improving patient consideration (Moc, 2020; Ingelman-Sundberg et al., 2023; Alajely et al., 2023) .

1.2 Role of Pharmacogenomics

Pharmacogenomics improves our understanding of how genetic polymorphisms influence medication response, which aids in drug discovery and customized therapy. This discipline aims to find biomarkers that predict patient reactions to medications by combining genetics and pharmacology. This is significant because it can increase therapy efficacy while minimizing negative effects. Pharmacogenomics allows for the classification of patients as responders or non-responders based on genetic traits such as polymorphisms in drug-metabolizing enzymes, which can then be used to inform tailored treatment approaches. Furthermore, incorporating pharmacogenomics into clinical trials and drug levels (Sharma et al., 2024; ElGarhy et al., 2024).

1.3 Importance in Drug Discovery

Due to safety and efficacity issues, shy clinical restatement stymies the development of new medicines. A considerable proportion (54) of investigational new drugs (INDs) fail clinical development, with safety enterprises counting for 17 and inadequate efficacity for 57. Pharmacogenomics can help medicine makers comprehend heritable diversity in drug response by incorporating inheritable data into medicine exploration and development, speeding up medicine development, and precluding remedial failures. This integration helps all stages of the clinical trial process, from natural target selection to patient recruiting. Next-generation sequencing (NGS) and other genomic advancements make pharmacogenomic exploration easier (Lauschke et al., 2024). Implicit gene studies, genome-wide association studies, and whole exome sequencing are three common genomic approaches used in medicine development. Pharmacogenomics can also reduce adverse medicine responses (ADRs), which are a leading cause of hospitalization and death. individualities may reply negatively to some medicines due to inheritable dissonances, and fetching these tendencies aids in developing safer medicine (Henderson et al., 2018). Individualized treatment is a pivotal element of pharmacogenomics, which involves modifying the remedy according to a case's genomic biographies to ameliorate treatment issues and patient compliance. This approach marks a change from generalist to more specific and technical care. Likewise, pharmacogenomics contributes in the discovery of innovative drugs for inheritable abnormalities, similar as targeted oncology treatments (Olivier & Williams-Jones, 2011). Pharmacogenomics can be used in a variety of remedial disciplines, and medicine repurposing grounded on inheritable information may reduce development time and costs.

1.4 Influence of Pharmacogenomics on Drug Discovery and Related Methodologies

Pharmacogenomics is revolutionizing drug discovery by utilizing genetic information to create safer and more effective treatments. This field uses genomic data to determine pharmacological targets, optimize leads, and customize therapies.

a. **Target Identification**

Pharmacogenomics identifies genetic differences associated with diseases, which can be exploited to find prospective therapeutic targets. Understanding the genetic foundation of diseases enables researchers to develop medications that particularly target these genetic variants, resulting in more effective treatments. For example, discovering mutations in the BRCA1 and BRCA2 genes has resulted in specific treatments for breast cancer.

b. **Lead Optimization**

Pharmacogenomics helps to refine lead molecules after they have been identified as possible therapeutic targets. This involves testing chemicals against genetic variations to determine their effectiveness and safety. High-throughput screening and bioinformatics technologies are utilized to assess how various genetic profiles react to these drugs, ensuring that the most promising leads are chosen for further development.

c. **Personalized Medicine**

Pharmacogenomics promotes the development of individualized therapeutics by adapting treatments to an individual's genetic makeup. This method aids in the prediction of drug reactions while minimizing undesirable effects. For example, genetic testing for CYP2C19 gene variants can guide the use of antiplatelet medications such as clopidogrel, ensuring adequate doses and lowering the risk of adverse responses.

Pharmacogenomics Methodologies

Pharmacogenomics uses a variety of approaches to determine how genetic differences influence drug responses. These strategies are critical for finding therapeutic targets, optimizing lead compounds, and creating tailored treatments.

a. **Genomic Data Analysis**

Genomic data analysis entails investigating genetic variants throughout the genome using modern sequencing technology. This includes the following:

i. Whole genome sequencing (WGS) is the process of sequencing the entire genome to find all genetic variants.
ii. Whole Exome Sequencing (WES) entails sequencing only the protein-coding areas of the genome that are most likely to influence medication response.
iii. Targeted sequencing focuses on genes or areas that are known to be involved in drug metabolism and response.

These procedures aid in the identification of genetic markers that determine how individuals respond to drugs, providing the conditions for individualized treatment.

b. **Bioinformatics**

Bioinformatics utilizes computational techniques to organize and understand huge genetic collections. Key features include:

i. Data management is storing and arranging large amounts of genetic data.
ii. Data analysis involves using computers and statistical approaches to discover patterns and correlations between genetic variants and medication reactions.
iii. Predictive Modeling: Creating models that forecast how genetic variants may affect drug efficacy and safety.

Bioinformatics is critical for understanding the complicated data provided by genomic investigations and discovering possible therapeutic targets.

c. **High-Throughput Screening**

High-throughput screening (HTS) is a technique for rapidly testing thousands of chemicals against genetic variations to uncover viable therapeutic candidates. This technique involves:

i. Compound Libraries: collections of thousands of chemical compounds that can be evaluated for biological activity.
ii. Assay development involves developing tests that can quantify the interaction between chemicals and genetic variations.
iii. Automation is the use of robotic devices to efficiently execute large-scale screenings.

HTS enables researchers to quickly identify chemicals that interact with specific genetic targets, hence accelerating the drug discovery process.

d. **Clinical Trials**

Pharmacogenomics clinical trials involve stratifying individuals based on genetic profiles in order to test the efficacy and safety of novel medications. This method includes:

i. Genetic screening is testing patients for certain genetic markers prior to enrolling them in clinical trials.
ii. Stratified trials include grouping patients based on their genetic profiles in order to customize treatments and enhance outcomes.

iii. Outcome Analysis: Assessing how genetic variants affect treatment responses and side effects.

Stratified clinical trials serve to guarantee that new treatments are both effective and safe for people with specific genetic profiles, opening the way for customized medicine. [14][15][16]

1.5 Global Impact and Relevance

Pharmacogenomics can improve therapeutic efficacy and safety, which is one of its key benefits. Healthcare practitioners can choose the most effective treatments and dosages by tailoring pharmaceuticals to their patient's genetic profiles, lowering the likelihood of adverse drug reactions. ADRs are a major cause of morbidity and mortality around the world, and pharmacogenomics provides a strategy to reduce these risks (Henderson et al., 2018). Personalized medicine can lead to more efficient healthcare systems by decreasing the need for trial and error in drug prescribing. Patients receive effective therapies faster, reducing the duration and cost of disease. This efficiency means lower healthcare expenses and better resource utilization. Pharmacogenomics is revolutionizing medication development by allowing researchers to identify patient subgroups that will benefit the most from novel treatments. This technique improves clinical trial success rates and speeds up the approval process for new pharmaceuticals. It also aids in the early detection of potential side effects throughout the drug development process, hence improving the overall safety of new treatments. Pharmacogenomics has the potential to reduce health inequities by providing information on how various populations respond to pharmaceuticals. Genetic variety among communities can cause varied pharmacological reactions, and recognizing these variances might help underrepresented groups get better treatment outcomes. This understanding is critical for establishing inclusive healthcare policies that address varied genetic backgrounds (Astras et al., 2020) . Figure 1 shows the global impacts and relevance of pharmacogenomics.

Figure 1. Global impacts and relevance of pharmacogenomics

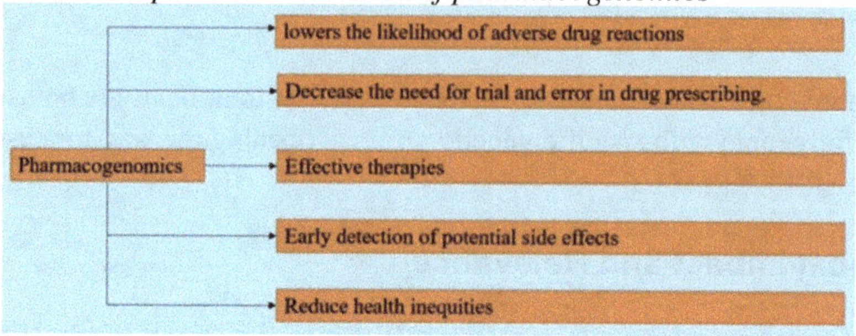

2. PHARMACOGENOMICS AROUND THE WORLD

2.1 North America

In North America, pharmacogenomics focuses on understanding how a person's inheritable composition influences their response to medicines. This field brings together pharmacology and genetics to produce personalized curatives that are more effective and have smaller negative goods. Inheritable testing examines DNA from a sample to find gene variants that impact drug metabolism. Genes generally examined include CYP450 enzymes, including CYP1A1, CYP1A2, CYP2A6, CYP2B6, CYP2C9, CYP2C19, CYP2D6, CYP2E1, CYP3A4, CYP3A5, and CYP4F2. The inheritable information is used to define medicines that are customized to an existent's inheritable profile, maximizing efficacity while avoiding adverse responses. Exemplifications include modifying warfarin tablets, opting for antidepressants, and deciding on chemotherapy treatments (Krebs & Milani, 2019).

The United States Food and Drug Administration (FDA) issues guidelines and concurrences for pharmacogenomic tests, as well as incorporating pharmacogenomic information into drug markers, while Health Canada follows an analogous nonsupervisory approach (Xu et al., 2024). Integrating inheritable data into electronic health records (EHRs) enables healthcare providers to pierce and use this information in clinical decision- timber (Sissung & Figg, 2022) . Hospitals, exploration associations, and inheritable testing pots work together to progress the field. Still, there are several issues, similar as sequestration enterprises, informed concurrence, indifferent access, the high expenditure of inheritable testing, and the need for ongoing education for healthcare providers.

Pharmacogenomic testing's clinical value is being validated as new gene-medicine relations are discovered. Individualized drug ways are decreasingly being used in clinical trials to estimate medicine responses rested on inheritable biographies. Organizations and support groups offer information and services to people witnessing pharmacogenomic testing, and patient participation is essential in addressing their inheritable knowledge and treatment ability (Abad-Santos et al., 2024) .

Pharmacogenomics has great eventuality for perfecting treatment results and customizing healthcare in North America, but it also confronts several practical and ethical problems that bear uninterrupted attention (Krebs & Milani, 2019).

2.2 Europe

Research in the field of pharmacogenomics across Europe covers a wide range of activities, with several key institutions and projects playing significant roles. Here's a look at some of the main entities and efforts involved.

Major Contribution and Innovations

Provides a general indication of the focus of this field by comparing the number of publications on pharmacogenetics or pharmacogenomics to the total number of publications on all themes, in these nations, using a classification based on the Scopus database. Above nations like Switzerland, Sweden, or Germany, where this discipline was first formed in Europe, are Greece, the Netherlands, and Spain on this ranking. The situation of France, which has roughly three times the research expenditure of Spain, is particularly pertinent.

Significant advancements in the pharmacogenomics of non-steroidal anti-inflammatory drugs (NSAIDs) have been made by Spanish groups. These include the identification of pharmaco-genes that determine the pharmacokinetics, the adverse effects, and the modulation of potential benefits for NSAIDs. A person with impaired metabolism is exposed to higher drug concentrations and is consequently more likely to have undesirable medication effects. Genetic variability in CYP2C activity is also linked to severe reactions to NSAIDs, such as gastrointestinal bleeding or drug hepatotoxicity (Magavern et al., 2022a) .

Regulatory Frameworks and Initiatives: A primary obstacle to the widespread use of PGx has been the death of data indicating enhanced effectiveness, safety, or economic viability. The position paper that goes with it contains examples of instances where there is proof, and this body of evidence is expanding. Most research to date indicates that PGx is a cost-effective solution. To win over government policymakers, longer-term evidence proving cost-effectiveness and favorable clinical outcomes are essential. Currently, recommendations exist for over 80 medications. Pharmaco- genetic variations are curated by PharmGKB. The majority are based on genetic research conducted in populations with Caucasian European heritage. Given the variety of ancestries seen throughout Europe and the admixing of people, interpreting variations of unknown significance (VUS) and reclassifying them as needed may prove difficult [21] [22].

Framework for Laws and Regulations: European legislation pertaining to the use of genetic testing in healthcare systems is diverse and intended more for diagnostic purposes than for PGx testing. As such, they apply to PGx testing specifically but are not limited to it. The framework for diagnostic genetic testing might be unduly restrictive in a PGx context, which would prevent PGx from being implemented.

EU's (European Medicines Agency) Position on PGx: Best practices for genetic testing analysis and results actionability are supported by the European Medicines Agency (EMA); product information comprises current PGx. Currently, rather than an extensive EMA commentary on PGx panel testing, individual drugs carry PGx data as endorsed by EMA licensure. The overview of product characteristics of 15% of all EMA-licensed pharmacological medicines includes PGx information (Lo et al., 2020).

2.3 Asia

Asians account for more than 60% of the global population and have a large pharmacogenetic variety, which influences how people respond to drugs. Asian communities exhibit a high prevalence of gene variations in enzymes like CYP2C19, which contribute to the metabolism of drugs, in addition to alleles like HLA-B15:02, which are linked to bad drug reactions (Ingelman-Sundberg et al., 2023). Individuals with the HLA-B15:02 allele, for example, are more prone to experience severe skin responses such as Stevens-Johnson syndrome after interacting with medicines like carbamazepine ((Jolly & Scaria, 2024).

The Clinical Pharmacogenetics Implementation Consortium (CPIC) and the United States Food and Drug Administration (FDA) provide guidelines based on research from various populations, but these recommendations may not necessarily include specific proposals for particular ethnic groups (Alajely et al., 2023). Even this, the Precision Medicine Initiative and other NIH-funded groups, such as the Pharmacogenomics Research Network (PGRN), are researching to incorporate pharmacogenomics into clinical practice globally.

Government policies and financial assistance have significant effects on drugs research and application in healthcare. Understanding and utilizing these genetic variants allows healthcare providers to modify medication therapy to improve safety and efficacy across varied populations, leading the way for modified medicine worldwide (Jolly & Scaria, 2024; Sosa-Macías et al., 2016).

2.4 Latin America

In the third session, D. Remírez from the Cuban Agency for Medicinal Products presented their regulatory guidelines for pharmacogenomics research and drug development. These include tight sample management, accurate biomarker recording, storage, and classification, as well as adherence to ethical requirements. The talk also discussed Cuba's efforts in pharmacogenetics research and bioproduct development within the country's regulatory framework. G. Moya of Argentina noted the lack of particular standards for handling samples and data in pharmacogenetics, instead relying on normal medical and biomedical research laws. Moving on to Ecuador, F. Hernández demonstrated the practical application of pharmacogenetics with a study on CYP2D6 genotype and phenotype, underlining the significance of following worldwide standards in research design and ethical committees.

The workshop explores deeper into regulatory concerns in Mexico under the Norm NOM-012-SSA3-2012, with a focus on ethical considerations such as risk management, informed consent, and privacy rights in clinical research involving humans. M. Fors discussed methodological challenges in pharmacogenetic clinical trials, emphasizing the importance of research design, the use of biostatistics and bioinformatics tools, and ethical/legal concerns in getting informed permission. C. Altamirano further noted that healthcare providers in Latin America rely on drug prospectuses and data sheets for pharmacogenetic information (Chu et al., 2022).

In terms of broad consequences, pharmacogenomics has the potential to customize drugs and improve treatment results by identifying high-risk patients and reducing adverse effects. However, integrating pharmacogenomic analysis into clinical practice remains challenging due to implementation constraints and various regulatory approaches around the world. Collaborative efforts, such as those sponsored by collaborations like the International Transporter Consortium (ITC), aim to promote drug interaction understanding and improve pharmacogenomic research standards by sharing resources and data. These initiatives highlight ongoing attempts to standardize pharmacogenomic data among regulatory bodies and improve its therapeutic value globally (Miteva-Marcheva et al., 2020).

2.5 Global Disparities in Access to Pharmacogenomic Technologies

Despite progressions in pharmacogenomics, admittance to these advances changes fundamentally overall due to financial, infrastructural, and instructive contrasts.

1. **Economic Barriers**

Significant expenses of sequencing and genotyping advances limit their adoption in low- and middle-income countries, confining the execution of pharmacogenomic testing.

2. **Infrastructural Challenges**

Many developing regions lack the necessary lab facilities, gear, and prepared faculty. Obsolete medical care IT frameworks further entangle the coordination of pharmacogenomic information into clinical practice.

3. **Educational and Awareness Gaps**

There is a significant lack of mindfulness and training about pharmacogenomics among medical care suppliers and patients in many regions of the planet. This hole upsets the reception and utilization of pharmacogenomic testing in personalized medication.

4. **Regulatory and Ethical Considerations**

Fluctuating guidelines and moral worries about hereditary testing and information protection obstruct to the reception of pharmacogenomic advancements. Resolving these issues is critical for fair access.

Efforts to Address Disparities

Global joint efforts and subsidizing plans to create the pharmacogenomic foundation and prepare in low and middle-income nations. Fitting administrative structures and advancing moral guidelines are likewise fundamental for worldwide integration (• (David et al., 2024; Jordan et al., 2023; Ingelman-Sundberg et al., 2023; Tyumentseva et al., 2023).

3. CASE STUDIES OF SUCCESSFUL PHARMACOGENOMIC APPLICATIONS

The case studies show how pharmacogenomics might improve drug safety and efficacy by personalizing therapies to patients' genetic profiles were depicted in Figure 2.

Figure 2. Successful pharmacogenomics applications

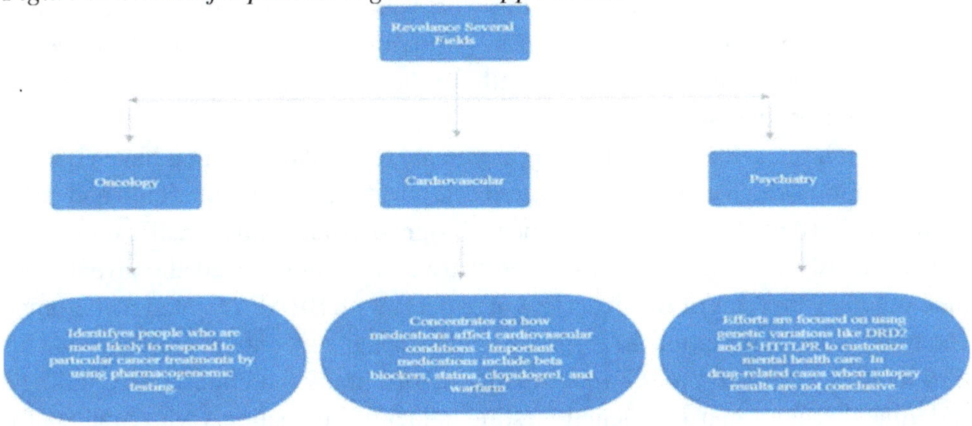

3.1 Personalized Medicine in Oncology

Personalized medicine in oncology is one of the most significant advances made possible by pharmacogenomics. Oncologists can greatly improve outcomes and reduce side effects by modifying cancer treatment to both the patient and the tumor's genetic profile. Here, we look at many case studies that demonstrate the efficacy of pharmacogenomic applications

Trastuzumab (Herceptin) for HER2-positive breast cancer -Trastuzumab (Herceptin) is one of the most well-known successes in customized oncology. HER2 is a protein that stimulates cancer cell development, and it is overexpressed in roughly 20% of breast tumors. Trastuzumab, a monoclonal antibody, inhibits the development of cancer cells by targeting the HER2 protein.

Case Study 1: A major clinical trial found that adding trastuzumab to chemotherapy increased survival rates for women with HER2-positive breast cancer. The study discovered that patients who received trastuzumab had a 52% increase in disease-free survival compared to those who received only chemotherapy. This tailored medicine has since become the standard treatment for HER2-positive breast cancer, demonstrating that the course of may alter cancer treatment by focusing on genetic markers.

Imatinib (Gleevec) and Chronic Myeloid Leukemia (CML) - CML is a white blood cell malignancy caused by the Philadelphia chromosome, resulting in the BCR-ABL gene fusion. Imatinib (Gleevec) was created to precisely inhibit the BCR-ABL tyrosine kinase, which causes uncontrolled cell division in CML [3].

Case Study 2: Imatinib transformed the treatment of CML. Clinical trials demonstrated that imatinib caused complete hematologic and cytogenetic responses in the vast majority of CML patients, changing what was formerly a deadly disease into a treatable chronic condition. The success of imatinib not only provided a targeted therapeutic option, but also paved the path for the development of further tyrosine kinase inhibitors in oncology.

BRAF Inhibitors for Melanoma- Approximately 50% of melanomas have a BRAF gene mutation, which causes uncontrolled cell proliferation. Vemurafenib and dabrafenib are BRAF inhibitors that have been designed to target this specific mutation.

Case Study 3: Clinical trials showed that vemurafenib exceeded conventional chemotherapy in terms of survival rates in patients with BRAF-mutated melanoma. The trials found that vemurafenib-treated patients had a median overall survival of 13.6 months, compared to 9.7 months for chemotherapy patients. These findings underline the significance of genetic testing in identifying patients who can benefit from tailored medicines, as well as the potential of pharmacogenomics in enhancing cancer treatment outcomes.

EGFR Inhibitors for Non-Small Cell Lung Cancer (NSCLC): Some individuals with NSCLC have mutations in the EGFR gene. EGFR inhibitors, such as gefitinib and erlotinib, were created to combat these mutations [31].

Case study 4: The research on gefitinib found that individuals with EGFR-mutant NSCLC had considerably greater response rates and longer progression-free survival than those receiving conventional treatment. Specifically, those treated with cancer had a progression-free survival of 10.8 months compared to 5.4 months for those treated with chemotherapy. This study highlighted the importance of genetic testing in guiding treatment decisions and revealed the efficacy of targeted medicines in improving patient outcomes [17].

These case studies highlight the revolutionary power of pharmacogenomics in oncology. Personalized medicine has greatly improved patient outcomes, minimized side effects, and created new pathways for focused therapy by discovering genetic abnormalities and adapting therapies accordingly. The continued integration of genetic testing into clinical practice, as well as ongoing research into novel genetic targets, promise to improve the precision and effectiveness of cancer treatments, ultimately leading to better prognoses and quality of life for cancer patients (Olivier & Williams-Jones, 2011; McDonough, 2021).

3.2 Cardiovascular Pharmacogenomics

Pharmacogenomics has produced significant improvements in cardiovascular care, resulting in more tailored and effective medicines. Clinicians can adapt medicines to individual patients by identifying genetic differences that influence medication metabolism, efficacy, and safety, decreasing side effects and increasing results. Here, we look at various case studies that demonstrate the efficacy of pharmacogenomic applications in cardiovascular pharmacogenomics (Roden, 2003; Božina et al., 2020).

Warfarin and VKORC1/CYP2C9 Genotyping - Warfarin, a popular anti-coagulant, has a narrow therapeutic window and high interpatient variability in dose requirements. Genetic differences in the VKORC1 and CYP2C9 genes have a significant impact on warfarin sensitivity and metabolism. VKORC1 influences the drug's target enzyme, whereas CYP2C9 has an impact on warfarin metabolism ((Roden, 2003; (Madan et al., 2015).

Case 1: An important study found that genotyping patients for VKORC1 and CYP2C9 polymorphisms before starting warfarin treatment greatly improved dose prediction accuracy. Patients on genotype-guided dose reached therapeutic INR (International Normalized Ratio) levels faster and saw fewer serious bleeding or thromboembolic events than those on standard dosing. This study resulted in the incorporation of genetic information into warfarin prescribing recommendations and demonstrated the therapeutic efficacy of pharmacogenomic testing in optimizing anticoagulant medication.

Clopidogrel and CYP2C19 Genotyping - Clopidogrel is a popular antiplatelet medicine used to prevent cardiovascular events in patients undergoing percutaneous coronary intervention (PCI) or those suffering from acute coronary syndromes. The CYP2C19 enzyme, which transforms the prodrug into the active form, influences clopidogrel's efficacy. Variants in the CYP2C19 gene can result in impaired metabolism and lower medication efficacy, raising the risk of severe cardiovascular events (Madan et al., 2015).

Case Study 2: According to the TRITON-TIMI 38 trial, patients with reduced-function CYP2C19 alleles have a considerably higher risk of serious adverse cardiovascular events when treated with clopidogrel than non-carriers. Following these findings, another research revealed that genotyping for CYP2C19 variations and utilizing alternative antiplatelet medications such as prasugrel or ticagrelor in poor metabolizers lowered the occurrence of adverse events. This method has been adopted into clinical practice, with guidelines advocating CYP2C19 genotyping to guide antiplatelet medication in high-risk patients (Abad-Santos et al., 2024).

Statins and SLCO1B1 Genotyping: Statins are commonly used to control cholesterol and decrease cardiovascular risk. However, statin-induced myopathy is a serious side effect that may limit their use. Variants in the SLCO1B1 gene, which encodes the liver transporter OATP1B1, have been linked to a higher risk of statin-induced myopathy, notably with simvastatin (Roden, 2003).

Case Study 3: The searching Collaboration Group performed a study which found an established connection between the SLCO1B1*5 allele and the risk of myopathy in simvastatin patients. Patients with this mutation had a much higher incidence of myopathy than those without it. Based on this research, clinical guidelines now urge SLCO1B1 genotyping before starting high-dose simvastatin therapy in order to identify people at risk and investigate alternate statins or lower doses. This genotyping contributes to the personalization of statin medication, improving safety and adherence (Roden, 2003; Božina et al., 2020; Madan et al., 2015).

ACE Inhibitors and ACE Genotyping- Angiotensin-converting enzyme (ACE) inhibitors are crucial in the treatment of hypertension and heart failure. The ACE gene insertion/deletion (I/D) polymorphism has been investigated to determine its effect on ACE inhibitor responsiveness. The D allele is linked to increased ACE levels and potentially differing treatment outcomes.

Case Study 4: Studies have found that patients with the DD genotype may enjoy more blood pressure decrease with ACE inhibitors than those with the II genotype. A meta-analysis of many studies indicated that ACE inhibitor medication was more successful in lowering cardiovascular events in people with the DD genotype. These data suggest that ACE genotyping could aid in the selection and administration of ACE inhibitors, improving treatment outcomes for hypertensive patients (Božina et al., 2020).

3.3 Pharmacogenomics in Psychiatry

Pharmacogenomics in psychiatry is transforming the treatment of mental illnesses by personalizing drugs to individual genetic profiles. This method improves treatment efficacy and reduces side effects while addressing the substantial diversity in patient reactions to psychiatric drugs. Several major studies and applications demonstrate the success and potential of pharmacogenomics in this field.

Antidepressants and the CYP2D6/CYP2C19 genotype - Antidepressants, particularly selective serotonin reuptake inhibitors (SSRIs) and tricyclic antidepressants (TCAs), have wide interindividual heterogeneity in therapeutic efficacy and adverse effects. The cytochrome P450 enzymes CYP2D6 and CYP2C19 play critical roles in the metabolism of several of these drugs. Genetic variations in these enzymes can classify people as poor, intermediate, extensive, or ultra-rapid metabolizers, which affects drug plasma levels and therapeutic outcomes. A meta-analysis of several trials found that patients with distinct CYP2D6 and CYP2C19 polymorphisms had different reactions to antidepressants. For example, poor CYP2D6 metabolizers who took nortriptyline, a TCA, had higher drug levels, which led to increased adverse effects, whereas ultra-rapid metabolizers had subtherapeutic levels and low efficacy. Incorporating genotyping into clinical practice enabled tailored dosage modifications or alternative pharmaceutical options, resulting in dramatically improved treatment outcomes. Clinical guidelines, such as those issued by the Clinical Pharmacogenetics Implementation Consortium (CPIC), increasingly propose using genotyping to guide antidepressant therapy.

Antipsychotics and the CYP2D6 Genotyping Antipsychotic medicines, which are used to treat disorders such as schizophrenia and bipolar disorder, show variability in efficacy and side effects due to CYP2D6 metabolism. Variants in the CYP2D6 gene can significantly change the metabolism of medicines such as risperidone and aripiprazole. In a study of risperidone-treated patients, those with poor CYP2D6 metabolizers had greater plasma concentrations of the drug, which resulted in increased side effects such as extrapyramidal symptoms and weight gain. In contrast, ultra-rapid metabolizers had lower medication levels, resulting in decreased efficacy. Genotyping for CYP2D6 variations allowed clinicians to tailor risperidone dosing or choose alternative antipsychotics, optimizing treatment while minimizing side effects. This method has been verified in clinical studies and is now part of psychiatric pharmacogenomic testing arrays [35] [36].

Mood Stabilizers and HLA-B*1502 Genotyping - Carbamazepine, a mood stabilizer used to treat bipolar disorder and epilepsy, has been linked to severe cutaneous adverse reactions (SCARs), such as Stevens-Johnson syndrome (SJS) and toxic epidermal necrolysis (TEN). The HLA-B*1502 allele considerably raises this risk, particularly among people of Asian heritage.A crucial investigation found a significant incidence of SJS/TEN in carbamazepine-treated patients carrying the HLA-B1502 allele. Regulatory organizations such as the FDA and the European Medicines Agency (EMA) have since suggested HLA-B1502 genotyping before beginning carbamazepine medication in Asian patients. This method has dramatically reduced the incidence of these life-threatening responses, emphasizing the importance of pharmacogenomics in improving drug safety in psychiatry (Getahun et al., 2024; Cai et al., 2020).

SSRIs and Serotonin Transporter Gene (SLC6A4) Genotyping: The serotonin transporter gene (SLC6A4) is critical to the effectiveness of SSRIs, the most commonly prescribed type of antidepressant. The SLC6A4 gene promoter polymorphism (5-HTTLPR) impacts serotonin reuptake and the responsiveness to SSRI therapy [2]. According to studies, those with the short (S) allele of 5-HTTLPR have lower response rates and greater side effects from SSRIs than people with the long (L) gene. By genotyping individuals for the S allele, physicians can examine alternate treatments or change SSRI doses to get better results. This genotyping approach is gradually being adopted in clinical practice, aiding in the personalization of antidepressants.

Pharmacogenomics is revolutionizing psychiatry by offering individualized treatment approaches that increase efficacy, reduce side effects, and improve patient outcomes. Healthcare professionals can better predict responses to psychiatric drugs by adding genetic testing into clinical practice, allowing them to adapt treatments to specific genetic profilesAs research advances, the potential for pharmacogenomics to refine and tailor psychiatric care grows, making precision medicine a more feasible objective in mental health (Morley & Hall, 2004; Caspar et al., 2021).

3.4 Other Notable Case Studies

Pharmacogenomics has achieved significant advances in fields other than oncology, cardiology, and psychiatry. Here, we look at important case studies from a variety of medical professions that show how genetic discoveries have improved treatments, reduced side effects, and better patient outcomes.

HIV Treatment and the HLA-B*5701 Genotyping- Abacavir, an antiretroviral medicine used to treat HIV, can produce severe hypersensitivity responses (HSRs) in certain people, which can be fatal. The HLA-B*5701 allele was found to be a significant predictor of these reactions.

Case Study 1: According to a landmark study, patients with the HLA-B5701 allele were significantly more likely to develop abacavir-induced HSR. Clinicians considerably lowered the incidence of HSR by performing routine HLA-B5701 screening prior to commencing abacavir treatment. This pharmacogenomic technique has been included into HIV treatment guidelines around the world, demonstrating how genetic testing can improve drug safety and patient outcomes (Getahun et al., 2024; Magavern et al., 2022a).

Thiopurines and TPMT Genomics in Leukemia - Thiopurine medicines, like mercaptopurine and azathioprine, are used to treat acute lymphoblastic leukemia (ALL) and autoimmune disorders. Thiopurine methyltransferase (TPMT) is an enzyme required for the metabolism of these medicines. TPMT genetic variants can cause differences in enzyme activity, which can have an impact on drug toxicity and effectiveness (Olivier & Williams-Jones, 2011).

Case Study 2: Research has indicated that patients with low or absent TPMT activity are more likely to experience severe myelosuppression when treated with conventional thiopurines. By genotyping for TPMT variations, doctors can alter thiopurine doses to reduce toxicity while preserving efficacy (Olivier & Williams-Jones, 2011) . This strategy has been especially successful in pediatric ALL, where customized dose based on TPMT genotype has improved treatment results while reducing side effects.

Warfarin and VKORC1/CYP2C9 Genotyping in Anticoagulation: Warfarin, a routinely prescribed anticoagulant, has a limited therapeutic window and high interindividual variability in dose requirements. Genetic differences in VKORC1 and CYP2C9 play an important role in warfarin sensitivity and metabolism (Roden, 2003).

Case Study 3: A clinical investigation found that genotyping patients for VKORC1 and CYP2C9 polymorphisms prior to starting warfarin medication improved dose prediction and reduced the incidence of adverse events such as bleeding or thrombosis. Patients who received genotype-guided treatment achieved therapeutic INR levels more quickly and with fewer problems. This personalized approach has been incorporated into clinical guidelines, with a focus on the use of pharmacogenomics in anticoagulant medication optimization (Roden, 2003).

Statins and SLCO1B1 Genotyping in Hypercholesterolemia - Although statins are commonly administered to control cholesterol and reduce cardiovascular risk, they can cause myopathy in some patients. Variants in the SLCO1B1 gene, which encodes the liver transporter OATP1B1, are linked to an increased risk of statin-induced myopathy, particularly with simvastatin.

Case Study 4: The SEARCH Collaborative Group discovered a substantial relationship between the SLCO1B1*5 allele and the risk of myopathy in simvastatin patients. Patients with this variation had a considerably increased risk of developing myopathy. Clinical guidelines now advocate SLCO1B1 genotyping before starting high-dose simvastatin therapy to identify high-risk patients and investigate alternate statins or lower dosages. This genotyping contributes to the personalization of statin medication, improving safety and adherence (Getahun et al., 2024; Roden, 2003).

Fluoropyrimidines and DPYD Genotyping in Cancer Treatment – Fluoropyrimidines, including 5-fluorouracil (5-FU) and capecitabine, are key chemotherapeutic medicines treating a variety of malignancies. However, they can cause severe toxicity in some patients, notably those with dihydropyrimidine dehydrogenase (DPD) impairment, which is caused by DPYD gene variations (Olivier & Williams-Jones, 2011).

Case Study 5: Clinical studies have shown that patients with DPYD mutations are more likely to experience severe and potentially fatal toxicities when given conventional fluoropyrimidine dosages. Clinicians can reduce toxicity risk by genotyping for DPYD mutations before starting treatment and adjusting doses or choosing alternate medications. This pharmacogenomic method has been integrated into cancer therapy procedures, thereby enhancing patient safety and treatment tolerance.

Tamoxifen and CYP2D6 Genotyping in Breast Cancer - Tamoxifen, a selective estrogen receptor modulator, is used to treat and prevent estrogen receptor-positive breast cancer. The effectiveness of tamoxifen is dependent on its conversion to the active metabolite endoxifen, which is predominantly mediated by CYP2D6.

Case Study 6: Research has found that patients with reduced-function CYP2D6 alleles have lower levels of endoxifen and worse clinical outcomes than those with normal CYP2D6 function. Genotyping for CYP2D6 polymorphisms enables clinicians to identify patients who may benefit from other treatments or tamoxifen dosage adjustments to ensure efficacy. This strategy is gradually being implemented in clinical practice, with the goal of tailoring breast cancer treatment to individual genetic profiles [34].

Antiepileptics and HLA-B*1502 genotyping in Epilepsy - Carbamazepine and other antiepileptic medicines (AEDs) can induce serious adverse drug reactions, such as Stevens-Johnson syndrome (SJS) and toxic epidermal necrolysis (TEN), in certain people. The HLA-B*1502 allele has been substantially linked to these reactions, notably in Asian populations.

Case study: In a critical trial, patients carrying the HLA-B1502 allele had a substantially increased chance of developing SJS/TEN when treated with carbamazepine. Routine HLA-B1502 testing prior to beginning carbamazepine therapy in at-risk patients has significantly reduced the prevalence of these serious events. This pharmacogenomic application exemplifies how genetic testing might avoid potentially fatal adverse medication responses while also improving patient safety in epilepsy therapy.

Abacavir with HLA-B*5701 in HIV treatment - Abacavir, an HIV nucleoside reverse transcriptase inhibitor, can produce severe hypersensitivity responses in certain people. The presence of the HLA-B*5701 allele is an excellent predictor of abacavir-induced HSR (Lauschke et al., 2024).

Case Study 7: The PREDICT-1 trial found that screening for HLA-B*5701 before starting abacavir therapy dramatically reduced the occurrence of HSR. This procedure has become a routine recommendation in HIV treatment guidelines, demonstrating how pharmacogenomics can improve drug safety and patient care in infectious diseases (Cai et al., 2020).

Pharmacogenomics has shown significant benefits in a wide number of medical specialties. By using genetic information to guide medicine selection and administration, healthcare providers can improve treatment efficacy, reduce side effects, and improve patient outcomes. Pharmacogenomics' impressive applications in HIV treatment, cancer therapy, pain management, and cardiovascular health demonstrate the technology's promise to change personalized care. As research uncovers additional genetic markers and their clinical implications, the incorporation of pharmacogenomics into ordinary practice is anticipated to grow, allowing for even more accurate and effective medicines tailored to individual patients.

4. TECHNOLOGICAL AND METHODOLOGICAL ADVANCES

4.1 High-Throughput Genomic Screening

Because there is a lack of regulation and defined rules, different testing facilities may use different applicable procedures and question genes/variants, which could produce conflicting results. Generally, the tested versions are chosen based on the drug class—such as anticoagulants, psychotropics, or antihypertensives or the indication. Pre-emptive testing of numerous variants is superior to a single gene-drug approach when viewed from an economic standpoint, especially when taking into account potential repurposing of results. Targeted genotyping, also known as PCR- and microarray-based targeted genotyping, has been widely used for pharmacogenetic profiling because of its cost-effectiveness, scalability, quick turnaround time, and high clinical usefulness. For example, the ubiquitous pharmacogenomics (U-PGx) program uses a PCR-based assay for pharmacogenetic profiling to demonstrate the therapeutic value of preemptive pharmacogenetic profiling.

Several factors may serve to emphasize the significance of PGx. Approximately 0.5% of adverse medication responses are fatal, and adverse drug reactions account for up to 12.8% of hospital admissions in the EU, representing a significant socio-economic cost. PGx could be used for patient stratification in clinical care and clinical studies since nearly every individual possesses at least one actionable sequence variant following the DPWG or CPIC recommendations. Despite this, most medications are still administered without taking pharmacogenetic information into account (Tong et al., 2021) .

With the development of third- and next-generation sequencing technology, the cost of genome sequencing has dropped dramatically; in 2003, it cost 2.7 billion USD to sequence the first human genome, a process that took 13 years; by 2019, however, the cost of sequencing an entire genome was less than $1,000.3–5. Three key tools are utilized to target discovery in PGx studies to adopt personalized medicine: next-generation sequencing (high throughput parallel sequencing), candidate gene analysis, and genome-wide association studies (GWASs). Six the candidate gene strategy begins with a gene that is thought to be important, whereas the first method, which is agnostic, finds a genome-wide signal to a locus. This is typically a gene that codes for the drug's protein targets or proteins involved in pharmacokinetics (Magavern et al., 2022b).

4.2 Bioinformatics and Computational Tools

Thanks to developments in big data analytics and high-performance computers, we can now leverage vast amounts of chemical, biological, and biomedical data to identify intricate genetic pathways underlying ADRs. However, even though it's now less expensive and quicker to get sequencing data, the analysis of that data remains the barrier. The two types of analysis that will be discussed in the next paragraphs are genotyping/pharmacogenomics variants/haplotypes and annotating them for ADR investigations. Modern bioinformatics pipelines and tools for these kinds of investigations, as well as newly developed machine learning methods for categorizing genomic variations into ADRs, were compiled.

Determining the impact of genetic variants on genes and disease is known as variant annotation.124 Generic variant annotation tools like VEP, SnpEff/SnpSift, Annovar, or Intervar can be used to acquire annotation for PGx variants. PGx gene annotation could be more helpful when utilizing specialized tools like PharmCAT, which was developed in accordance with CPIC recommendations. These recommendations prescribe treatments based on the relationship between genetics and phenotype (Ingelman-Sundberg et al., 2023).

4.3 Integration of Multi Omics Data

The Deep DRK Model

They created Deep DRK, a deep learning model that integrates data from several sources, different cancer kinds, and different chemical components to predict medication response. Deep neural networks (DNNs) were trained for drug response prediction using an integrated model of the interaction partners of a CCL and an anticancer medication, using a kernel-based technique.

Kernel Technique: They created numerous similarity matrices of both CCLs and anticancer medications using kernel approaches to combine heterogeneous data. The similarity matrices between each of the available CCLs were determined for each sort of characteristic separately (Sathyanarayanan et al., 2023).

Integrating GWAS with Omics

Most of the important variations found in the GWAS are located in non-coding areas. This calls into question how the GWAS data should be functionally interpreted and clinically applied. The need for large sample numbers and the inability to identify causal variations are two more drawbacks. Mapping the functional pathways of the GWAS-associated SNPs has been demonstrated to be facilitated by integrating genomics data with other omics information. It makes sense that if an SNP linked to an illness is also linked to other molecular characteristics, such as DNA methylation or gene expression, then both the SNP and the related omics trait are probably important for understanding the disease process. Owing to the effectiveness of GWAS in treating complex conditions, including psychiatric disorders, several techniques and readily available software tools are regularly produced (Miao et al., 2020).

4.4 The Role of AI and Machine Learning in Shaping Pharmacogenomics

Arising advancements like artificial intelligence (AI) and machine learning (ML) are reforming pharmacogenomics by empowering more exact and productive examination of hereditary information. These headways are vital for understanding hereditary varieties that impact drug reactions and for creating customized medication.

Enhancing Data Analysis

AI and ML algorithms can process immense measures of hereditary and clinical information to uncover examples and relationships that are challenging to physically recognize. For example, profound learning has been utilized to anticipate drug reactions by examining complex datasets, including genomic and clinical data.

Identifying Genetic Variants

AI strategies are fundamental for distinguishing novel hereditary variations related to drug reactions. By examining the enormous scope of genomic data, these algorithms can pinpoint explicit hereditary markers that impact how individuals metabolize and respond to medications. This is crucial for creating designated treatments and further developing medication adequacy.

Personalized Medicine

Artificial intelligence and machine learning are vital to progressing customized medication. These innovations empower the production of prescient models that forecast an individual's response to a specific drug based on their genetic profile. This approach helps tailor medicines to individual patients, improving remedial results and limiting unfriendly impacts.

Drug Discovery and Development

In drug Discovery, artificial intelligence and ML are utilized to recognize new medication targets and anticipate the viability and security of potential medication competitors. These innovations can recreate what different hereditary varieties mean for drug communications, accelerating the advancement of new treatments.

Clinical Applications

artificial intelligence-driven devices are being coordinated into clinical practice to help independent direction. For instance, pharmacogenomic testing panels powered by artificial intelligence furnish clinicians with experiences in the best treatment choices for patients given their hereditary cosmetics. This coordination enhances drug choice and dosing and prompts better quiet consideration (Cilluffo et al., 2021; Kalinin et al., 2018; Lin et al., 2020).

Case Study 1: Profound Learning for Patient Separation

Specialists at the College of Michigan utilized profound learning calculations to order patients in light of their hereditary profiles and clinical chronicles. This strategy revealed new administrative variations in non-coding districts of the genome, which are fundamental for grasping medication reactions. The review demonstrated the way that profound learning could foresee customized drug reactions, advancing prescription determination and dosing.

Case Study 2: AI for Variation Grouping

A concentration in Human Genomics presented an AI-based procedure to group pharmacogenomic variations. Utilizing calculations like arbitrary woods, the analysts arranged new variations because of their anticipated effect on protein capability. This approach upgraded the precision of anticipating what hereditary variations mean for drug digestion, helping with the improvement of customized treatments.

Case Study 3: Man-Made Intelligence in Accuracy Psychiatry

In the field of accuracy psychiatry, man-made intelligence, and AI have been utilized to assess illness status and medication medicines. By coordinating multiomics and neuroimaging information, scientists made models that anticipate patient reactions to mental prescriptions. This approach has worked on the comprehension of how hereditary and natural variables impact drug viability in emotional wellness medicines (• (Lin et al., 2020; Pandi et al., 2021; Jarvis et al., 2022).

4.5 Applications of Recent Advancements in Gene Editing Tools in Pharmacogenomic

Late headways in quality-altering innovations, particularly CRISPR/Cas frameworks, significantly affect pharmacogenomics by empowering exact hereditary alterations. These instruments are vital for understanding how hereditary varieties impact drug reactions and for creating customized medication.

a) **CRISPR/Cas9 and Its Variations**

Upgraded forms of CRISPR/Cas9, like high-devotion Cas9, have further developed precision by lessening askew impacts. This accuracy is fundamental for concentrating on unambiguous hereditary variations that influence drug digestion and reaction.

b) **Base Editing and Prime Editing**

Base Editing and Prime Editing offer significantly more prominent accuracy. Base editors can change over one DNA base to one more without causing twofold strand breaks, while prime editors take into account more intricate hereditary adjustments. These instruments are significant for making models to concentrate on the effect of hereditary variations on drug adequacy and harmfulness.

c) **Drug Disclosure and Advancement**

Gene editing tools are utilized to recognize and approve new medication targets. By making models with explicit hereditary alterations, specialists can see what these progressions mean for drug reaction, prompting the revelation of new helpful targets. These apparatuses additionally assist with creating customized treatments custom-made to a person's hereditary profile, further developing medication viability and lessening unfriendly impacts (Xu et al., 2024; Garcia-Agundez et al., 2023).

5. ETHICAL, LEGAL, AND SOCIAL IMPLICATIONS

5.1 Patient Privacy and Data Security

The use of genetic data in pharmacogenomics and drug discovery raises significant concerns about patient privacy and data security. Ensuring the confidentiality, integrity, and availability of genetic data is essential to prevent genetic discrimination and stigma. Robust security measures must be implemented to prevent data breaches and unauthorized access to genetic information. This includes encryption, access controls, and audit trails. Additionally, patients must be informed about how their genetic data will be used, shared, and protected. Clear policies and procedures for data management and security must be established and communicated to patients. This includes transparency about data sharing, storage, and retention. Patients must also be informed about their rights regarding their genetic data, including the right to access, correct, or delete their data [49]. Figure 3 gives the overall idea of patient privacy and data security.

Figure 3. Patient privacy and data security

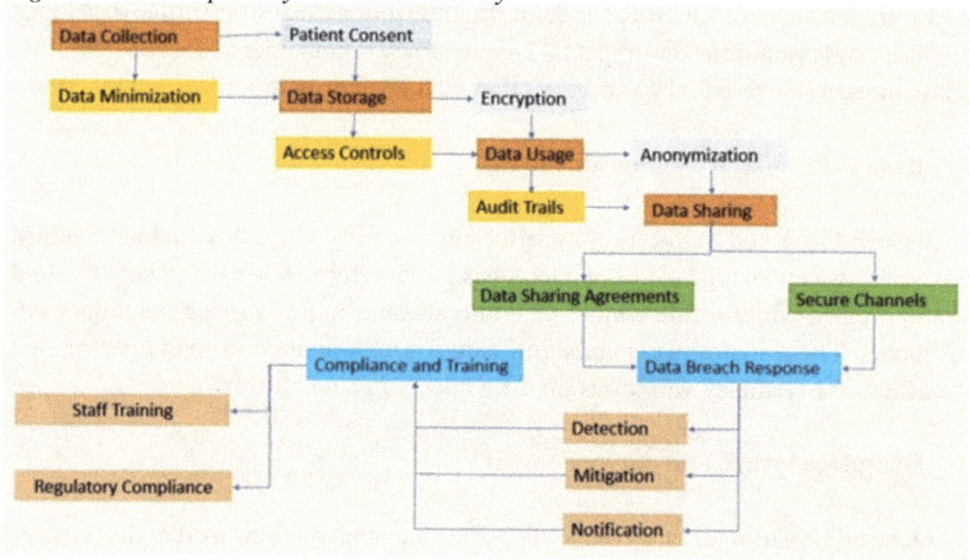

5.2 Informed Consent and Genetic Counselling

Informed consent is a critical aspect of pharmacogenomics and drug discovery. Patients must be fully informed about the risks and benefits of genetic testing, including the potential for genetic discrimination. Adequate genetic counseling and support must be provided to patients to help them understand the implications of genetic testing. This includes counseling on the risks and benefits of genetic testing, the limitations of genetic testing, and the potential for genetic discrimination. Furthermore, informed consent must be obtained from patients before genetic testing and data sharing. Clear and concise language must be used to explain the purpose and scope of genetic testing, and patients must be allowed to ask questions and provide consent. Patients must also be informed about their right to withdraw consent at any time (Hewitt, 2018) .

5.3 Equity and Access to Pharmacogenomic Advances

Pharmacogenomics and drug discovery raise important questions about equity and access to genetic testing and interpretation. Ensuring equal access to genetic testing and interpretation for diverse populations is essential to address health disparities and genetic determinants of health. Inclusive research and development of pharmacogenomic tests must be promoted to ensure that genetic data are representative of diverse populations. This includes ensuring diverse participant recruitment in clinical trials and using diverse population data in test development. Furthermore, genetic testing and interpretation must be made accessible and affordable for all patients, regardless of their race, ethnicity, or socioeconomic status. This includes ensuring that genetic testing is covered by insurance and that patients have access to genetic counseling and support (Callier et al., 2016).

5.4 International Regulatory Harmonization

The global nature of pharmacogenomics and drug discovery requires international regulatory harmonization. Harmonizing regulations and standards for pharmacogenomic testing and data sharing is essential to ensure compliance with international ethical and legal frameworks. Clear guidelines and standards must be established for the use of genetic data in research and clinical practice. This includes guidelines for data sharing, storage, and retention, as well as standards for genetic testing and interpretation. Furthermore, international collaboration and data sharing in pharmacogenomics research must be facilitated to promote global advances in personalized medicine. Regulatory harmonization must balance individual privacy and autonomy with the need for data sharing and collaboration. This includes ensuring that data-sharing agreements are in place and that data are de-identified and anonymized to protect patient privacy (Chenoweth et al., 2020).

6. CHALLENGES AND OPPORTUNITIES

6.1 Scientific and Technical Challenges

Pharmacogenomics and drug discovery face significant scientific and technical challenges. Figure 4 gives an overview of the challenges in pharmacogenomics. One of the main challenges is the complex relationship between genetic variants and drug response, which can be influenced by multiple genetic and environmental factors. This complexity makes it difficult to predict drug response and toxicity and requires the development of sophisticated analytical and computational tools. Additionally, the high dimensionality of genetic data and the limited understanding of gene-environment interactions pose significant analytical and interpretational challenges. Furthermore, standardization and validation of genetic tests are essential to ensure accuracy and reliability, but this process can be time-consuming and resource-intensive.

Moreover, the lack of standardized protocols for genetic testing and data analysis, and the limited availability of high-quality genetic data, pose significant challenges to the development of personalized medicine. Additionally, the need for advanced bioinformatic tools and expertise to analyze and interpret large amounts of genetic data poses a significant challenge to researchers and clinicians (Shabaruddin et al., 2015).

Figure 4. Scientific and technical challenges

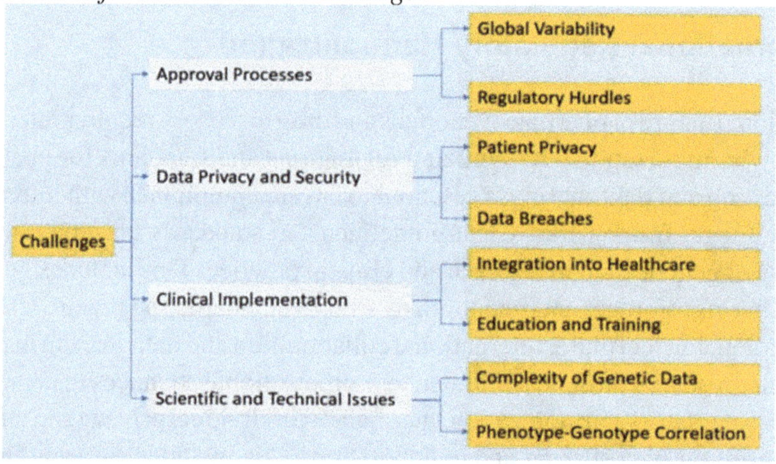

6.2 Economic and Funding Issues

Pharmacogenomics and drug discovery also face significant economic and funding challenges. Genetic testing and data analysis can be expensive, and the cost of developing and commercializing new drugs is high. Moreover, reimbursement for genetic tests is often limited, and funding for pharmacogenomics research is insufficient. These economic barriers can limit access to genetic testing and personalized medicine, particularly for underserved populations.

Moreover, the high costs of drug development and commercialization can limit the availability of new drugs and therapies. The cost of developing a new drug can range from hundreds of millions to billions of dollars, and the process can take several years to complete. This can limit the availability of new drugs and therapies, particularly for rare diseases or diseases that affect small populations (Klein et al., 2017).

The development and application of pharmacogenomic techniques meet numerous economic and funding challenges, affecting their accessibility and integration into healthcare systems globally.

High Costs of Technology

One of the primary obstacles is still the cost of high-throughput sequencing and genotyping technology. Even with the decline in costs, they continue to be a major financial burden particularly for low and middle-income nations this respects the use of pharmacogenomic testing in some areas. High throughput sequencing and genotype technologies continue to be prohibitively expensive. Because of this, pharmacogenomic testing is not as widely used in these areas.

Limited Research Funding

Pharmacogenomic study often lacks investment, especially in underdeveloped nations. Institutions and governments frequently place a higher priority on short-term medical needs than on long-term genetic research expenditures. Large-scale research required to verify pharmacogenomic biomarkers and create novel treatments is hampered by this lack of funding.

Infrastructural Deficiencies

The infrastructure required to facilitate pharmacogenomic research and application is lacking in many areas. This includes a lack of skilled staff, obsolete equipment, and inadequate lab space. Even well-funded projects may face difficulties in the absence of adequate infrastructure.

Insurance and Reimbursement Issues

Pharmacogenomic testing is not always covered by insurance, with many plans not including this type of testing. Pharmacogenomic services are not widely adopted because patients and healthcare professionals are deterred from using them by this lack of payment.

Investment Hesitancy

Pharmacogenomics investment is frequently resisted by investors, particularly in areas with less developed healthcare infrastructure. The apprehension stems from the perceived elevated risk and unpredictability of the return on investment, thereby impeding innovation and delaying the advancement of novel technology (Shukla, 2020; Sadee et al., 2023).

6.3 Policy and Regulatory Hurdles

Pharmacogenomics and drug discovery must navigate complex policy and regulatory hurdles. Currently, there is a lack of clear regulations and guidelines for genetic testing and data sharing, which can create uncertainty and inconsistency. Moreover, international regulatory frameworks are often inconsistent, which can create challenges for global collaboration and data sharing.

Furthermore, concerns about data privacy and security are paramount, and regulations must balance individual autonomy with public health needs. The use of genetic data raises significant ethical concerns, including the potential for genetic discrimination and the need for informed consent. Regulations must ensure that genetic data are protected from unauthorized access and misuse, while also allowing for the sharing of data for research and clinical purposes.

Finally, limited awareness and education among healthcare providers can create barriers to adoption and implementation. Many healthcare providers may not have the necessary knowledge or training to interpret and use genetic data in clinical practice, which can limit the adoption of personalized medicine (Buchanan et al., 2017; Katara & Yadav, 2019).

6.4 Real-World Case Studies on Pharmacogenomics Implementation and Patient Outcomes

Case Study 1: Comprehensive Medication Management in Medicare Advantage

Coriell Life Sciences directed a study on a pharmacogenomics-improved complete prescription administration (PGx + CMM) program including 5,288 Medicare Advantage members aged 65 and older. The program included hereditary testing, at-home saliva collection, and customized drug activity plans. The outcomes were impressive, showing huge upgrades in drug safety and medical services results, with $37 million saved in direct clinical charges for more than 32 months. This study highlights the capability of pharmacogenomics to decrease medical care expenses and improve patient results.

Case Study 2: Pharmacogenomic Testing in Pediatric Care

A review distributed in the JAMA Network Open investigated the execution of pharmacogenomic testing in a pediatric tertiary care setting. The examination included entire genome sequencing and designated genotyping to recognize drug-gene interactions in children. Coordinating pharmacogenomic information into clinical practice prompted better prescription administration and less antagonistic medication responses, featuring the worth of pharmacogenomics in pediatric care.

Case Study 3: Pharmacogenomics in Geriatric Populations

Joseph Jarvis and colleagues directed research on the execution of pharmacogenomics in a Federal health care Benefit populace. The review found a 15% decrease in clinic days, a 7% decrease in emergency room visits, and a 2% decrease in physician visits, bringing about general expense reserve funds of $315 per member each month. These discoveries show the capability of pharmacogenomics to develop results further and lessen medical care costs among more established grown-ups (Bienfait et al., 2022; Sukri et al., 2022).

6.5 Future Opportunities and Directions

Despite these challenges, pharmacogenomics and drug discovery offer significant future opportunities and directions. Personalized medicine and targeted therapies can improve patient outcomes and reduce adverse reactions. Integration of pharmacogenomics with artificial intelligence and machine learning can enhance predictive analytics and drug discovery. Global collaboration and data sharing can accelerate research and development, and increased investment in pharmacogenomics research can address unmet medical needs.

Moreover, innovative business models and reimbursement strategies can improve access to genetic testing and personalized medicine. The development of new business models, such as pay-for-performance and value-based pricing, can align the incentives of drug developers and payers with the needs of patients. Additionally, the development of new reimbursement strategies, such as bundled payments and risk-sharing agreements, can improve access to genetic testing and personalized medicine.

Overall, the future of pharmacogenomics and drug discovery holds much promise for improving human health and well-being. However, addressing the scientific, technical, economic, and regulatory challenges will be critical to realizing this promise [58] [60]. Figure 5 depicts the future opportunities and directions of pharmacogenomics.

Figure 5. Future opportunities and directions

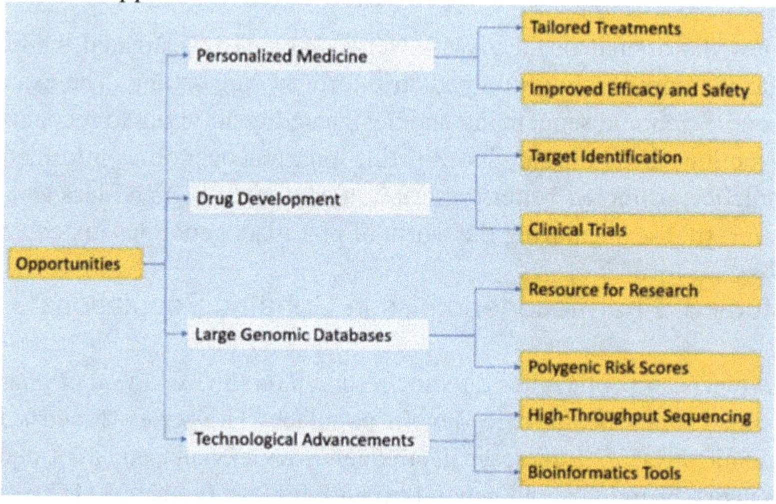

7. GLOBAL COLLABORATIONS AND PARTNERSHIPS

7.1 International Consortia and Networks

International consortia and networks play a pivotal role in advancing pharmacogenomics and drug discovery. These collaborative initiatives bring together researchers, institutions, and organizations from different countries. Their purpose is multifaceted:

- **Facilitate Data Sharing:** By pooling resources and expertise, international networks enable efficient data sharing. Researchers can access diverse datasets, enhancing the robustness of pharmacogenomic studies.
- **Cross-Disciplinary Collaboration:** These networks foster collaboration across disciplines—genomics, clinical medicine, ethics, and policy. Such cross-pollination accelerates scientific progress.
- **Global Challenges:** Pharmacogenomics faces complex global challenges, including diverse patient populations, varying regulatory landscapes, and ethical considerations. International consortia provide a platform to address these challenges collectively [61].

7.2 Public-Private Partnerships (PPPs)

Public-private partnerships are critical for bridging gaps in pharmacogenomics research and drug development as they lead to:

- **Resource Synergy:** Combining public (academic, nonprofit, or government) and private (for-profit) entities leverages complementary strengths. Funding, infrastructure, and expertise are pooled.
- **Accelerating Innovation:** PPPs expedite drug discovery by streamlining processes. Collaborators share risks, costs, and knowledge, leading to faster breakthroughs.
- **Patient Impact:** Successful PPPs translate into improved patient outcomes. For example, integrating pharmacogenomics into Medicare populations has reduced hospital stays and emergency visits [62].

7.3 Case Studies of Successful Collaborations

Examining real-world examples sheds light on effective collaboration strategies:

- **Medicare Advantage Population:** Implementing pharmacogenomics reduced healthcare utilization and costs. This case demonstrates the tangible benefits of collaboration [63].
- **Drug Development Trajectory:** Analyzing successful drug classes reveals patterns. For instance, understanding how PCSK9 inhibitors or PDE5 inhibitors emerged provides insights for future collaborations [64].
- **Corporate Employee Health:** Self-insured employers adopting pharmacogenomics enhance workforce health. By investing in genetic testing, they improve employee well-being and productivity [65].

8. FUTURE OUTLOOK

8.1 Trends and Innovations on the Horizon

The future of pharmacogenomics and drug discovery is promising, with several trends and innovations on the horizon. Artificial intelligence (AI) and machine learning (ML) will revolutionize pharmacogenomics, enabling predictive analytics and personalized medicine. AI-powered algorithms will analyze vast amounts of genetic data, identifying patterns and correlations that inform drug development and treatment strategies. Single-cell genomics and epigenomics will provide new insights into cellular heterogeneity and drug response, enabling researchers to develop targeted therapies that address specific cell types and mechanisms. Gene editing technologies like CRISPR/Cas9 will enable precision medicine and targeted therapies, allowing for the precise modification of genes to treat genetic diseases.

Additionally, the integration of pharmacogenomics with other omics technologies, such as proteomics and metabolomics, will provide a more comprehensive understanding of the complex relationships between genes, proteins, and metabolites. This will enable the development of personalized therapies that take into account an individual's unique molecular profile (Erceg, 2024).

8.2 Potential Impact on Global Health

The potential impact of pharmacogenomics on global health is significant. Personalized medicine will improve drug efficacy and reduce adverse reactions, enhancing patient outcomes and quality of life. Targeted therapies will address unmet medical needs in diverse populations, reducing health disparities and improving health equity. Pharmacogenomics will enable precision public health, informing health policy and disease prevention strategies. Global collaborations will accelerate research and development, addressing health disparities and improving health outcomes worldwide.

By integrating pharmacogenomics into healthcare, we can reduce healthcare costs, improve patient quality of life, and enhance the overall efficiency of healthcare systems. Pharmacogenomics will also enable the development of personalized prevention strategies, allowing individuals to take proactive steps to prevent disease (Crisafulli et al., 2019).

8.3 Strategies for Advancing Pharmacogenomics in Drug Discovery

Several strategies are essential to advance pharmacogenomics in drug discovery. We must integrate pharmacogenomics into drug development pipelines, incorporating genetic data and biomarkers into the drug development process. Standardized frameworks for genetic data sharing and analysis must be developed, facilitating collaboration and knowledge sharing across industries and borders.

Global collaborations and knowledge sharing must be fostered, bringing together researchers, clinicians, and industry experts to advance pharmacogenomics research and translation. Investment in education and training programs for healthcare providers is critical, ensuring that clinicians are equipped to interpret and apply genetic data in clinical practice.

Addressing ethical and regulatory challenges will ensure equity and access to pharmacogenomic testing and personalized medicine, promoting responsible innovation and translation. This includes ensuring that genetic data are protected from unauthorized access and misuse and that individuals have the right to make informed decisions about their genetic data.

Overall, the future of pharmacogenomics and drug discovery holds much promise for improving human health and well-being. By leveraging trends and innovations, addressing global health challenges, and implementing strategies for advancing pharmacogenomics, we can create a future where personalized medicine is the norm, and individuals receive the right treatment at the right time (Elgarhy et al., 2024).

9. CONCLUSION

Pharmacogenomics is reshaping the landscape of drug discovery and improvement by merging pharmacology and genomics to make customized, effective, and safe drugs. This section has investigated the worldwide points of view on pharmacogenomics, underlining the commitments of key regions, research organizations, regulatory frameworks, technological advancements, and the ethical, legitimate, and social difficulties related to this field. North America, with its driving foundations like the NIH and a solid administrative system through the FDA, has secured itself as a trailblazer in pharmacogenomics research. Europe, with commitments from associations like Biobank and Germany's DZHK, and administrative oversight from the EMA, keeps on assuming a huge part. Asia is quickly turning into a key part, with nations like China, Japan, and India putting vigorously into creative exploration and development. Technological headways are vital to the advancement of pharmacogenomics, empowering the improvement of customized treatment choices. Notwithstanding, the field faces huge moral, lawful, and social difficulties. Guaranteeing patient protection and information security is basic, given the delicate idea of hereditary data and the requirement for worldwide information sharing. Despite these difficulties, the future of pharmacogenomics holds incredible commitment. Worldwide consortia and coordinated efforts are fundamental for cultivating information sharing, normalizing rehearses, and advancing helpful examination. These endeavors feature the capability of cooperative undertakings to defeat snags and advance the field. Laying out worldwide norms and strategies to guarantee fair admittance to pharmacogenomic headways is critical for the overall reconciliation of customized medication.

Pharmacogenomics is set to alter drug discovery and advancement by giving customized treatment choices based on individual hereditary profiles. Proceeded with worldwide joint effort, mechanical development, and a solid obligation to moral guidelines will be imperative in understanding the maximum capacity of pharmacogenomics, prompting work on persistent results and more successful medical care arrangements around the world.

REFERENCES

Abad-Santos, F., Aliño, S. F., Borobia, A. M., García-Martín, E., Gassó, P., Maroñas, O., & Agúndez, J. A. (2024). Developments in pharmacogenetics, pharmacogenomics, and personalized medicine. *Pharmacological Research*, 200, 107061. DOI: 10.1016/j.phrs.2024.107061 PMID: 38199278

Antol, D. D., Schwartz, R., Caplan, A., Casebeer, A., Erwin, C. J., Shrank, W. H., & Powers, B. W. (2022, October). Comparison of health care utilization by Medicare Advantage and traditional Medicare beneficiaries with complex care needs. [). American Medical Association.]. *JAMA Health Forum*, 3(10), e223451–e223451. DOI: 10.1001/jamahealthforum.2022.3451 PMID: 36206006

Astras, G., Papagiannopoulos, C. I., Kyritsis, K. A., Markitani, C., & Vizirianakis, I. S. (2020). Pharmacogenomic testing to guide personalized cancer medicine decisions in private oncology practice: A case study. *Frontiers in Oncology*, 10, 521. DOI: 10.3389/fonc.2020.00521 PMID: 32411592

Bienfait, K., Chhibber, A., Marshall, J. C., Armstrong, M., Cox, C., Shaw, P. M., & Paulding, C. (2022). Current challenges and opportunities for pharmacogenomics: Perspective of the Industry Pharmacogenomics Working Group (I-PWG). *Human Genetics*, 141(6), 1165–1173. DOI: 10.1007/s00439-021-02282-3 PMID: 34081195

Božina, N., Vrkić Kirhmajer, M., Šimičević, L., Ganoci, L., Mirošević Skvrce, N., Klarica Domjanović, I., & Merćep, I. (2020). Use of pharmacogenomics in elderly patients treated for cardiovascular diseases. *Croatian Medical Journal*, 61(3), 147–158. DOI: 10.3325/cmj.2020.61.147 PMID: 32378381

Buchanan, A., Califano, A., Kahn, J., McPherson, E., Robertson, J., & Brody, B. (2017). Pharmacogenetics: ethical issues and policy options. In *Genetics and Gene Therapy* (pp. 327-341). Routledge. DOI: 10.4324/9781315254517-12

Butnariu, A., Samasca, G., & Lupan, I. (2015). Ethical implications in pharmacogenetics and pharmacogenomics. *Medical Connections.*, 10, 41–44.

Cai, W., Liu, Z., Miao, L., & Xiang, X. (Eds.). (2020). *Pharmacogenomics in precision medicine: from a perspective of ethnic differences*. Springer. DOI: 10.1007/978-981-15-3895-7

Callier, S. L., Abudu, R., Mehlman, M. J., Singer, M. E., Neuhauser, D., Caga-Anan, C., & Wiesner, G. L. (2016). Ethical, legal, and social implications of personalized genomic medicine research: Current literature and suggestions for the future. *Bioethics*, 30(9), 698–705. DOI: 10.1111/bioe.12285 PMID: 27767224

Caspar, S. M., Schneider, T., Stoll, P., Meienberg, J., & Matyas, G. (2021). Potential of whole-genome sequencing-based pharmacogenetic profiling. *Pharmacogenomics*, 22(3), 177–190. DOI: 10.2217/pgs-2020-0155 PMID: 33517770

Charnysh, E., Pal, S., Reader, J. M., Uhlmann, W. R., McCain, S., Sanghavi, K., Blasco, D., Brandt, R., Feero, W. G., Ferber, R., Giri, V. N., Hendy, K., Prince, A. E. R., Lee, C., Roberts, J. S., Crumpler, N., Leader, A., Mathews, D., Ryan, K., & Saidi, F. (2024). Health care utilization and behavior changes after workplace genetic testing at a large US health care system. *Genetics in Medicine*, 26(8), 101160. DOI: 10.1016/j.gim.2024.101160 PMID: 38733246

Cheng, F., Ma, Y., Uzzi, B., & Loscalzo, J. (2020). Importance of scientific collaboration in contemporary drug discovery and development: A detailed network analysis. *BMC Biology*, 18(1), 1–9. DOI: 10.1186/s12915-020-00868-3 PMID: 33050894

Chenoweth, M. J., Giacomini, K. M., Pirmohamed, M., Hill, S. L., van Schaik, R. H., Schwab, M., Shuldiner, A. R., Relling, M. V., & Tyndale, R. F. (2020). Global pharmacogenomics within precision medicine: Challenges and opportunities. *Clinical Pharmacology and Therapeutics*, 107(1), 57–61. DOI: 10.1002/cpt.1664 PMID: 31696505

Chu, X., Prasad, B., Neuhoff, S., Yoshida, K., Leeder, J. S., Mukherjee, D., Taskar, K., Varma, M. V. S., Zhang, X., Yang, X., & Galetin, A. (2022). Clinical implications of altered drug transporter abundance/function and PBPK modeling in specific populations: An ITC perspective. *Clinical Pharmacology and Therapeutics*, 112(3), 501–526. DOI: 10.1002/cpt.2643 PMID: 35561140

Cilluffo, G., Fasola, S., Ferrante, G., Malizia, V., Montalbano, L., & La Grutta, S. (2021). Machine learning: An overview and applications in pharmacogenetics. *Genes*, 12(10), 1511. DOI: 10.3390/genes12101511 PMID: 34680905

Crisafulli, C., Romeo, P. D., Calabrò, M., Epasto, L. M., & Alberti, S. (2019). Pharmacogenetic and pharmacogenomic discovery strategies. *Cancer Drug Resistance (Alhambra, Calif.)*, 2(2), 225. PMID: 35582724

David, S. P., Devine, B., Palaniappan, L., & Houwink, E. J. (2024). State of the science of pharmacogenomics implementation in healthcare systems and communities. *Frontiers in Pharmacology*, 15, 1463384. DOI: 10.3389/fphar.2024.1463384 PMID: 39114360

Davis, A. M., Engkvist, O., Fairclough, R. J., Feierberg, I., Freeman, A., & Iyer, P. (2021). Public-private partnerships: Compound and data sharing in drug discovery and development. *SLAS Discovery*, 26(5), 604–619. DOI: 10.1177/2472555220982268 PMID: 33586501

Elgarhy, F. M., Borham, A., Alziny, N., AbdElaal, K. R., Shuaib, M., Musaibah, A. S., Hussein, M. A., & Abdelnaser, A. (2024). From Drug Discovery to Drug Approval: A Comprehensive Review of the Pharmacogenomics Status Quo with a Special Focus on Egypt. *Pharmaceuticals (Basel, Switzerland)*, 17(7), 881. DOI: 10.3390/ph17070881 PMID: 39065732

Erceg, D. (2024). Pharmacogenomics in Drug Research and Development. In *Pharmacogenomics in Clinical Practice* (pp. 439–461). Springer International Publishing.

Fadya, M. (2024). From Drug Discovery to Drug Approval: A Comprehensive Review of the Pharmacogenomics Status Quo with a Special Focus on Egypt. *Pharmaceuticals (Basel, Switzerland)*, 17(7), 881–881. DOI: 10.3390/ph17070881 PMID: 39065732

Garcia-Agundez, A., Garcia-Martin, E., & Eickhoff, C. (2023). The Potential of Machine-learning in Pharmacogenetics, Pharmacogenomics and Pharmacoepidemiology: Volume II. *Frontiers in Pharmacology*, 14, 1277561. DOI: 10.3389/fphar.2023.1277561 PMID: 37711174

Getahun, K. A., Angaw, D. A., Asres, M. S., Kahaliw, W., Petros, Z., Abay, S. M., Yimer, G., & Berhane, N. (2024). The Role of Pharmacogenomics Studies for Precision Medicine Among Ethiopian Patients and Their Clinical Implications: A Scoping Review. *Pharmacogenomics and Personalized Medicine*, 17, 347–361. DOI: 10.2147/PGPM.S454328 PMID: 38974617

Ginsburg, G. S. (2019). A global collaborative to advance genomic medicine. *American Journal of Human Genetics*, 104(3), 407–409. DOI: 10.1016/j.ajhg.2019.02.010 PMID: 30849327

Henderson, L. M., Claw, K. G., Woodahl, E. L., Robinson, R. F., Boyer, B. B., Burke, W., & Thummel, K. E. (2018). P450 pharmacogenetics in indigenous North American populations. *Journal of Personalized Medicine*, 8(1), 9. DOI: 10.3390/jpm8010009 PMID: 29389890

Hewitt, J. E. (2018). The ethical, legal and regulatory issues associated with pharmacogenomics: Systematically quantifying the literature. *Journal of Law and Medicine*, 25(3), 782–793. PMID: 29978667

Ingelman-Sundberg, M., Nebert, D. W., & Lauschke, V. M. (2023). Emerging trends in pharmacogenomics: From common variant associations toward comprehensive genomic profiling. *Human Genomics*, 17(1), 105. DOI: 10.1186/s40246-023-00554-9 PMID: 37996916

Ingelman-Sundberg, M., Nebert, D. W., & Lauschke, V. M. (2023). Emerging trends in pharmacogenomics: From common variant associations toward comprehensive genomic profiling. *Human Genomics*, 17(1), 105. DOI: 10.1186/s40246-023-00554-9 PMID: 37996916

Jarvis, J. P., Peter, A. P., Keogh, M., Baldasare, V., Beanland, G. M., Wilkerson, Z. T., Kradel, S., & Shaman, J. A. (2022). Real-world impact of a pharmacogenomics-enriched comprehensive medication management program. *Journal of Personalized Medicine*, 12(3), 421. DOI: 10.3390/jpm12030421 PMID: 35330421

Jolly, B., & Scaria, V. (2024). Ethnic differences in pharmacogenomic variants: A south Asian perspective. *Pharmacogenomics*, 25(4), 171–174. DOI: 10.2217/pgs-2024-0026 PMID: 38511426

Jordan, I. K., Sharma, S., & Mariño-Ramírez, L. (2023). Population pharmacogenomics for health equity. *Genes*, 14(10), 1840. DOI: 10.3390/genes14101840 PMID: 37895188

Kalinin, A. A., Higgins, G. A., Reamaroon, N., Soroushmehr, S., Allyn-Feuer, A., Dinov, I. D., Najarian, K., & Athey, B. D. (2018). Deep learning in pharmacogenomics: From gene regulation to patient stratification. *Pharmacogenomics*, 19(7), 629–650. DOI: 10.2217/pgs-2018-0008 PMID: 29697304

Katara, P., & Yadav, A. (2019). Pharmacogenes (PGx-genes): Current understanding and future directions. *Gene*, 718, 144050. DOI: 10.1016/j.gene.2019.144050 PMID: 31425740

Klein, M. E., Parvez, M. M., & Shin, J. G. (2017). Clinical implementation of pharmacogenomics for personalized precision medicine: Barriers and solutions. *Journal of Pharmaceutical Sciences*, 106(9), 2368–2379. DOI: 10.1016/j.xphs.2017.04.051 PMID: 28619604

Krebs, K., & Milani, L. (2019). Translating pharmacogenomics into clinical decisions: Do not let the perfect be the enemy of the good. *Human Genomics*, 13(1), 1–13. DOI: 10.1186/s40246-019-0229-z PMID: 31455423

Lauschke, V. M., & Ingelman-Sundberg, M. (2020). Emerging strategies to bridge the gap between pharmacogenomic research and its clinical implementation. *NPJ Genomic Medicine*, 5(1), 9. DOI: 10.1038/s41525-020-0119-2 PMID: 32194983

Lauschke, V. M., Zhou, Y., & Ingelman-Sundberg, M. (2024). Pharmacogenomics beyond single common genetic variants: The way forward. *Annual Review of Pharmacology and Toxicology*, 64(1), 33–51. DOI: 10.1146/annurev-pharmtox-051921-091209 PMID: 37506333

Lei, X., & Liyin, L. (2024, August). Qiutang, Wang., Bing, Pan., Leilei, Zheng., Zheng, Lin. (2024). Effect of pharmacogenomic testing on the clinical treatment of patients with depressive disorder: A randomized clinical trial. *Journal of Affective Disorders*, 359, 117–124. Advance online publication. DOI: 10.1016/j.jad.2024.05.063

Lin, E., Lin, C. H., & Lane, H. Y. (2020). Precision psychiatry applications with pharmacogenomics: Artificial intelligence and machine learning approaches. *International Journal of Molecular Sciences*, 21(3), 969. DOI: 10.3390/ijms21030969 PMID: 32024055

Lo, C., Nguyen, S., Yang, C., Witt, L., Wen, A., Liao, T. V., Nguyen, J., Lin, B., Altman, R. B., & Palaniappan, L. (2020). Pharmacogenomics in Asian subpopulations and impacts on commonly prescribed medications. *Clinical and Translational Science*, 13(5), 861–870. DOI: 10.1111/cts.12771 PMID: 32100936

Madan, A., Walker, C. R., Weinstein, B., & Fowler, J. C. (2015). Pharmacogenomics in practice: A case report of personalized inpatient psychiatric care. *Pharmacogenomics*, 16(5), 433–439. DOI: 10.2217/pgs.15.9 PMID: 25916515

Magavern, E. F., Kaski, J. C., Turner, R. M., Drexel, H., Janmohamed, A., Scourfield, A., Burrage, D., Floyd, C. N., Adeyeye, E., Tamargo, J., Lewis, B. S., Kjeldsen, K. P., Niessner, A., Wassmann, S., Sulzgruber, P., Borry, P., Agewall, S., Semb, A. G., Savarese, G., & Caulfield, M. J. (2022). The role of pharmacogenomics in contemporary cardiovascular therapy: A position statement from the European Society of Cardiology Working Group on Cardiovascular Pharmacotherapy. *European Heart Journal. Cardiovascular Pharmacotherapy*, 8(1), 85–99. DOI: 10.1093/ehjcvp/pvab018 PMID: 33638977

Magavern, E. F., Kaski, J. C., Turner, R. M., Drexel, H., Janmohamed, A., Scourfield, A., Burrage, D., Floyd, C. N., Adeyeye, E., Tamargo, J., Lewis, B. S., Kjeldsen, K. P., Niessner, A., Wassmann, S., Sulzgruber, P., Borry, P., Agewall, S., Semb, A. G., Savarese, G., & Caulfield, M. J. (2022). Challenges in cardiovascular pharmacogenomics implementation: A viewpoint from the European Society of Cardiology Working Group on Cardiovascular Pharmacotherapy. *European Heart Journal. Cardiovascular Pharmacotherapy*, 8(1), 100–103. DOI: 10.1093/ehjcvp/pvab063 PMID: 34463331

McDonough, C. W. (2021). Pharmacogenomics in cardiovascular diseases. *Current Protocols*, 1(7), e189. DOI: 10.1002/cpz1.189 PMID: 34232575

. Miao, L., Zhang, J., Yi, L., & Huang, S. (2020). The ethical, legal, and regulatory issues associated with pharmacogenomics. *Pharmacogenomics in Precision Medicine: From a Perspective of Ethnic Differences*, 219-239.

Miteva-Marcheva, N. N., Ivanov, H. Y., Dimitrov, D. K., & Stoyanova, V. K. (2020). Application of pharmacogenetics in oncology. *Biomarker Research*, 8(1), 32. DOI: 10.1186/s40364-020-00213-4 PMID: 32821392

Moc, C. (2020). Pharmacogenomics: An evolving clinical tool for precision medicine. *Cleveland Clinic Journal of Medicine*, 87(2), 91–99. DOI: 10.3949/ccjm.87a.19073 PMID: 32015062

. Mohammad, Alajely, S., Ms., Subasri, G., Meena, Singh., K., Ilango. (2023). Pharmacogenomics -a genome based drug tailoring. 256-265. DOI: 10.58532/V2BS25P2CH5

Morley, K. I., & Hall, W. D. (2004). Using pharmacogenetics and pharmacogenomics in the treatment of psychiatric disorders: Some ethical and economic considerations. *Journal of Molecular Medicine*, 82(1), 21–30. DOI: 10.1007/s00109-003-0496-x PMID: 14598045

Olivier, C., & Williams-Jones, B. (2011). Pharmacogenomic technologies: A necessary" luxury" for better global public health? *Globalization and Health*, 7(1), 1–13. DOI: 10.1186/1744-8603-7-30 PMID: 21864366

Pandi, M. T., Koromina, M., Tsafaridis, I., Patsilinakos, S., Christoforou, E., van der Spek, P. J., & Patrinos, G. P. (2021). A novel machine learning-based approach for the computational functional assessment of pharmacogenomic variants. *Human Genomics*, 15(1), 1–13. DOI: 10.1186/s40246-021-00352-1 PMID: 34372920

Pirmohamed, M. (2023). Pharmacogenomics: Current status and future perspectives. *Nature Reviews. Genetics*, 24(6), 350–362. DOI: 10.1038/s41576-022-00572-8 PMID: 36707729

Pirmohamed, M. (2023). Pharmacogenomics: Current status and future perspectives. *Nature Reviews. Genetics*, 24(6), 350–362. DOI: 10.1038/s41576-022-00572-8 PMID: 36707729

. Pooja, Sharma., Ritu, Rani., Ajeet, Singh., Amar, Pal, Singh. (2024). Pharmacogenomics. International journal of pharmaceutics and drug analysis, 79-85. DOI: 10.47957/ijpda.v12i1.580

Rita, N.-Á. (2023). Pharmacogenomics in clinical trials: An overview. *Frontiers in Pharmacology*, 14, 1247088. Advance online publication. DOI: 10.3389/fphar.2023.1247088 PMID: 37927590

Roden, D. M. (2003). Cardiovascular pharmacogenomics. *Circulation*, 108(25), 3071–3074. DOI: 10.1161/01.CIR.0000110626.24310.18 PMID: 14691022

Sadee, W., Wang, D., Hartmann, K., & Toland, A. E. (2023). Pharmacogenomics: Driving personalized medicine. *Pharmacological Reviews*, 75(4), 789–814. DOI: 10.1124/pharmrev.122.000810 PMID: 36927888

Sathyanarayanan, A., Mueller, T. T., Moni, M. A., Schueler, K., Baune, B. T., Lio, P., & Xicota, L. (2023). Multi-omics data integration methods and their applications in psychiatric disorders. *European Neuropsychopharmacology*, 69, 26–46. DOI: 10.1016/j.euroneuro.2023.01.001 PMID: 36706689

Shabaruddin, F. H., Fleeman, N. D., & Payne, K. (2015). Economic evaluations of personalized medicine: Existing challenges and current developments. *Pharmacogenomics and Personalized Medicine*, •••, 115–126. DOI: 10.2147/PGPM.S35063 PMID: 26309416

Shukla, R. (2020). *Pharmacogenomics: overview, applications, and recent developments*. Drug Design-Novel Adv Omics Field Appl.

. Sosa-Macías, M., Teran, E., Waters, W., Fors, M. M., Altamirano, C., Jung-Cook, H., ... & LLerena, A. (2016). Pharmacogenetics and ethnicity: relevance for clinical implementation, clinical trials, pharmacovigilance and drug regulation in Latin America.

Sukri, A., Salleh, M. Z., Masimirembwa, C., & Teh, L. K. (2022). A systematic review on the cost effectiveness of pharmacogenomics in developing countries: Implementation challenges. *The Pharmacogenomics Journal*, 22(3), 147–159. DOI: 10.1038/s41397-022-00272-w PMID: 35319010

Tong, H., Phan, N. V., Nguyen, T. T., Nguyen, D. V., Vo, N. S., & Le, L. (2021). Review on databases and bioinformatic approaches on pharmacogenomics of adverse drug reactions. *Pharmacogenomics and Personalized Medicine*, 14, 61–75. DOI: 10.2147/PGPM.S290781 PMID: 33469342

Tristan, M. (2022). Pharmacogenomics Testing in Phase I Oncology Clinical Trials: Constructive Criticism Is Warranted. *Cancers (Basel)*, 14(5), 1131–1131. DOI: 10.3390/cancers14051131 PMID: 35267440

Tyumentseva, M., Tyumentsev, A., & Akimkin, V. (2023). CRISPR/Cas9 landscape: Current state and future perspectives. *International Journal of Molecular Sciences*, 24(22), 16077. DOI: 10.3390/ijms242216077 PMID: 38003266

Wang, Y., Yang, Y., Chen, S., & Wang, J. (2021). DeepDRK: A deep learning framework for drug repurposing through kernel-based multi-omics integration. *Briefings in Bioinformatics*, 22(5), bbab048. DOI: 10.1093/bib/bbab048 PMID: 33822890

. Xiang, X., & Yuan, Y. (2020). Application of pharmacogenomics in drug discovery and development. *Pharmacogenomics in Precision Medicine: From a Perspective of Ethnic Differences*, 257-276.

Xu, F., Zheng, C., Xu, W., Zhang, S., Liu, S., Chen, X., & Yao, K. (2024). Breaking genetic shackles: The advance of base editing in genetic disorder treatment. *Frontiers in Pharmacology*, 15, 1364135. DOI: 10.3389/fphar.2024.1364135 PMID: 38510648

Chapter 6

Introduction to Genomic-Driven Drug Discovery

Arli Aditya Parikesit
https://orcid.org/0000-0001-8716-3926
Indonesia International Institute for Life Sciences, Indonesia

ABSTRACT

This chapter provides a comprehensive overview of the role of genomics in accelerating the drug development process. It highlights the significant advancements in large-scale population collections and whole genome sequencing approaches, which have significantly influenced multiple stages across the lifespan of a drug discovery program. The chapter emphasizes the increasing importance of genomic data in target identification, target prioritization, and tractability, as well as predicting outcomes from pharmacological perturbations. It also discusses integrating genomic technologies with other omics approaches, such as proteomics and metabolomics, to understand disease mechanisms better and identify potential therapeutic targets. The integration of bioinformatics is pivotal in harnessing the power of genomic data for drug discovery endeavors. This chapter delves into the computational approaches and analytical frameworks that enable extracting meaningful insights from vast genomic datasets. Furthermore, it examines the contributions of bioinformatics in the field of pharmacogenomics.

INTRODUCTION

The field of drug discovery has been undergoing a significant transformation thanks to the advent of genomics. Genomics, the study of an organism's complete set of DNA, including all of its genes, has provided a new dimension to understanding disease mechanisms and developing targeted therapies (Butler, 2001; Galperin &

DOI: 10.4018/979-8-3693-6597-7.ch006

Koonin, 2000). Thus, integrating genomics into drug discovery has catalyzed the development of precision medicine. Precision medicine, or personalized medicine, tailors medical treatment to the individual characteristics of each patient's genetic profile (Fernald et al., 2011; Valeska et al., 2019). This approach contrasts sharply with the traditional "one-size-fits-all" method, where disease treatment and prevention strategies are developed for the average person, with less consideration for individual variations (Naithani et al., 2021). The genomic revolution has enabled researchers to dissect the genetic underpinnings of various diseases at an unprecedented resolution, revealing the complex interplay of genes and their contributions to disease susceptibility and drug response (Kabadi et al., 2020).

Furthermore, genomics-driven drug discovery has been instrumental in the emergence of pharmacogenomics, the branch of genetics concerned with determining an individual's likely response to therapeutic drugs (Shukla, 2020; Thorn et al., 2010). This field has the potential to significantly reduce the trial-and-error approach of drug prescribing, minimize adverse drug reactions, and increase the efficacy of treatments. By analyzing a patient's genome, scientists can predict which drugs and dosages would be most effective and safe for each individual (Akhtari et al., 2018; Lal et al., 2007; Yuliwulandari et al., 2019). As we move forward, the convergence of genomics with other disciplines, such as bioinformatics and systems biology are expected to refine drug discovery processes further, leading to a new era of medicine where treatments are not only reactive but also predictive and preventive (Liu et al., 2017; Rastogi, 2008; Slezak, 2003; Tan et al., 2010; Zhang et al., 2010). The future of medicine looks promising, with genomics at the helm, steering us toward more effective and personalized healthcare solutions. The translational frameworks for cardiovascular drug discovery in the clinic have been developed, as well for potentially many other diseases (Spreafico et al., 2020a; Z. Yang et al., 2024) .

UNDERSTANDING GENOMICS-DRIVEN DRUG DISCOVERY

Genomics-driven drug discovery is a process that leverages the vast amount of genetic data available to identify novel therapeutic targets. The approach is rooted in the analysis of genome-wide association studies (GWAS), which correlate genetic variations with disease traits (Pasmant et al., 2011; Teh et al., 2016). By understanding these correlations, researchers can pinpoint specific genes that may contribute to disease and target them with precision therapies. Biobanks, large collections of biological samples for research, play a crucial role in genomics-driven drug discovery (Levine et al., 2023). They provide the necessary scale of data to validate new drug targets (Hukerikar et al., 2024). Integrating diverse datasets from biobanks with other omics data, such as proteomics and metabolomics, is crucial for a comprehensive understanding of disease pathways. Artificial Intelligence (AI) and Machine Learning (ML) are increasingly being utilized to expedite the drug discovery process (Blanco-Gonzalez et al., 2022). These technologies can analyze large datasets more efficiently than traditional methods, identifying patterns and insights that might take humans much longer to recognize (Blaschke et al., 2020). AI/ML can also predict how different genetic profiles will respond to potential drugs, which is invaluable for personalized medicine.

Functional genomics investigates the roles of genes and their products in cellular processes. It aims to understand how genetic variations influence disease phenotypes (Conesa & Götz, 2008; Galperin & Koonin, 2000; Kabadi et al., 2020; Mattick, 2005). One powerful approach is CRISPR-Cas9 gene editing, which allows researchers to modify specific genes in cells or model organisms selectively (Atçeken et al., 2022; Latella et al., 2016). Scientists can identify potential therapeutic targets by perturbing genes and observing the resulting effects. For example, if a gene knockout improves disease outcomes in a cellular model, that gene becomes a candidate for drug development(Budiani et al., 2015). Functional genomics also involves transcriptomics, proteomics, and metabolomics studies to unravel complex interactions within biological systems.

Network pharmacology integrates data from various sources, including genomics, proteomics, and chemical databases, to construct interaction networks (Syahruni et al., 2023). These networks reveal connections between genes, proteins, and small molecules. Researchers can identify drug targets by analyzing these networks and predict potential off-target effects (Afdhal et al., 2020; Diansyah et al., 2021). Polypharmacology, the study of drugs that interact with multiple targets, has gained prominence (Anighoro et al., 2014; Bolognesi, 2019; Ryszkiewicz et al., 2023). Instead of focusing on a single target, polypharmacological approaches recognize that diseases often involve intricate networks of molecular interactions. Computational methods, such as ligand-based and structure-based drug design, aid in identifying compounds that modulate multiple targets simultaneously.

The Impact of Genomics on Disease Understanding and Treatment

Genomics, the study of an organism's complete set of DNA, has significantly impacted our understanding and treatment of various diseases (Aworunse et al., 2018). By analyzing individuals' genetic makeup, researchers have identified genetic variations that contribute to the development of diseases, leading to more effective and personalized treatments (Carrillo-Vázquez et al., 2013; Yan et al., 2012). Here are some diseases where genomics has played a crucial role:

1. Cancer

Genomic studies have transformed the way we understand and treat cancer. By sequencing the genomes of cancer cells, scientists can identify specific mutations that drive the growth of tumors. This has led to the development of targeted therapies that specifically attack cancer cells with certain genetic profiles, sparing healthy cells and reducing side effects. Mutations in the BRCA1 gene significantly increase the risk of developing hereditary breast cancer. Genomic data analysis has allowed researchers to identify specific BRCA1 mutations consistently present in individuals with breast cancer, leading to more personalized treatment approaches.

Moreover, the upregulated and downregulated CT45A1, GCGR, TPTE, ABCC2 and PI16 genes in Lung Carcinoma have been annotated in epigenetic research (Shemuel et al., 2023). Thus, the protein-protein interaction-based approach is also instrumental in canceromics research (Margaretha et al., 2023). Based on extensive pharmacogenomics studies, transcription factors protein is widely known to be a suitable target for cancer drugs (Gunawan et al., 2024).

2. Heart Disease

Genomics has also made strides in the field of cardiovascular diseases. For instance, identifying genetic markers associated with heart disease risk has enabled the development of preventive strategies and treatments tailored to individual genetic profiles. There has been a quite recent development in ANGPTL3 Inhibition for Cardiovascular Disease. The research involving the sequencing of exomes from patients highlighted the role of angiopoietin-related protein 3 (ANGPTL3) inactivation in protecting against cardiovascular disease (Bini et al., 2023; Luo et al., 2024). This has opened the door for the development of therapies targeting ANGPTL3 to prevent cardiovascular events

3. Rare Diseases

Mutations in single genes cause many rare diseases. Genomics has enabled the identification of these mutations, which has improved diagnosis and opened up new avenues for treatment (Heerlie et al., 2024). Conditions such as Huntington's disease and Duchenne Muscular Dystrophy are examples of genomics enhancing our understanding and management.

4. Infectious Diseases

The genomic analysis of pathogens has improved our ability to track disease outbreaks, understand their evolution, and develop vaccines. The COVID-19 pandemic is a prime example where genomic sequencing of the SARS-CoV-2 virus has been critical in monitoring the spread of the virus and identifying variants of concern (Aditya Parikesit, 2023; Parikesit et al., 2022). Moreover, zoonotic diseases such as rabies, which is rampant in animals, could be transmitted to humans as well. Genomics technology has been deployed to design an immunoinformatics-based vaccine for rabies, although it is yet to reach the wet laboratory phase(A. N. M. Ansari et al., 2024; Asia et al., 2024).

5. Chronic Diseases

Chronic diseases such as diabetes and asthma have also been impacted by genomics. Understanding the genetic factors contributing to these conditions can lead to better risk assessment, prevention, and personalized treatment plans. The discovery of gain- and loss-of-function mutations in the PCSK9 gene, which are linked to altered levels of low-density lipoprotein cholesterol (LDL-C), has led to the development of PCSK9 inhibitors. These drugs can lower LDL-C levels and reduce the risk of cardiovascular events, representing a major advancement in the management of hypercholesterolemia. Moreover, the role of long noncoding (lnc) RNA has also been elucidated in the progression of diabetes (Fachrul et al., 2018). Lastly, based on extensive genomics studies, it is known that the PTP1B protein is a feasible target for diabetes drugs (Pratama et al., 2023).

The integration of genomics into healthcare has the potential to revolutionize medicine by enabling more precise and personalized treatments. As research advances, we expect genomics to play an even greater role in improving health outcomes for various diseases. The future of genomics in medicine is promising, and it holds the key to unlocking new treatments tailored to each individual's genetic makeup, offering hope for many who suffer from genetic disorders and diseases.

The Role of Bioinformatics in Genomic-Based Drug Discovery

Bioinformatics is pivotal in leveraging genomic data for drug discovery and development(Aditya Parikesit, 2023; Parikesit, 2018, 2021). The advent of high-throughput sequencing technologies has generated vast amounts of genomic data, including whole genomes, exomes, and transcriptomes. Bioinformatics provides the computational tools and analytical frameworks to extract meaningful insights from this data deluge, enabling a genomic-driven approach to drug discovery (Arabnia & Tran, 2015; Ouzounis, 2012; Roos, 2001).

One of the primary applications of bioinformatics in genomic-based drug discovery is identifying and validating potential drug targets(Jiang & Zhou, 2005). By analyzing genomic data from diseased and healthy individuals, researchers can pinpoint genetic variations, such as single nucleotide polymorphisms (SNPs), insertions, deletions, and structural variations, that are associated with specific diseases or phenotypes(Hukerikar et al., 2024). These genetic alterations can then be mapped to their corresponding proteins, which may serve as potential drug targets (Xia, 2017).

Bioinformatics tools, such as sequence alignment algorithms, variant callers, and pathway analysis software, are employed to identify and prioritize these targets based on their biological relevance, druggability, and potential for modulation. Additionally, integrating genomic data with other omics data, such as transcriptomics, proteomics, and metabolomics, can provide a more comprehensive understanding of the molecular mechanisms underlying disease pathways, further refining target selection (Katara, 2013; Nobile, 2023; Xia, 2017).

Once potential drug targets have been identified, bioinformatics plays a crucial role in the design and screening of drug candidates (Gunawan et al., 2024; LORELL et al., 2024). Computational approaches, such as molecular docking, virtual screening, and structure-based drug design, can predict the binding affinities and interactions between small molecules and their target proteins (Valeska et al., 2019; Wooller et al., 2017).

Bioinformatics tools can also optimize drug candidates by predicting their physicochemical properties, such as solubility, permeability, and metabolic stability, which are essential for drug efficacy and safety(Cheng et al., 2012; Norinder & Bergström, 2006; H. Yang et al., 2019). These computational methods can significantly reduce the time and cost associated with traditional drug discovery processes by prioritizing the most promising drug candidates for further experimental validation (Wooller et al., 2017; Xia, 2017).

Bioinformatics is also instrumental in pharmacogenomics, which aims to understand how genetic variations influence an individual's drug response (Jiang & Zhou, 2005; Thorn et al., 2010). Bioinformatics can identify genetic markers associated with drug efficacy, toxicity, and adverse reactions by analyzing genomic data from patient populations. This information can guide the development of personalized medicine approaches, where drug therapies are tailored to an individual's genetic makeup, improving treatment outcomes and minimizing adverse effects (Katara, 2013; Nobile, 2023). The real-word-data from the patient's genomic expression already incorporated by pharmaceutical industry for their drug development pipeline, and this is a great opportunity to leverage this field even further (Zhao et al., 2022).

CHALLENGES

Challenges and ethical considerations must be addressed despite the promise of genomics-driven drug discovery. One of the main challenges is the need for a framework that can handle evidence from cross-population GWAS meta-analysis. Additionally, using genomic information raises ethical questions regarding privacy and the potential for genetic discrimination. Establishing guidelines for the ethical use of genomics in drug discovery is imperative (Hongladarom, 2006; Wilkinson, 2008). Moreover, the exisiting clinical trials framework should be expanded to incorporate the latest iteration of the regulation from the relevant regulatory bodies (Zeggini et al., 2019) .

Besides ethical issues, new technologies have also started to be implemented. One is CRISPR-based gene editing (A. NM. Ansori et al., 2023). This approach has already been implemented in malaria disease surveillance and is undergoing extensive development as a point-of-care (POC) diagnostics tool (Parikesit et al., 2023). However, the widespread use of CRISPR-based diagnostics is not assured at this point due to the establishment of existing tools such as RT-PCR-based detection.

While GWAS and other genomic studies can identify numerous disease-susceptibility genes, translating these findings into effective clinical treatments remains challenging. Researchers must validate potential drug targets and estimate their potential side effects, which often fails promising clinical trials (Sonehara & Okada, 2021). In this regard, Genomics-driven drug discovery is resource-intensive, requiring significant financial investment and access to advanced technologies. The high costs associated with sequencing, data storage, and computational analysis can be prohibitive for many research institutions (Alvarellos et al., 2023; Spreafico et al., 2020b). There are already successful biomarker-driven cancer drug designs, such as vemurafenib, erlotinib, and gefitinib (Serelli-Lee et al., 2022) . However, challenges remain for the development for other diseases.

The vast amount of genomic data generated from high-throughput sequencing technologies presents a significant challenge in data management and integration (Cook et al., 2016; Grapov et al., 2018). Researchers must handle and analyze complex datasets from various sources, including genomics, transcriptomics, proteomics, and metabolomics, to draw meaningful conclusions (Alvarellos et al., 2023; Spreafico et al., 2020b). The complexity is added to many genome-wide association studies (GWAS) that lack diversity in their cohorts, which can lead to biased results and spurious disease associations. This limitation hinders the generalizability of findings across different populations and may leave out large groups of patients who still need treatment (Alvarellos et al., 2023).

CONCLUSION

Genomics-driven drug discovery represents a paradigm shift in the development of new drugs. It offers the potential for more effective and personalized treatments, transforming the landscape of clinical medicine. As the field continues to evolve, it will undoubtedly bring forth innovative solutions to some of our most pressing medical challenges. Integrating bioinformatics in drug discovery is a testament to the power of interdisciplinary collaboration. By harnessing the vast amounts of genomic data available, researchers can pinpoint potential drug targets with unprecedented precision. This approach accelerates the identification of viable candidates and enhances the understanding of disease mechanisms at a molecular level. As computational methods become more sophisticated, the predictive accuracy of drug efficacy and safety profiles will improve, further revolutionizing the field. For those interested in exploring this topic further, the rest of the book will provide overviews of the role of omics in drug discovery and the management of diverse data sets. Those topics can serve as a gateway to understanding and participating in this exciting and rapidly advancing field.

ACKNOWLEDGMENTS

We thank the Research Institute and Community Service (LPPM) of the Indonesia International Institute for Life Sciences for the opportunity to develop this research article. We also thank the Community Language Center of the Indonesia International Institute for Life Sciences for their excellent proofreading of our manuscript.

REFERENCES

Aditya Parikesit, A. (2023). Introductory Chapter: Current Landscape of Animal and Human Virus Treatments. In Parikesit, A. A. (Ed.), *Antiviral Strategies in the Treatment of Human and Animal Viral Infections as*. IntechOpen., DOI: 10.5772/intechopen.111733

Afdhal, D., Ananta, K. W., & Hartono, W. S. (2020). Adverse Drug Reactions Prediction Using Multi-label Linear Discriminant Analysis and Multi-label Learning. *2020 International Conference on Advanced Computer Science and Information Systems (ICACSIS)*, 69–76. DOI: 10.1109/ICACSIS51025.2020.9263166

Akhtari, J., Tafazoli, A., Mehrad-Majd, H., & Mahrooz, A. (2018). Nanovehicle-based Small Interfering RNA (siRNA) Delivery for Therapeutic Purposes: A New Molecular Approach in Pharmacogenomics. *Current Clinical Pharmacology*, 13(3), 173–182. DOI: 10.2174/1574884713666180709152610 PMID: 29992895

Alvarellos, M., Sheppard, H. E., Knarston, I., Davison, C., Raine, N., Seeger, T., Prieto Barja, P., & Chatzou Dunford, M. (2023). Democratizing clinical-genomic data: How federated platforms can promote benefits sharing in genomics. *Frontiers in Genetics*, 13, 1045450. Advance online publication. DOI: 10.3389/fgene.2022.1045450 PMID: 36704354

Anighoro, A., Bajorath, J., & Rastelli, G. (2014). Polypharmacology: Challenges and opportunities in drug discovery. In *Journal of Medicinal Chemistry* (Vol. 57, Issue 19, pp. 7874–7887). DOI: 10.1021/jm5006463

Ansori, A. N. M., Antonius, Y., Susilo, R. J. K., Hayaza, S., Kharisma, V. D., Parikesit, A. A., Zainul, R., Jakhmola, V., Saklani, T., Rebezov, M., & Ullah, Md. E., Maksimiuk, N., Derkho, M., & Burkov, P. (2023). Application of CRISPR-Cas9 genome editing technology in various fields: A review. In *Narra J* (Vol. 3, Issue 2, pp. e184–e184). DOI: 10.52225/narra.v3i2.184

Ansori, A. N. M., Chandara, N., Sucipto, T. H., Kharisma, V. D., Murtadlo, A. A. A., Kolesnik, E., Rebezov, M., Parikesit, A. A., Wiradana, P. A., Widhiantara, I. G., Sahadewa, S., Durry, F. D., Jakhmola, V., & Zainul, R. (2024). Computational analysis of Indonesian rabies vaccine enhanced by myeloid differentiation factor 88 (Myd88) genetic adjuvant. *Journal of Medicinal and Pharmaceutical Chemistry Research*, 6(9), 1327–1339. DOI: 10.48309/JMPCR.2024.448250.1145

Arabnia, H. R., & Tran, Q. N. (2015). *Emerging trends in computational biology, bioinformatics, and system biology: Algorithms and software tools*. Elsevier. https://books.google.co.id/books?id=VuGoBAAAQBAJ

Asia, S., Immunoinf, A., Amin, A. M., Nur, A., Ansori, M., & Kharisma, V. D. (2024). T-Cell Epitope Vaccine Prediction Analysis Targeting Phosphoprotein (P) Rabies Virus Based on the Presence of HLA-I Alleles A, B, and C Loci Throughout T-Cell Epitope Vaccine Prediction Analysis Targeting Phosphoprotein (P) Rabies Virus Based on the. *Research Journal of Pharmacy and Technology*, 17(May), 2001–2008. DOI: 10.52711/0974-360X.2024.00317

Atçeken, N., Yigci, D., Ozdalgic, B., & Tasoglu, S. (2022). CRISPR-Cas-Integrated LAMP. *Biosensors (Basel)*, 12(11), 1035. Advance online publication. DOI: 10.3390/bios12111035 PMID: 36421156

Aworunse, O. S., Adeniji, O., Oyesola, O. L., Isewon, I., Oyelade, J., & Obembe, O. O. (2018). Genomic Interventions in Medicine. *Bioinformatics and Biology Insights*, 12, 1177932218816100. DOI: 10.1177/1177932218816100 PMID: 30546257

Bini, S., Tramontano, D., Minicocci, I., Di Costanzo, A., Tambaro, F., D'Erasmo, L., & Arca, M. (2023). How ANGPTL3 Inhibition Will Help Our Clinical Practice? In *Current Atherosclerosis Reports* (Vol. 25, Issue 1). DOI: 10.1007/s11883-022-01076-w

Blanco-Gonzalez, A., Cabezon, A., Seco-Gonzalez, A., Conde-Torres, D., Antelo-Riveiro, P., Pineiro, A., & Garcia-Fandino, R. (2022). *The Role of AI in Drug Discovery: Challenges, Opportunities, and Strategies*. https://doi.org//arxiv.2212.08104DOI: 10.48550

Blaschke, T., Arús-Pous, J., Chen, H., Margreitter, C., Tyrchan, C., Engkvist, O., Papadopoulos, K., & Patronov, A. (2020). REINVENT 2.0: An AI Tool for de Novo Drug Design. *Journal of Chemical Information and Modeling*, 60(12), 5918–5922. DOI: 10.1021/acs.jcim.0c00915 PMID: 33118816

Bolognesi, M. L. (2019). Harnessing Polypharmacology with Medicinal Chemistry. *ACS Medicinal Chemistry Letters*, 10(3), 273–275. Advance online publication. DOI: 10.1021/acsmedchemlett.9b00039 PMID: 30891125

Budiani, A., Nugroho, I. B., Sari, D. A., Palupi, I., & Putranto, R. A. (2015). CRISPR / Cas9-mediated Knockout of An Oil Palm Defense-related Gene to The Pathogenic Fungi Ganoderma boninense. *Indonesian Journal of Biotechnology*, 24(1), 1–10. DOI: 10.22146/ijbiotech.52170

Butler, D. (2001). Genomics. Are you ready for the revolution? *Nature*, 409(6822), 758–760. DOI: 10.1038/35057400 PMID: 11236971

Carrillo-Vázquez, J. P., Chimal-Vega, B., Zamora-López, B., Marchat, L. A., Benítez-Cardoza, C. G., Reyes-López, C. A. S., & Zamorano-Carrillo, A. (2013). Structural Consequences Of The Polymorphism Q223r In The Human Leptin Receptor: A Molecular Dynamics Study. *American Journal of Agricultural and Biological Sciences*, 8(3), 239–248. DOI: 10.3844/ajabssp.2013.239.248

Cheng, F., Li, W., Zhou, Y., Shen, J., Wu, Z., Liu, G., Lee, P. W., & Tang, Y. (2012). AdmetSAR: A comprehensive source and free tool for assessment of chemical ADMET properties. *Journal of Chemical Information and Modeling*, 52(11), 3099–3105. DOI: 10.1021/ci300367a PMID: 23092397

Conesa, A., & Götz, S. (2008). Blast2GO: A comprehensive suite for functional analysis in plant genomics. *International Journal of Plant Genomics*, 619832, 1–12. Advance online publication. DOI: 10.1155/2008/619832 PMID: 18483572

Cook, C. E., Bergman, M. T., Finn, R. D., Cochrane, G., Birney, E., & Apweiler, R. (2016). The European Bioinformatics Institute in 2016: Data growth and integration. *Nucleic Acids Research*, 44(D1), D20–D26. DOI: 10.1093/nar/gkv1352 PMID: 26673705

Diansyah, M. R., Annisa, A., & Kusuma, W. A. (2021). Analysis using top-k skyline query of protein-protein interaction reveals alpha-synuclein as the most important protein in Parkinson's disease. *Indonesian Journal of Biotechnology*, 26(4), 197–205. DOI: 10.22146/ijbiotech.63023

Fachrul, M., Utomo, D. H., & Parikesit, A. A. (2018). lncRNA-based study of epigenetic regulations in diabetic peripheral neuropathy. *In Silico Pharmacology*, 6(1), 7. DOI: 10.1007/s40203-018-0042-8 PMID: 30607320

Fernald, G. H., Capriotti, E., Daneshjou, R., Karczewski, K. J., & Altman, R. B. (2011). Bioinformatics challenges for personalized medicine. *Bioinformatics (Oxford, England)*, 27(13), 1741–1748. DOI: 10.1093/bioinformatics/btr295 PMID: 21596790

Galperin, M. Y., & Koonin, E. V. (2000). Who's your neighbor? New computational approaches for functional genomics. *Nature Biotechnology*, 18(6), 609–613. DOI: 10.1038/76443 PMID: 10835597

Grapov, D., Fahrmann, J., Wanichthanarak, K., & Khoomrung, S. (2018). Rise of Deep Learning for Genomic, Proteomic, and Metabolomic Data Integration in Precision Medicine. *OMICS: A Journal of Integrative Biology*, 22(10), 630–636. DOI: 10.1089/omi.2018.0097 PMID: 30124358

Gunawan, A., Nadia, J., & Parikesit, A. A. (2024). The Molecular Docking Approach Toward Targeting Transcription Factors in Prostate Cancer. In *A Closer Look at Cancer Biomarkers* (p. Chapter 3). Nova Science Publishers. https://www.researchgate.net/publication/380032683_The_Molecular_Docking_Approach_Toward_Targeting_Transcription_Factors_in_Prostate_Cancer

Heerlie, D. M., Margaretha, F., Fugaha, D. R., & Parikesit, A. A. (2024). The Effects and Treatments for Usher Syndrome: A Review. *Indonesian Journal of Life Sciences*, 6(1), 68–82. DOI: 10.54250/IJLS.V6I01.184

Hongladarom, S. (2006). Ethics of bioinformatics: A convergence between bioethics and computer ethics. *Asian Biotechnology and Development Review*, 9(1), 37–44.

Hukerikar, N., Hingorani, A. D., Asselbergs, F. W., Finan, C., & Schmidt, A. F. (2024). Prioritising genetic findings for drug target identification and validation. *Atherosclerosis*, 390, 117462. Advance online publication. DOI: 10.1016/j.atherosclerosis.2024.117462 PMID: 38325120

Jiang, Z., & Zhou, Y. (2005). Using bioinformatics for drug target identification from the genome. *American Journal of Pharmacogenomics*, 5(6), 387–396. DOI: 10.2165/00129785-200505060-00005 PMID: 16336003

Kabadi, A., McDonnell, E., Frank, C. L., & Drowley, L. (2020). Applications of Functional Genomics for Drug Discovery. In *SLAS Discovery* (Vol. 25, Issue 8). DOI: 10.1177/2472555220902092

Katara, P. (2013). Role of bioinformatics and pharmacogenomics in drug discovery and development process. *Network Modeling and Analysis in Health Informatics and Bioinformatics*, 2(4), 225–230. DOI: 10.1007/s13721-013-0039-5

Lal, S., Wong, Z. W., Jada, S. R., Xiang, X., Shu, X. C., Ang, P. C. S., Figg, W. D., Lee, E. J. D., & Chowbay, B. (2007). Novel SLC22A16 polymorphisms and influence on doxorubicin pharmacokinetics in Asian breast cancer patients. *Pharmacogenomics*, 8(6), 567–575. DOI: 10.2217/14622416.8.6.567 PMID: 17559346

Latella, M. C., Di Salvo, M. T., Cocchiarella, F., Benati, D., Grisendi, G., Comitato, A., Marigo, V., & Recchia, A. (2016). In vivo Editing of the Human Mutant Rhodopsin Gene by Electroporation of Plasmid-based CRISPR/Cas9 in the Mouse Retina. *Molecular Therapy. Nucleic Acids*, 5, e389. DOI: 10.1038/mtna.2016.92 PMID: 27874856

Levine, K. S., Leonard, H. L., Blauwendraat, C., Iwaki, H., Johnson, N., Bandres-Ciga, S., Ferrucci, L., Faghri, F., Singleton, A. B., & Nalls, M. A. (2023). Virus exposure and neurodegenerative disease risk across national biobanks. *Neuron*, 0(0), 1086–1093.e2. Advance online publication. DOI: 10.1016/j.neuron.2022.12.029 PMID: 36669485

Liu, Y., Bian, C., Huang, Y., You, X., & Shi, Q. (2017). Recent advances in bioinformatics for fish genomics. In *Bioinformarics* (p. 11). Avid Science., https://www.pacb.com/publications/recent-advances-in-bioinformatics-for-fish-genomics/

Lorell, J., Ivana, K., Tanoto, J. P. W., Michael, M., Jescika, S., Gautama, N. R. P., Tjoa, O., Alina, K., Parikesit, A. A., & Sutanto, F.LORELL. (2024). In silico testing of C9H12ClNO2 and C6H5Cl2NO as derivatives of acetaminophen using molecular docking method. *Notulae Scientia Biologicae*, 16(1), 11632–11632. DOI: 10.55779/nsb16111632

Luo, F., Das, A., Khetarpal, S. A., Fang, Z., Zelniker, T. A., Rosenson, R. S., & Qamar, A. (2024). ANGPTL3 inhibition, dyslipidemia, and cardiovascular diseases. In *Trends in Cardiovascular Medicine* (Vol. 34, Issue 4). DOI: 10.1016/j.tcm.2023.01.008

Margaretha, F., Subagja, M., Putri, A. D., Salam, C. Q., Valenthenardo, L., Adhiwijaya, P. K., Kaitlyn, G. Z., & Parikesit, A. (2023). Comprehensive analysis of protein-protein interactions (PPIs) with structure prediction program for breast cancer determination. *Berkala Penelitian Hayati*, 29(3), 87–94. DOI: 10.23869/bphjbr.29.3.20232

Mattick, J. S. (2005). The functional genomics of noncoding RNA. *Science*, 309(5740), 1527–1528. DOI: 10.1126/science.1117806 PMID: 16141063

Naithani, N., Sinha, S., Misra, P., Vasudevan, B., & Sahu, R. (2021). Precision medicine: Concept and tools. In *Medical Journal Armed Forces India* (Vol. 77, Issue 3). DOI: 10.1016/j.mjafi.2021.06.021

Nobile, C. (2023). Bioinformatics in drug discovery and development: Methods and applications. *Pre QC. Revised*, 12. Advance online publication. DOI: 10.35841/bmcr

Norinder, U., & Bergström, C. A. S. (2006). Prediction of ADMET properties. In *ChemMedChem* (Vol. 1, pp. 920–937). DOI: 10.1002/cmdc.200600155

Ouzounis, C. A. (2012). Rise and demise of bioinformatics? promise and progress. *PLoS Computational Biology*, 8(4), e1002487. Advance online publication. DOI: 10.1371/journal.pcbi.1002487 PMID: 22570600

Parikesit, A. A. (2018). Introductory Chapter: The Contribution of Bioinformatics as Blueprint Lead for Drug Design. In Glavic, I. (Ed.), *Molecular Insight of Drug Design* (p. 7). InTech., DOI: 10.5772/intechopen.79664

Parikesit, A. A. (2021). Introductory Chapter: The Emerging Corner of the Omics Studies for Rational Drug Design. In Parikesit, A. A. (Ed.), *Drug Design: Novel Advances in the Omics Field and Applications* (p. 4). IntechOpen., DOI: 10.5772/intechopen.95544

Parikesit, A. A., Hermantara, R., Kevin, G., & Sidhartha, E. (2023). Designing hybrid CRISPR-Cas12 and LAMP detection systems for treatment-resistant Plasmodium falciparum with in silico method. *Narra J*, 3(3), e301. DOI: 10.52225/narra.v3i3.301 PMID: 38455618

Parikesit, A. A., Nurdiansyah, R., Parikesit, A. A., Nurdiansyah, R., Aditya Parikesit, A., & Nurdiansyah, R. (2022). *Introductory Chapter: Current Perspective of COVID-19 Drugs. COVID-19 Drug Development - Recent Advances*. New Perspectives and Applications., DOI: 10.5772/intechopen.105537

Pasmant, E., Sabbagh, A., Vidaud, M., & Bieche, I. (2011). ANRIL, a long, noncoding RNA, is an unexpected major hotspot in GWAS. *The FASEB Journal*, 25(2), 444–448. Advance online publication. DOI: 10.1096/fj.10-172452 PMID: 20956613

Pratama, R. A., Astina, J., & Parikesit, A. A. (2023). In Silico Antidiabetic Study of Phenolic Compounds from Banana (Musa spp.) Peel Targeting PTP1B Protein. *Journal of Tropical Biodiversity and Biotechnology*, 8(3), 83124. DOI: 10.22146/jtbb.83124

Rastogi, P. (2008). *Bioinformatics Methods And Applications: Genomics Proteomics And Drug Discovery 3Rd Ed.* Prentice-Hall Of India Pvt. Limited. http://books.google.co.id/books?id=YP1rEFxFgDcC

Roos, D. S. (2001). COMPUTATIONAL BIOLOGY: Bioinformatics—Trying to Swim in a Sea of Data. *Science*, 291(5507), 1260–1261. DOI: 10.1126/science.291.5507.1260 PMID: 11233452

Ryszkiewicz, P., Malinowska, B., & Schlicker, E. (2023). Polypharmacology: promises and new drugs in 2022. In *Pharmacological Reports* (Vol. 75, Issue 4). DOI: 10.1007/s43440-023-00501-4

Serelli-Lee, V., Ito, K., Koibuchi, A., Tanigawa, T., Ueno, T., Matsushima, N., & Imai, Y. (2022). A State-of-the-Art Roadmap for Biomarker-Driven Drug Development in the Era of Personalized Therapies. In *Journal of Personalized Medicine* (Vol. 12, Issue 5). DOI: 10.3390/jpm12050669

Shemuel, J., Angelique, P., Josephine, E., Fugaha, D. R., Gabriela, V., Alhusain, S., & Parikesit, A. A. (2023). ANALISIS GEN KOMPARATIF KARSINOMA SEL SKUAMOSA PARU-PARU ANTARA INDIVIDU MEROKOK DAN TIDAK MEROKOK [Comparative Gene Analysis of Squamous Cell Lung Carcinoma Between Smoking and Non-smoking Individuals]. *Berita Biologi*, 22(3), 291–304. DOI: 10.55981/beritabiologi.2023.908

Shukla, R. (2020). *Pharmacogenomics: Overview*. Applications, and Recent Developments., DOI: 10.5772/intechopen.93737

Slezak, T. (2003). Comparative genomics tools applied to bioterrorism defence. *Briefings in Bioinformatics*, 4(2), 133–149. DOI: 10.1093/bib/4.2.133 PMID: 12846395

Sonehara, K., & Okada, Y. (2021). Genomics-driven drug discovery based on disease-susceptibility genes. *Inflammation and Regeneration*, 41(1), 8. Advance online publication. DOI: 10.1186/s41232-021-00158-7 PMID: 33691789

Spreafico, R., Soriaga, L. B., Grosse, J., Virgin, H. W., & Telenti, A. (2020a). Advances in Genomics for Drug Development. *Genes*, 11(8), 1–19. DOI: 10.3390/genes11080942 PMID: 32824125

Spreafico, R., Soriaga, L. B., Grosse, J., Virgin, H. W., & Telenti, A. (2020b). Advances in Genomics for Drug Development. *Genes*, 11(8), 1–19. DOI: 10.3390/genes11080942 PMID: 32824125

Syahruni, R., Umar, A. H., Rahman, H. N., & Kusuma, W. A. (2023). Exploration of Annona muricata (Annonaceae) in the Treatment of Hyperlipidemia through Network Pharmacology and Molecular Docking. *Sains Malaysiana*, 52(3), 899–939. DOI: 10.17576/jsm-2023-5203-17

Tan, T. W., Tong, J. C., Khan, A. M., de Silva, M., Lim, K. S., & Ranganathan, S. (2010). Advancing standards for bioinformatics activities: Persistence, reproducibility, disambiguation and Minimum Information About a Bioinformatics investigation (MIABi). *BMC Genomics*, 11(SUPPL. 4), S27. DOI: 10.1186/1471-2164-11-S4-S27 PMID: 21143811

Teh, C.-K., Ong, A.-L., Kwong, Q.-B., Apparow, S., Chew, F.-T., Mayes, S., Mohamed, M., Appleton, D., & Kulaveerasingam, H. (2016). Genome-wide association study identifies three key loci for high mesocarp oil content in perennial crop oil palm. *Scientific Reports*, 6(1), 19075. DOI: 10.1038/srep19075 PMID: 26743827

Thorn, C. F., Klein, T. E., & Altman, R. B. (2010). Pharmacogenomics and bioinformatics: PharmGKB. *Pharmacogenomics*, 11(4), 501–505. DOI: 10.2217/pgs.10.15 PMID: 20350130

Valeska, M. D., Adisurja, G. P., Bernard, S., Wijaya, R., Aldino, M., & Parikesit, A. A. (2019). The Role of Bioinformatics in Personalized Medicine: Your Future Medical Treatment. *Cermin Dunia Kedokteran*, 46(12), 785–788. DOI: 10.5281/zenodo.4460835

Wilkinson, S. (2008). "Eugenics talk" and the language of bioethics. *Journal of Medical Ethics*, 34(6), 467–471. DOI: 10.1136/jme.2007.021592 PMID: 18511622

Wooller, S. K., Benstead-Hume, G., Chen, X., Ali, Y., & Pearl, F. M. G. (2017). Bioinformatics in translational drug discovery. *Bioscience Reports*, 37(4), 20160180. DOI: 10.1042/BSR20160180 PMID: 28487472

Xia, X. (2017). Bioinformatics and Drug Discovery. *Current Topics in Medicinal Chemistry*, 17(15), 1709–1726. DOI: 10.2174/1568026617666161116143440 PMID: 27848897

Yan, W., Cui, J., Wang, Y., Zhang, X., Yang, F., Chen, E. F., Wang, M., & Yang, J. (2012). Comparative Analysis of *UGT1A9* Genetic Polymorphisms between Chinese Han and Tibetan Populations. *American Medical Journal*, 3(2), 194–202. DOI: 10.3844/amjsp.2012.194.202

Yang, H., Lou, C., Sun, L., Li, J., Cai, Y., Wang, Z., Li, W., Liu, G., & Tang, Y. (2019). AdmetSAR 2.0: Web-service for prediction and optimization of chemical ADMET properties. *Bioinformatics (Oxford, England)*, 35(6), 1067–1069. DOI: 10.1093/bioinformatics/bty707 PMID: 30165565

Yang, Z., Zhao, J. V., Qi, Y., Deng, X., Ji, Z., & Liu, J. (2024). A translational framework of genoproteomic studies for cardiovascular drug discovery. *Npj Cardiovascular Health 2024 1:1, 1*(1), 1–12. DOI: 10.1038/s44325-024-00015-9

Yuliwulandari, R., Prayuni, K., Susilowati, R. W., Subagyo, S., Soedarsono, S. M., Sofro, A. S., Tokunaga, K., & Shin, J. G. (2019). NAT2 slow acetylator is associated with anti-tuberculosis drug-induced liver injury severity in Indonesian population. *Pharmacogenomics*, 20(18), 1303–1310. DOI: 10.2217/pgs-2019-0131 PMID: 31699005

Zeggini, E., Gloyn, A. L., Barton, A. C., & Wain, L. V. (2019). Translational genomics and precision medicine: Moving from the lab to the clinic. *Science*, 365(6460), 1409–1413. DOI: 10.1126/science.aax4588 PMID: 31604268

Zhang, L., Zhang, X., Ma, Q., Ma, F., & Zhou, H. (2010). Transcriptomics and Proteomics in the Study of H1N1 2009. In *Genomics, Proteomics and Bioinformatics* (Vol. 8, Issue 3, pp. 139–144). Elsevier. DOI: 10.1016/S1672-0229(10)60016-2

Zhao, X., Iqbal, S., Valdes, I. L., Dresser, M., & Girish, S. (2022). Integrating real-world data to accelerate and guide drug development: A clinical pharmacology perspective. *Clinical and Translational Science*, 15(10), 2293–2302. DOI: 10.1111/cts.13379 PMID: 35912537

Chapter 7
Precision Medicine in View of Genomic Biomarker Discovery

Abhidha Kohli
Department of Biotechnology, Noida Institute of Engineering and Technology, Greater Noida, India

Aryan Kushwaha
Department of Biotechnology, Noida Institute of Engineering and Technology, Greater Noida, India

Sonia Verma
Department of Biotechnology, Noida Institute of Engineering and Technology, Greater Noida, India

Arpit Gupta
Department of Biotechnology, Noida Institute of Engineering and Technology, Greater Noida, India

Neeraj Agarwal
https://orcid.org/0000-0002-4395-4301
Department of Biotechnology, Noida Institute of Engineering and Technology, Greater Noida, India

Ashwini Kumar
https://orcid.org/0000-0002-4696-4652
School of Basic Sciences and Research, Sharda University, Greater Noida, India

ABSTRACT

Precision Medicine encompasses decision making about prevention, diagnosis, and treatment of a particular disease in a patient tailored based on the information derived from a patient's genetic profile, environment, and lifestyle. It is patient-specific based on their specific clinical and biological characteristics with minimum side effects. With advancements in DNA sequencing technology, this concept gained attention among scientists. The projects like Human Genome Project and International HapMap Project provided deep insight into human genomics and information about specific disease-associated gene variants. Moreover, pharmacogenetics and pharmacogenomics have helped in understanding molecular basis

DOI: 10.4018/979-8-3693-6597-7.ch007

of diseases and corresponding therapeutic strategy designed for individuals for a particular drug and the effect of multiple variations in genome on drug treatment, helping biomarker identification, influencing pathogen and drug effects on body, establishing the need for genomic biomarkers discovery for designing patient-specific treatment modalities for a specific disease.

1. INTRODUCTION TO PRECISION MEDICINE

The term "Precision medicine" describes the approach of treating a disease with more clarity by using genomic and other technologies for enabling precise targeting of disease subgroups and development of more accurate treatment strategies with new therapies (Ashley, 2016). Although definitions of precision medicine vary, it can be understood as the approach that uses diagnostic tools like genomics, epigenomics, pharmacogenomics and pharmacokinetics to develop targeted treatments for treating each patient based on their genetic, biomarkers, or psychosocial characteristics (Ramaswami et al., 2018).

Precision medicine is an emerging approach of therapy that takes into account variability in genes, lifestyle, and environment of an individual for disease treatment and prevention. This approach is suggested to help researchers and doctors as it increases the accuracy of diagnosis and prognosis of a disease and hence help in accurately strategizing the treatment and prevention of a specific disease in a specific group of people (Hossain, 2023). Briefly, precision medicine can be defined as an approach that help researchers and doctors to develop tailored drug treatment for any disease in an individual by using their genetics, environment, and lifestyle (Sisodiya, 2021).

Precision medicine is developed on the basis of patient stratification, also known as novel taxonomy that is derived utilizing large scale data that includes clinical, lifestyle, genetic, and biomarker information and not merely on the basis of traditional "signs and symptoms" approach (König et al., 2017). It helps to predict suitable treatments that work best for specific patients. It improvises the understanding of the underlying mechanisms of various diseases and helps to prevent, diagnose and treat a wide range of diseases (Dzau and Ginsburg, 2016). It uses integration of electronic medical records into patient care, allowing physicians and researchers to have easier access to medical information (Sitapati et al., 2017) and aims to maximize effectiveness of treatment while minimizing adverse effects (characteristics (Ramaswami et al., 2018).

Therefore, it is important to promote and develop this concept for certain potential benefits which includes promising aspects of improving health and health care, aiming to create new therapies characteristics, to learn and comprehend about disease processes, and to potentially take preventive measure before a disease arises its characteristics (Ramaswami et al., 2018). Additionally, the goal of precision medicine has been consistent with the communicable diseases with identification of causative organisms and generation of data repositories about infectious and resistant organisms. Moreover, the noncommunicable diseases have had treatment implications leading to a shift in the accepted way of potential therapeutic interventions that could maximize disease treatment and minimize adverse events characteristics (Ramaswami et al., 2018). These studies have facilitated **basket** and **umbrella** trials (discussed later in this chapter) for drug discovery for various cancers based on precision oncology studies.

1.1. Definition and Concept

As per National Health Service (NHS) England, precision medicine help delivers four important P's related to an individual's health i.e., prevention and prediction of a disease, precise diagnoses, personalized or tailored therapeutic interventions, and participation of the patient. These P's can help determine the risk of an individual developing an illness and possible health choices that can be made be it medicines, lifestyle or diet-related to ensure individual-specific health care is being delivered (NHS England). Although this term "precision medicine" is relatively new, the concept has been a part of health care for several years. The evolution of how the traditional medicine has evolved to now paving the way for potential precision medicine is shown in **table 1**.

Table 1. The evolution of medicine from traditional to precision approach.

Evolution from "Traditional" to "Precision" approach of medicine (König IR et al., 2017)	
Previously only "personalized" medicine-Physicians treated patients on a personal level	Now, new tailored biomedical information about the genetic, epigenetic profile of the individual adds substantial information and make it "Precision" medicine
Restricted to "Signs and Symptoms" approach	Promises to shed this limitation and tailor it with the right medicine, at the right time, in a the right dosage as per an individual
Worked on "one-size-fits-all" concept	Offers customized approach for diagnosis, prognosis, and treatment of a patient-specific disease treatment
Example: ATLAS, NSABP B20 trials for HR+ breast cancer (Hensing et al., 2020)	Examples: Basket and umbrella trials for precision oncology-based cancer studies

An example where precision medicine has been known to be playing its role is blood transfusion where the type of blood is matched of the donor with the receiver to ensure risk free transfusion and there can be numerous other such examples;

however, the role of precision medicine in everyday health is limited and needs intervention and expansion in the near future harnessing information procured from underlying technologies like genomics, transcriptomics, proteomics, metabolomics, pharmacogenetics, pharmacogenomics (Hasanzad et al., 2022). Pharmacogenetics deals with the interactions of drug with relatively limited number of genes. However, pharmacogenomics is the study involving the effect of genes on the person's response to certain medications or in other words interactions of drug with entire genome (Hasanzad et al., 2022). It is a relatively new field that combines the science of drugs i.e., pharmacology and the study of genes and their functions i.e., genomics that helps in developing the treatments which are designed according to a person's genetic variations.

1.2. Historical Background

The idea of personalized medicine is ingrained by the fact that individuals possess refined and unique characteristics at the molecular, environmental, physiological exposure and behavioral levels which are variable and specific for different individuals, thus when it comes to the treatment that is required to be provided against any disease, it should be tailored to these nuanced and unique characteristics. This idea seemed feasible through the use of emerging technologies such as proteomics, DNA sequencing, wireless health monitoring devices, and imaging protocols which have showed great inter-individual variation in disease processes at molecular level (Goetz and Schork, 2018). In the year 2015, Barack Obama mentioned in his speech at the State of the Union address about the virtues of precision medicine as a potential initiative to revolutionize individual health and treat disease. Historically, the advent of precision medicine goes back to 2700 BC starting from the epidemic of malaria to Hippocratic "personalized" medicine. This evolution gained momentum from the first idea of biomarkers in 1950s; however, the term "personalized medicine" appears as a scientific term only in the 1990s even with new technologies of sequencing and generation of huge data repositories and still struggles to find its way in application for diagnosis, prediction and targeted therapy (Visvikis-Siest et al., 2020).

Around a century ago, an English physician Archibald Garrod, studied numbers of metabolic diseases including albinism, alkaptonuria, pentosuria and cystinuria. While working on Alkaptonuria, he observed that some members of families having alkaptonuria showed measurably outlying values for certain biochemical assays as compared to those members who did not have alkaptonuria e.g., from urine. This made him conclude that alkaptonuria was due to specific alteration in metabolism among affected individuals, which was later proven correct. Thus Garrod, while considering other diseases like alkaptonuria mentioned his belief that these diseases are important examples of variation of chemical behavior which are present everywhere up to certain extent and no two individuals can possess same bodily condition and neither their chemical processes can be performed in identical way. This belief of Gerrod represents that human are widely different from each other in terms of their metabolism, and this difference could explain the presence of overt and visible phenotypic difference among individuals, such as their varying reactions to various conditions that they are exposed to, like diseases, and the way they manifest diseases (Goetz and Schork, 2018).

With the traditional methods of diagnosis, prognosis and treatment, doctors design therapies to treat large groups of people with same or even similar disease adopting the one-size-fits-all framework. Despite many variable factors like age, sex, weight, a doctor prescribes the same treatment to any patient that he/she prescribes to any other patient with the same disease. However, not everyone responds to the treatment in a similar way as anticipated. Some individuals respond very well to the treatment but others may show allergic response. This led to finding treatments that suits the individual and thus gave birth to the idea of personalized medicine (Goetz and Schork, 2018).

1.3. Key Principles

The basic principle behind developing precision medicine is to develop drugs and treatment strategies based on the information available about biomarkers and genomic, epigenomic profiling of the patient that leads to increase in accuracy of drug target and treatment efficiency, which eventually increases the chances of clinical success and drug approval. Precision medicine can assist decision-making process of whether or not to select a particular drug for the drug target, and to understand the response of different individuals to that particular drug in both nonclinical and clinical settings. It also helps in deciding appropriate doses of drugs to show its maximum efficiency, and to understand drug's mode-of-action (Cook et al., 2018).

Basically, the inter-individual variations in response to drug treatment may be resulting from genetic polymorphisms, epigenetic factors, environmental factors like lifestyle, diet, prior drug treatments or microbiome and disease-related or demographic factors like age, sex, bodyweight. The identification of these factors or predictors in drug pharmacology i.e., in both pharmacokinetic and pharmacodynamic processes is essential for guiding the process of drug selection and dosing to achieve personalized medicine (Kantae et al., 2017).

In the pharmaceutical and drug discovery, the development of patient-specific disease biomarkers plays an important role by helping researchers and doctors identify a mechanism of action of a particular drug, its toxicity, efficiency and its effect on different individuals. Biomarkers are used to check the development status of disease, its course, effect on individual, therapy that needs to be provided to the individual that they can assist in selecting the primary treatment before diagnosis of disease and help in monitoring the therapy and selecting new therapy for the treatment of disease (Bodaghi et al., 2023).

1.4. Real World Clinical Challenges and Industry Trends

The real-world clinical challenges in implementation of precision medicine lies in the organizational, technological and people perspective as they hold decision-makers and other stake holders of the healthcare systems. The organizational, people and technological challenges include data quality or incomplete data of patients; issues of trust, data access, cost of drugs, regulations; and security of patient data and awareness, respectively (Grimberg et al., 2021). Genetic testing for patients is challenges in clinical settings due to heterogeneity, insufficiency of data, dearth of policies aligning standards, approval, lack of real-world evidences and reimbursement. Therefore, the development of more sophisticated biomarkers would reduce the gap of lack of role of real-world evidences which would eventually improve informed treatment decision making and deployment of targeted precision treatments (Horgan et al., 2021). The challenge of combined analysis of genomic and non-genomic patient data pertaining to their symptoms, history, lifestyle has been suggested to be solved utilizing the tool of artificial intelligence for tailored diagnosis and prognosis. One such example of an AI tool used for real-time detection of food-born illness is FINDER (Johnson et al., 2021). Interpretation and management of healthcare data challenge is dealt through advancements in translational research via development of electronic health record archiving systems and "real-world data (RWD)" as integrated data sources for harvesting data in real life contexts which may be collected through randomized clinical trials or non-interventional methodology (normal clinical practice) (De Maria Marchiano et al., 2021).

In light of promoting precision medicine and taking it to individuals it is essential to shift mindset to use AI-based methods for drug discovery and development. A comprehensive industry analysis shows that as compared to leading technology-based companies such as Microsoft, Google, the pharmaceutical companies are still maturing to catch up the pace of integration of AI and R&D. However, hopes are high as leading companies like Johnson & Johnson and Novartis have foot forwarded on digitization, developing competency and literacy for AI and producing AI-based services and products in healthcare (Hartl et al., 2021).

For integration of precision medicine into everyday clinical practice a systemic effort from all stakeholders is desirable. Healthcare providers, insurers, policy makers, employers, as well as patients and their families need to have a keen understanding of the fundamental advancement of how personalized medicine is being practiced and delivered. The diverse ethical, legal, and societal issues revolving around the concept must be understood and acceptable to the patient before consenting for the treatment. Clinical decisions should be well guided and supported by established effective healthcare infrastructure and data management system ensuring usefulness and user-friendliness.

2. GENOMICS IN PRECISION MEDICINE

Through basic and clinical research, it is evident that molecular biomarkers play an important role in facilitating prognosis, diagnosis of a disease and selection of targeted therapies for patients as many biomarkers can be tested in parallel and thus can be essentially playing a critical role in designing precision medicine by improving identification of patients having favorable response to particular drug, phenomena called as disease stratification. Therefore, the techniques for molecular biomarkers detection must be accurate, relatively easy to conduct, and should be highly reproducible (Dirks et al., 2016).

2.1. Genomic and Epigenomic Profiling

Currently, the techniques underplay for discovering molecular biomarkers primarily include genomic and epigenomic profiling. Genomic profiling includes changes in the DNA sequence itself, such as mutations and rearrangements, and epigenomic profiling includes changes in the epigenome, represented by molecules and structures associated with the DNA such as DNA methylation and post-translational histone modifications (Dirks et al., 2016). It has been observed that numerous diseases are caused by alterations(mutations) of epigenetic patterns, including changes in DNA methylation, post-translational histone modifications, or chromatin structure such as mutations in epigenetic enzymes is frequently observed in cancer (Vogelstein et al., 2013), neurological disorders like Parkinson's, Alzheimer's and Huntington's disease (Urdinguio et al., 2009), type 1 diabetes (Miao et al., 2008), intellectual disability (Bjornsson, 2015) and autoimmune diseases like rheumatoid arthritis (Jin et al., 2008; Karouzakis et al., 2009; Liu et al., 2013). These changes carry important information regarding the associated disease and thus act as a biomarker. Therefore, to detect such changes in patterns (biomarkers) genome-wide epigenomic profiling is implemented. Some of the main genome-wide profiling technologies include (i) Bisulfite sequencing for determining DNA methylation patterns at base-pair resolution, (ii) ChIP-Seq for identification of protein binding sites on the genome, (iii) DNaseI-Seq/ATAC-Seq for profiling open chromatin, and (iv) 4C-Seq and HiC-Seq for determining the spatial organization of chromosomes (Dirks et al., 2016). The epigenome includes the chromatin state of a cell i.e., when genetic material is in condensed state, including the genomic localization of modifications of DNA and histones, transcription factor binding sites, and 3D DNA structure. Thus, the epigenome regulates and controls which part of genome will be functional and which will be inactive. Similarly, they are found to regulate the occurrence and development of disease in individuals. Thus, to get an insight into the epigenetic control, it is required to be comprehensively studied at genome wide scale using epigenomic profiling. In this chapter, we have made sincere efforts to summarize how these epigenomic profiling helped biomarker discovery (Dirks et al., 2016).

2.2. Prediction and Prevention of Disease

Researchers utilize individual-specific genomic profiling information to study their traits and inherited risks of major diseases. In several cases, patients are unaware of the presence of disease-causing potential genetic variants in their genome. For individuals to predict the risk of cardiovascular diseases, they may undergo polygenic risk scores (PRS) assessment which can help them recognize genetic variants, that can even be fatal, before appearance of any symptoms (Abraham et al., 2021). This information helps doctors to design a tailored treatment and therapy for that patient against cardiovascular diseases. Particularly, for rheumatoid arthritis, prediction models are mostly based on abnormalities of biomarkers, and elevations in the levels of anti-citrullinated protein/peptide antibodies (ACPA) and rheumatoid factor (RF), including other factors such as genetic and environmental factors, symptoms of inflammatory arthritis (IA), additional markers and findings from imaging (Mahler et al., 2020). Furthermore, the application of artificial intelligence (AI) and high-performance computing can help predict disease risks with greater accuracy on the basis of the available multidimensional clinical and biological datasets. The AI derived precision medicine provides opportunity for the clinicals to specifically tailor early therapeutic interventions for each individual (Subramanian et al., 2020). This is how genomic profiling aids prediction and prevention of disease. Scientists are working towards improvement and application of polygenic scores to help patients with disease like breast cancer (Mavaddat et al., 2019), cardiovascular disease (Abraham G et al., 2021), atrial fibrillation (Gladding et al., 2020).

2.3. Precision Medicine Spanning Complex Diseases

Apart from oncology, precision medicine can be applicable to variety of medical fields like cardiology, neurology and even further. Genetic testing used for cardiology to predict, prognose and prevent inherited heart problems and risk assessment and thus could be beneficial in avoiding life-threatening situations (Singla, 2024). Moreover, precision medicine has paved its path into allergic illnesses by understanding and analyzing large datasets of allergic endotypes; however, it is still in infancy owing to their limitation of characterization of allergic endotypes, understanding of interrelationship of allergic multi-morbidities, environmental impacts, genetic risks, integration into 'omics', development of patient cohorts, and planning refined clinical trials (Proper et al., 2021). Certain mental illnesses such as major depressive disorder (MDD) are being dealt with at single-patient level using multi-omics approaches by utilizing the combination of technologies for diagnosis, stratification, prognosis and discovery of specific biomarkers that will be responsive to the treatment with the help of AI for improved assessment and treatment of MDD (Stolfi et al., 2024).

Furthermore, the underlying hereditary reasons for neurological-based diseases can be deeply understood utilizing genetic testing methods and potentially making it possible to tailor therapies addressing patients on a case-to-case basis (Singla, 2024). For example, genetic test results may well indicate the risk of Alzheimer's disease and can be dealt with precision therapy, but limited by the practical challenge of first stigma around such genetic test results especially relating to mental illnesses and second third parties asking disclosure of health information for those seeking health insurance or any coverage. Such "compulsory disclosures" contradict the ability of an individual or healthcare provider to keep the confidentiality of the information (Brothers and Rothstein, 2015). Therefore, the question arises, are too many stakeholders or participants in structuring the personalized medicine for dealing with the challenges of information confidentiality possible to be reduced for imparting a carefree and efficient personalized therapy?

3. BIOMARKERS IN PRECISION MEDICINE

The National Academy of Sciences/National Research Council's Committee in the US defined biological markers as signaling indicators which are responsible for any activity in a biological system. Primarily, they classified them into three types including those of effect, exposure, and susceptibility. Biomarker development plays a critical important role in area of precision medicine and further helps in development of targeted drugs and designing therapy against specific disease thereby promoting human health improvement and health care cost reduction. As precision medicine sets on developing treatment and therapies for an individual or subgroups of patients based on disease specific biomarkers, several analytical tools are used for identification of these biomarkers including genomics, proteomics, molecular imaging, next generation sequencing and metabolomics (Slikker, 2018). These and several other approaches like reactivating immune checkpoints for immune-based therapy have made it possible to search for new biomarkers that are responsible for major diseases like cancer (Kato et al., 2017). Several cancer biomarkers have been discovered using these FDA approved approaches such as Alectinib, Ceritinib and Crizotinib targeted drugs that act on biomarker *ALK* for diagnoses of non-small cell lung cancer with approximate response rates of 60% to 70%. (Shumei Kato et al., 2017). Additionally, it has been reported that diagnostic and prognostic markers are being investigated and reported such as urine biomarkers (nuclear matrix protein 22, Uro Vysion, microRNA, epigenetic marker, Mutation) for early detection of bladder cancer and molecular profiling of fluid and tissue are to be used an earlier detection and prognosis prediction as well as drug sensitivity (Kojima et al., 2017). Furthermore, techniques like high-throughput sequencing and repositories of data show that biomarkers can even be RNA molecules derived from body fluids like serum, saliva, urine for diagnosis of cancer, or monitoring tumor progression or predicting a therapy response (Xi et al., 2017). In their editorial by Peterson in 2020, they reported that biomarkers are currently limited relating to mental illnesses apart from certain candidates identified as diagnostic biomarkers for detecting Alzheimer's disease such as Ab42 and total tau levels in the cerebrospinal fluid. However, it would be beneficial to direct funds for investigating risk, diagnostic, predictive, treatment response, and safety biomarkers for mental illnesses given the available technologies for measuring biomarkers (Peterson, 2020) and psychosomatic nature of most illnesses (Nisar et al., 2018).

Therefore, discovery of biomarkers is important for individualizing the medicine for the purpose of disease screening, diagnosis, monitoring the ongoing development and application of therapy, developing prognostic indicators, and targeted drugs for individual or patient subgroup in order to predict and treat adverse drug reactions; and for their pharmacodynamic and dose-response studies (Aronson and Ferner, 2017). An ideal clinical biomarker should be simpler and less expansive to measure and can be measured repeatedly over a short period of time. Some examples of biomarkers for individualizing drug therapy include measuring HbA_{1c}, thyroxine/TSH levels, and blood pressure for diabetes mellitus, thyroid disease and hypertension, respectively (Aronson and Ferner, 2017). As there is only 10%–30% of drug response rate of the patients with same cancer type owing to possible change in multiple types of potential genes at genetic or epigenetic levels, the challenge for precision medicine programs is to identify biomarkers using omics data alone or in combination with environmental or lifestyle factors (Wang et al., 2017).

3.1. Biomarker Types, Role, and Limitations in Precision Medicine

An online document named as "Biomarkers, EndpointS, and other Tools (BEST)" was formed by the joint conference of US Food and Drug Administration (FDA) and National Institutes of Health (NIH) to establish appropriate definitions for biomarkers in contextual usage. The BEST for framed to ensure enabling efficiency, speed, and precision in strategic diagnostic and therapeutic development of technologies and benefitting as well as implementing public health policies (Califf et al., 2018). The **table 2** below shows different type of biomarkers stating their function and their potential role in precision medicine with some known examples; however, these examples are yet to be explored for individualized therapeutic trials.

Table 2. Types of biomarkers and their potential role in precision medicine.

S.No.	Biomarker types	Function	Potential role in precision medicine	Example	References
1	Diagnostic biomarkers	Responsible for the presence of disease	Detects disease presence or identifies the individual with disease subtype	Prostate-specific antigen is a diagnostic biomarker for detecting prostate cancer	(Califf, 2018; Makarov et al., 2009)
2	Monitoring biomarkers	Reveal changes in response to the treatment with due time	Can be measured regularly to check the status of disease and effect of treatment on specific disease	Hemoglobin, blood pressure, LDL cholesterol, in case of HIV infection, CD4 count is observed for monitoring	(Califf, 2018; Ford et al., 2015)
3	Pharmacodynamic/ response biomarkers	The level of this biomarker changes when exposed to a medical product or environmental agent	Can be efficiently used for checking the effect of drug in phase 1 trials. Drug doses are defined using alterations in this biomarker phase 1, however, validation is required.	Epigenetic markers targeting HDAC assess epigenetic modification in response to treatment such as DNA acetylation and measures Acetylated histone 3 in phase I/ II trials	(Califf, 2018; Gainor et al., 2014)
4	Predictive biomarkers	It shows if an individual will eventually have a favourable or unfavourable response to medication	Predicts if an individual or group of individuals will be more likely to experience a favorable or unfavorable effect when they are exposed to medication or environmental agent	Cancer patients with HER2 receptor are highly likely to respond to treatment with Herceptin	(Califf, 2018; Mitri et al., 2012)
5	Prognostic biomarkers	Represents disease progression and likelihood of recurrence	Identifies the chances of disease recurrence, or progress of disease development in patients or in medical condition of interest	Ki67, Metallothionein I and II are higher expressing biomarkers that are associated with worse prognosis of melanoma	(Califf, 2018; Weinstein et al., 2014)

continued on following page

Table 2. Continued

S.No.	Biomarker types	Function	Potential role in precision medicine	Example	References
6	Safety biomarker	Measures the presence or extent of toxicity of a medical intervention	Measures before or after medication to check for toxicity.	Stress-related hormones for mitigating safety risk and gastrin as a drug-induced stomach toxicity safety biomarker (Schomaker et al., 2019)	(Califf, 2018; Schomaker et al., 2019)
7	Susceptibility/risk biomarkers	Indicates the chances of an individual of potentially developing a disease or medical condition	Useful for epidemiological estimation of the risk of occurrence of disease in an individual.	Gene expression of metallothionein in peripheral blood lymphocytes as a susceptibility biomarker for cadmium-induced kidney damage	(Califf, 2018; Nordberg, 2010)

Although identification of potential biomarkers has led to opportunities for effective treatment therapies, they are restricted by the limitations of transference to clinical practice owing to various factors pertaining to structure and science. In order to be approved for clinical usage, biomarkers need to be specific, sensitive and reproducible and should be confirmed and validated through large volume of specimens. The pitfall is to deal with the lacking quality of biomarker validation strategies, making selection before initiation of discovery phase, powerful-built of analytical techniques to be used in clinical trials. Commencing from the patient stratification (selection for likely therapeutic success), to efficacy (approval), differentiation (selecting best and safe drug for each patient), toxicity (monitoring and avoiding potential toxic effects), screening (patients' recruitment for trials) up to prognosis (prediction of likely disease course), the clinical markers have to face limitations at each step. The limitations can be attributable to factors such as smaller sample size for analysis, unavailability of medical history of patients, unmatched age and sex of case and controls, limited proteomic and metabolomic data of patients, the strict regulations to be abide by for selection, collection, storage, handing, analysis of samples and interpretation of data. (Drucker and Krapfenbauer, 2013). In 2022, a thorough analysis by Bing Yuan has shown that the precision medicine research progress has not been quite satisfactory and has raised some valid scientific concerns. These concerns are as follows:

> -There are more than 55000 diseases and are caused by multiple factors; however, currently only precision medicine works for single target diseases which are only a few. Although biomarkers greatly reduce the number of individual diseases still the scale is too large to accommodate plethora of diseases that require attention.
>
> -The side effects of personalized medicine have remained unavoidable.

-Dealing with different biomarkers corresponding to the same treatment and relationship between them.

-Patients suffer from diseases attributable to more than one biomarker; however targeted drugs are studied individually; personalized medicine needs to answer multiple diseases in same patient attributable to multiple biomarkers, which is often the case.

-If and whether improving biomarker abnormalities actually improve patient's illness. (Yuan, 2022).

4. TECHNIQUES AND TOOLS

As precision medicine depend on large repositories of data of population level for determining appropriate individualistic treatment, there is a need to have technologies which can signal dynamic diseased states, can continuously optimize therapeutic intervention and can be modulated for selection of drug and its dose. Notably, enabling biomarker discovery, creating sensing and diagnostic platforms, and devising new methods of drug delivery contribute to the development of precision medicine using data from genetic and environmental sources, medical records, clinical sources of "big data" and from artificial intelligence and machine learning, environmental and social factors (Naithani et al., 2021) and integrating them to develop precision therapy (Ho et al., 2020)

In addition, various tools that have been suggested to be contributing to achievement of precision medicine include big data, artificial intelligence, "omics," pharmaco-omics, environmental, social, behavioral factors which encompass clinical and omics based biomarkers; genomics, proteomics, transcriptomics, metabolomics (Naithani et al., 2021; Ota et al., 2021), and microbiomics areas, which comprises various sequencing, profiling and network association studies; and studying about the suitability of particular drug to a particular sub-population (Naithani et al., 2021). Integration of profiles of different "omes" like genome, epigenome, proteome, transcriptome, metabolome, antibodyome of an individual provides valuable information for monitoring health, prevention and developing precision medicine (Chen et al., 2013). Therefore, efforts are required to develop the necessary analytical infrastructure for effectively generating, analyzing, and annotating multi-omics data to promote individualized decision-making for precision medicine (Ota et al., 2021). Furthermore, an interesting easy to use, public, free to use web-server tool, DBPOM ((http://www.dbpom.net/) is available with no requirement for login, which integrates pharmaco-omics data of cancer precision medicine and has the capacity to provide calculations for more than 10,000 small molecules and drugs and drug combinations on cancer patients and their targeted biological functions (Liu et al.,

2023). Such tools and high throughput technologies are a framework for future development of precision-based therapies.

5. CHALLENGES AND FUTURE DIRECTIONS

5.1. Challenges

The primary challenge with precision medicine is to generate huge amount of data and analyze it which puts economic as well as burden of labor and must ensure technical expertise for ensuring error-free data generation. This data being individualistic has to be secured and hence encompass ethical challenges which will be discussed in the next section in this chapter. Data also needs to be speedily generated in order to make relevant decisions in acute care settings by health care professionals. The infrastructure in terms of capacity building in terms of training of health care workers, and availability of high-quality artificial intelligence, machine leaning and laboratory equipment tools is another challenge that precision medicine treatments face and population health depends upon (Naithani et al., 2021a). Above all, precision medicine requires collaboration with different scientific groups to work on large scale data and is beneficial for patients but has yet to reach its high (Liu et al., 2019).

Additionally, it is difficult and complex to deal with huge data from various platforms like proteomics genomics metabolomics and others and expensive machines are used for the data collection which cannot be managed to be made available in rural areas (Davis et al., 2018). Furthermore, the accessibility of products and services in rural setups is meagre and patients are sometimes reluctant or unable to travel to other health centers. The availability of geneticists, molecular pathologists, and genetic counsellors is also limited in rural settings. Lack of high-speed internet impacts the educational awareness about e-learning and e-health programs thus challenging the adoption of precision medicine in rural setups (Pritchard et al., 2017).

5.1.1. Economic Implications

One of the major implications that precision medicine is on the economic evaluation and sustenance of tailored drugs for their development. The antibiotic drugs tend to develop resistance over time and thus impact the costs of development, fundings, healthcare costs and global economy. These implications significantly affect pharmaceutical companies, healthcare systems and holistically impact social ecosystems. An increase in healthcare costs requires patients to extend hospital stays, opt for complex treatments and sometimes even go for costly second line antibiotics further burdening the limit of available financial resources which could rather be utilized for critical healthcare purposes (Founou et al., 2017). On the patient front, because of prolonged sickness, decrease in work productivity and antibiotic resistance even leading to disability or death results in economic losses of family income as well as workforce output. Developing antibiotics is a cost- and time-intensive process. The pharmaceutical companies get financially discouraged to research and develop tailored-drugs due to scientific, regulatory, and financial barriers spanning the following reasons: one for their limited market and second short returns on investments in antibiotics as they are required only for short durations unlike for chronic diseases for which extended medication is required (Dutescu and Hillier, 2021). The changing dynamics of market of drugs owing to challenges faced by newer drugs in reaching the market owing to production and distribution issues causes rise in the prices of existing drugs adversely affecting healthcare as well as patients. These challenges are implicating towards innovative precision medicine-based treatments such as phage therapy and monoclonal antibodies-based treatments for drug development costs and market competition (Muteeb et al., 2023).

5.1.2. Cost-Benefit Analysis, Cost-Effectiveness of Drugs

A significant challenge faced by pharmaceutical companies lies in aligning the economic incentives and the cost and risks involved in the development of drugs. Bringing in new drugs to the market becomes challenging for companies as it is limited by profit margins and short periods of treatment at the same time presenting technical difficulties of development of resistance, complications of clinical trials and challenging reimbursement regulations (Muteeb et al., 2023). In general, the investment curve takes a bell-shaped parabola through the process of drug development moving through the phases of exponential growth post-registration, competition-induced decline phase and expiration of patent term validity for most drugs; however, the specific nature of the curve may vary for each different drug. Under continuous surveillance of drug's benefit-risk assessment and resulting safety profile, the drugs undergo life cycle of entering and retreating from the market through the process of selection and adaptation and are replaced by newer alternatives. Many factors play role through these stages such as effectiveness, side effects, contradicting use, limited application, and adverse reactions. The changes in this drug life cycle dynamics impacts the pricing of the drug which is primarily based on the sale volumes of the pharmaceutical companies.

The greater the difficulty in achieving higher and sustainable sale volumes, the prices of the drugs are to be increased to maintain the revenue. For example, a drug called imatinib, used for treating chronic myeloid leukemia got its cost tripled in spite of huge volumes being sold after it was allowed and approved by US FDA for repurposing from $31,930 in 2005 to $118,000 in 2015 by Novartis despite research costs for new indication being included in its initial price. In the US, another drug sofosbuvir (Sovaldi) for treating hepatitis C infections has a price of $84,000 for 12-week treatment while it is $1000 per pill causing health plans to refuse routine coverage of this drug despite its effective ability to prevent the ultimate requirement of liver transplant. However, the Sovaldi was found unsustainable in Spain and Latvia and alternative combination treatment of ledipasvir/sofosbuvir (Harvoni) was marketed by Gilead Sciences. The prices of many drugs such as ivacaftor for cystic fibrosis, pyrimethamine from Turing Pharmaceuticals, Mylan's EpiPen, cholesterol-lowering Evolocumab has become a critical concern for patients, health insurers, pharmaceutical companies as well as governments (Gronde et al., 2017). From the patient's perspective, high costs of drugs, difficulties pertaining to insurance coverage, and high co-pays (like for testing) rather than the main targeted therapy which is mainly oral, becomes the critical challenges in bringing home the personalized medicine and making it individualistic (Hamilton et al., 2021). Therefore, critiques and governments have started to ponder over regulations for balancing the intent of pharmaceutical companies from only-maximize-profit to do what is beneficial for all the stakeholders (Gronde et al., 2017).

5.2. Future Directions

The technological process is expected to be continued for precision medicine through certain programmes such as the Precision Medicine Initiative and the 100,000 Genomes Project anticipating a better learned and organized healthcare information system that can effectively analyze molecular and health record data for prevention, detection and treatment of a disease (Love-Koh et al., 2018). The three types of technological advancements in precision medicine that are likely to become widespread in clinical practice over the next decade include "omics"-based biomarkers; complex artificial intelligence-based algorithms; and digital health applications (Love-Koh et al., 2018). The successful integration of development of required screening tools, pharmacogenomics programming, educational programming, digital medical records, strong clinical decision making tools putting together for NorthShore University HealthSystem has set an example for other communities to follow where genetic variations of over 4000 patients have been identified utilizing the medical management system which could be beneficial for generating any required potential precision therapy based on the data (Awwad et al., 2024). Genomic biomarkers have already been playing crucial roles in precision medicine; however, proteomic and metabolomic biomarkers are also evolving as two new biomarkers and are expected evolve in the next 5 years (Love-Koh et al., 2018).

With the progress in the field of precision medicine, certain complex algorithms are expected to be AI-based in near future i.e., AI Biocomputing. For example, Sapientia, which will utilize the clinical phenotyping along with genome sequencing for making therapy choices (Love-Koh et al., 2018). Furthermore, certain apps like "My heart counts" that records and analyses data on activity, risk factors and hematology, provides suggestions on improving heart health, collects data and does the analysis. The number of such apps are expected to drastically increase in the next decade assisting doctors and expanding individualistic healthcare (Love-Koh et al., 2018).

6. ETHICAL AND LEGAL CONSIDERATIONS

Although precision medicine greatly promises to a tailored approach of treating individuals based on their genomic, epigenomic, environmental, lifestyle and specific information, ethical and legal concerns arise owing to big amount of data that is required to be carefully secured and dealt with. It becomes challenging to ensure the privacy of patients, having informed consent for research and therapy development and no genetic discrimination of patients based on their genetic information (Mezinska et al., 2021), particularly in case where a polygenic risk assessment is executed without attention to macrolevel environmental factors (Parker, 2024). The famous case of *Heneritta Lacks* is an example of the informed consent implication of precision medicine where an American African lady in 1951 received medical treatment for cervical cancer when her cells were procured without her consent and later observed to be highly robust (Gold, 1985). These cells are now extensively used in medical research and referred as HeLa cells significantly contributing to research advancement since many years; however, there was no notification to Lacks or her family regarding utilization of her cancerous cells for research purpose and scientific intentions were clarified many years later and the case ignited debates and discussions about informed consent and rights of patients and research ethics. Certain legal implications of precision medicine are also a concern as there is larger amount of genomic information involved and the professional responsibilities of clinicians and doctors about reporting the information derived from whole genome sequencing (sometimes sensitive) for a patient remain uncertain (Brothers and Rothstein, 2015).

6.1. Ethical Concerns and Data Privacy Issues

As medicine is an individual's health related phenomena, development of precision medicine needs to acknowledge and consider ethical concerns with an approach that is personal, empathetic and efficient and has to involve factors which are more than just advancements of technology (Singla, 2024). For example, genetic profiling in cardiology offers promises; however, they are ethically challenged by privacy of genetic data, informed consent and comparable accessibility of medicines and genetic testing facilities. Further, for neurological treatments, the genetic testing and counselling has ethical concerns of helping individuals and their families understand the possible consequences of test results. A plethora of genetic and molecular data that is generated for precision medicine has to be carefully considered regarding its accessibility and privacy as it the sensitive patient's information and there lies a challenge of protecting potential data breaching (Singla, 2024). Therefore, it is essential to adhere to regulations and policies governing ethical concerns such as ealth Insurance Portability and Accountability Act (HIPAA) in the U.S. and the General Data Protection Regulation (GDPR) in the EU for ensuring patient's trust remains earned and treatment moves in compliance with the legal requirements (Udegbe et al., 2024). The use of AI in healthcare also poses risk of breach of data privacy, may also reflect biases in training data used for AI, and difficulty in making patients understand how their data will be used. The potential solutions for AI-driven ethical concerns in advancements of precision medicine include encryption techniques to tackle unauthorized patient data access, development of diversified and yet representative training datasets and ensuring implementation of transparent process of consent (Yogeshappa, 2024).

Other important ethical concerns apart from data privacy include informed consent (to make patients understand the possibility of unexpected outcomes or incidental discoveries from individualistic data); careful management and handling of patient data by researchers and healthcare providers regarding genetic information integrity, transparency and its sharing; decision of whether, how and when to share patient's results if pertaining to any accidental discoveries or findings from their data; prevention of any discrimination at work, for insurance or others based on the content of genetic data; consent for utilizing patient data for research purposes; promoting knowledge, awareness and education regarding precision medicine where healthcare providers and institutions are obligated to play their role. (Singla, 2024).

6.2. Legal Developments and Regulatory Frameworks

The ethical and legal concerns and developments are closely associated and solve issues with complementarity as it is not only essential to determine a situation to be ethical wrong or right but also to be liked and accepted by the law and abided by the policies (Alvarez et al., 2022). Hence, regulations and policies are developed for a safer, stable, and sustainable development of the new concept of precision medicine ruling the medicine to be tailored individualistically. Research in precision medicine has paralleled developments of choices of laws by the states pertaining to human subjects, the specimens and their treatment, genetic testing, genetic discrimination, right to access of information and various others. These legal regulations and frameworks include policies and rules about common rules of using human subjects, exemptions if any from research, consent for using clinical specimens, property rights followed by destruction of specimen, and many others as well as repercussions of non-obedience of the law (Beskow and Wolf, 2022).

7. CASE STUDIES IN PRECISION MEDICINE

Despite faster identification of potential biomarkers through "omics" based technologies for precision medicine, there remains a dearth of statistical power and reproducibility of results for which "information commons" are proposed that will have data spanning from patient's medical history to demographics, molecular measurement to disease outcome data. On the basis of this reservoir, patient stratification can be appropriately conducted into new or specific taxa as per specific or group of identified biomarkers (Vargas and Harris, 2016). The two trial designs named as "Umbrella" and "Basket" as discussed hereafter have helped efficient and effective drug development by speeding up, innovation in science, as well as completion of trails.

7.1. Basket Trials

Basket trials design test the effect of single or multiple durgs on one or more mutations in multiple cancer types for example, STARTRK-2, Imatinib basket trial, BRAF+, NCI Match. Cancer biology has evolved into distinct cancer histologies that have common genetic and immune phenotypic traits. The participants with the presence of the targeted feature are assigned one or more therapies and followed up for clinical outcomes. Generally, basket trials are open-label, single-arm trials that are conducted in phase II setting. They overcome the requirement of conducting separate trials for individual histologies. This methodology includes the study of rare tumor types, which can be included as baskets. The prevalence of cancers can be known even for rare tumors through their statistical evaluations of outcomes. The unique responses corresponding to individual tumor types may trigger pathways pertaining to rapid expansion and/or for achieving accelerated regulatory approval. However, the robust responses corresponding to multiple tumor types may trigger achievement of pathways for tissue-agnostic regulatory review. The trialists choose basket trials implementation expecting that the targeted feature is specific for delineation of patients for whom the specific drug is efficacious in a manner that goes beyond tumor histology for the selected tumor types (https://www.ncbi.nlm.nih.gov/sites/books/NBK458878/).

The Bayesian statistical modeling based on the multisource exchangeability (data coming from different sources) model can be used for quantification and understanding the differences between various types of tumors (tumor-heterogeneity) and assess how well treatments work against tumors across different tissues (tissue-agnostic efficacy evidence). Evaluating the likelihood by Bayesian posterior decision making, arising from evaluating the posterior probability that a treatment is effective for each tumor type, we can make informed decisions either at the final analysis or during interim checks (periodically check in between the study treatment not working) to stop enrolling patient in tumor types that are not showing positive results. This can help develop appropriate stratification of patients according to tumor type thereby assisting potential precision medicine therapy generation (Hobbs et al., 2022).

7.2. Umbrella Trials

Umbrella trial design test the impact of different drugs on different mutations in a single type of cancer for example Lung-MAP, BATTLEI-SPY2, FOCUS4. After the advent of lung tumor-associated EGFR mutations, a plethora of oncogenic mutations were identified that further classified lung cancer into molecular subsets; however, the challenge was to match them against the targeted drug. The Lung-MAP umbrella clinical trials were launched to meet the therapeutic need of squamous cell carcinomas which make 30% of NSCLC. The objective of Lung-MAP trials was to develop drugs against unique genotypes using broad-based next generation sequencing, achieve faster molecular testing times and accelerate the process of approval of new biomarker. The trials aided quick identification and testing of new targeted treatments by choosing appropriate patient-population, designing adaptable protocols and encouraging broader participation (https://www.ncbi.nlm.nih.gov/sites/books/NBK458878/).

Apart from basket and umbrella trials, platform trials are also contributing to uncover the perspective of precision medicine for diseases other than tumors. Some of the master protocols that are in place working to find therapeutic solutions for antibiotic resistance pathogenic infections and neurological and mental illnesses have been covered here under.

7.3. ADAPT protocol

For evaluation of therapies targeting antibiotic-resistant pathogens at multiple sites of infection in a body the Antibiotic Resistance Leadership group works using master ADAPT protocol using adaptive methods and technology (Woodcock and LaVange, 2017).

7.4. PREVAIL II

With the availability of more experimental therapies, the clinical trial under the Partnership for Research on Ebola Virus in Liberia II (PREVAIL II) evaluates various therapies for disease from Ebola virus under flexible Bayesian adaptive design (Woodcock and LaVange, 2017).

7.5. DIAN-TU

The Dominantly Inherited Alzheimer Network Trial (DIAN-TU) evaluated gantenerumab and solanezumab treatments for patients with early-onset of Alzheimer's disease that is caused by genetic mutation (Woodcock and LaVange, 2017; Duan et al., 2024).

7.6. Trials for Schizophrenia

The Clinical Antipsychotic Trials of Intervention Effectiveness (CATIE) sponsored by the National Institute of Mental Health works on comparing unconventional antipsychotic agents against conventional treatments for schizophrenia (Woodcock and LaVange, 2017).

8. CONCLUSION

This chapter reviews precision medicine in light of genomic biomarker development and discovery. Precision medicine aiming to achieve individualistic tailored treatment based on a person's genetic makeup has only progressed slowly owing to the challenges and limitation on all different fronts be it scientific, ethical, structural, or regulatory. Though with the advancements of AI and machine learning-based technologies assisting "omics-technologies" in terms of handling, structuring, accessibility of large amount of patient-specific, sensitive personal medical data, the concept of personalizing the medication for prevention and treatment of diseases is deemed plausible. The pharmaceutical industries are picking up the trend to evolve with advent in new "omics-technologies" to disseminate, empower, and develop more AI-driven employee base to eventually gain momentum in R&D of biomarkers or agents which can potentially be put for clinical trial for personalized treatments for diseases. What limits the growth of personalized medicine at scientific front are the biomarker validation, building robust analytical techniques and sample size. At the regulatory front, it is limited by the mandatory standard protocols for trials and challenges of toxicity, screening, and prognosis. The lack of ethical reforms addressing issues of patient data privacy and protection from breach as well as legal framework genetic testing, discrimination, specimen usage also concerns that require attention. This chapter summarizes few case studies spanning different diseases that are in place from the perspective of precision medicine. Finally, to bring up novel system of healtcare into place a collaborative efforts and systemic thinking is required from researchers to healthcare practitioners, isurers, policy makers, to the patients themselves.

REFERENCES

Abraham, G., Rutten-Jacobs, L., & Inouye, M. (2021). Risk prediction using polygenic risk scores for prevention of stroke and other cardiovascular diseases. *Stroke*, 52(9), 2983–2991. DOI: 10.1161/STROKEAHA.120.032619 PMID: 34399584

Alvarez, M. J. R., Griessler, E., & Starkbaum, J. (2022). Ethical, Legal and Social Aspects of Precision Medicine. In *Precision Medicine in Clinical Practice* (pp. 179–196). Springer Nature Singapore. DOI: 10.1007/978-981-19-5082-7_11

Aronson, J. K., & Ferner, R. E. (2017). Biomarkers—A general review. *Current Protocols in Pharmacology*, 76(1), 9–23. DOI: 10.1002/cpph.19 PMID: 28306150

Ashley, E. A. (2016). Towards precision medicine. *Nature Reviews. Genetics*, 17(9), 507–522. DOI: 10.1038/nrg.2016.86 PMID: 27528417

Awwad, O., Ahram, M., Coperchini, F., & Jalil, M. A. (2024). Precision medicine: Recent advances, current challenges and future perspectives. *Frontiers in Pharmacology*, 15, 1439276. DOI: 10.3389/fphar.2024.1439276 PMID: 38910892

Beskow, L. M., & Wolf, L. E. (2022). "Choice of law" in precision medicine research. *American Journal of Human Genetics*, 109(8), 1347–1352. DOI: 10.1016/j.ajhg.2022.06.009 PMID: 35931047

Bjornsson, H. T. (2015). The Mendelian disorders of the epigenetic machinery. *Genome Research*, 25(10), 1473–1481. DOI: 10.1101/gr.190629.115 PMID: 26430157

Bodaghi, A., Fattahi, N., & Ramazani, A. (2023). Biomarkers: Promising and valuable tools towards diagnosis, prognosis and treatment of Covid-19 and other diseases. *Heliyon*, 9(2), e13323. DOI: 10.1016/j.heliyon.2023.e13323 PMID: 36744065

Brothers, K. B., & Rothstein, M. A. (2015). Ethical, legal and social implications of incorporating personalized medicine into healthcare. *Personalized Medicine*, 12(1), 43–51. DOI: 10.2217/pme.14.65 PMID: 25601880

Califf, R. M. (2018). Biomarker definitions and their applications. *Experimental Biology and Medicine*, 243(3), 213–221. DOI: 10.1177/1535370217750088 PMID: 29405771

Chen, R., & Snyder, M. (2013). Promise of personalized omics to precision medicine. *Wiley Interdisciplinary Reviews. Systems Biology and Medicine*, 5(1), 73–82. DOI: 10.1002/wsbm.1198 PMID: 23184638

Cook, J. C., Wu, H., Aleo, M. D., & Adkins, K. (2018). Principles of precision medicine and its application in toxicology. *The Journal of Toxicological Sciences*, 43(10), 565–577. DOI: 10.2131/jts.43.565 PMID: 30298845

Davis, A. A., McKee, A. E., Kibbe, W. A., & Villaflor, V. M. (2018). Complexity of delivering precision medicine: Opportunities and challenges. *American Society of Clinical Oncology Educational Book*, 38(38), 998–1007. DOI: 10.1200/EDBK_200279 PMID: 30231318

De Maria Marchiano, R., Di Sante, G., Piro, G., Carbone, C., Tortora, G., Boldrini, L., Pietragalla, A., Daniele, G., Tredicine, M., Cesario, A., Valentini, V., Gallo, D., Babini, G., D'Oria, M., & Scambia, G. (2021). Translational research in the era of precision medicine: Where we are and where we will go. *Journal of Personalized Medicine*, 11(3), 216. DOI: 10.3390/jpm11030216 PMID: 33803592

Dirks, R. A., Stunnenberg, H. G., & Marks, H. (2016, December). Genome-wide epigenomic profiling for biomarker discovery. *Clinical Epigenetics*, 8(1), 1–7. DOI: 10.1186/s13148-016-0284-4 PMID: 27895806

Drucker, E., & Krapfenbauer, K. (2013). Pitfalls and limitations in translation from biomarker discovery to clinical utility in predictive and personalised medicine. *The EPMA Journal*, 4(1), 1–10. DOI: 10.1186/1878-5085-4-7 PMID: 23442211

Duan, X. P., Qin, B. D., Jiao, X. D., Liu, K., Wang, Z., & Zang, Y. S. (2024). New clinical trial design in precision medicine: Discovery, development and direction. *Signal Transduction and Targeted Therapy*, 9(1), 57. DOI: 10.1038/s41392-024-01760-0 PMID: 38438349

Dutescu, I. A., & Hillier, S. A. (2021). Encouraging the development of new antibiotics: Are financial incentives the right way forward? A systematic review and case study. *Infection and Drug Resistance*, 14, 415–434. DOI: 10.2147/IDR.S287792 PMID: 33574682

Dzau, V. J., & Ginsburg, G. S. (2016). Realizing the full potential of precision medicine in health and health care. *Journal of the American Medical Association*, 316(16), 1659–1660. DOI: 10.1001/jama.2016.14117 PMID: 27669484

Ford, N., Meintjes, G., Pozniak, A., Bygrave, H., Hill, A., Peter, T., Davies, M.-A., Grinsztejn, B., Calmy, A., Kumarasamy, N., Phanuphak, P., deBeaudrap, P., Vitoria, M., Doherty, M., Stevens, W., & Siberry, G. K. (2015). The future role of CD4 cell count for monitoring antiretroviral therapy. *The Lancet. Infectious Diseases*, 15(2), 241–247. DOI: 10.1016/S1473-3099(14)70896-5 PMID: 25467647

Founou, R. C., Founou, L. L., & Essack, S. Y. (2017). Clinical and economic impact of antibiotic resistance in developing countries: A systematic review and meta-analysis. *PLoS One*, 12(12), e0189621. DOI: 10.1371/journal.pone.0189621 PMID: 29267306

Gainor, J. F., Longo, D. L., & Chabner, B. A. (2014). Pharmacodynamic biomarkers: Falling short of the mark? *Clinical Cancer Research*, 20(10), 2587–2594. DOI: 10.1158/1078-0432.CCR-13-3132 PMID: 24831281

Gladding, P. A., Legget, M., Fatkin, D., Larsen, P., & Doughty, R. (2020). Polygenic risk scores in coronary artery disease and atrial fibrillation. *Heart Lung and Circulation*, 29(4), 634–640. DOI: 10.1016/j.hlc.2019.12.004 PMID: 31974023

Goetz, L. H., & Schork, N. J. (2018). Personalized medicine: Motivation, challenges, and progress. *Fertility and Sterility*, 109(6), 952–963. DOI: 10.1016/j.fertnstert.2018.05.006 PMID: 29935653

Gold, M. (1985). *A Conspiracy of Cells: One woman's immortal legacy-and the medical scandal it caused*. State University of New York Press.

Grimberg, F., Asprion, P. M., Schneider, B., Miho, E., Babrak, L., & Habbabeh, A. (2021). The real-world data challenges radar: A review on the challenges and risks regarding the use of real-world data. *Digital Biomarkers*, 5(2), 148–157. DOI: 10.1159/000516178 PMID: 34414352

Gronde, T. V. D., Uyl-de Groot, C. A., & Pieters, T. (2017). Addressing the challenge of high-priced prescription drugs in the era of precision medicine: A systematic review of drug life cycles, therapeutic drug markets and regulatory frameworks. *PLoS One*, 12(8), e0182613. DOI: 10.1371/journal.pone.0182613 PMID: 28813502

Hamilton, J. G., Banerjee, S. C., Carlsson, S. V., Vera, J., Lynch, K. A., Sar-Graycar, L., Martin, C. M., Parker, P. A., & Hay, J. L. (2021). Clinician perspectives on communication and implementation challenges in precision oncology. *Personalized Medicine*, 18(6), 559–572. DOI: 10.2217/pme-2021-0048 PMID: 34674550

Hartl, D., de Luca, V., Kostikova, A., Laramie, J., Kennedy, S., Ferrero, E., Siegel, R., Fink, M., Ahmed, S., Millholland, J., Schuhmacher, A., Hinder, M., Piali, L., & Roth, A. (2021). Translational precision medicine: An industry perspective. *Journal of Translational Medicine*, 19(1), 245. DOI: 10.1186/s12967-021-02910-6 PMID: 34090480

Hasanzad, M., Sarhangi, N., Ehsani Chimeh, S., Ayati, N., Afzali, M., Khatami, F., Nikfar, S., & Aghaei Meybodi, H. R. (2022). Precision medicine journey through omics approach. *Journal of Diabetes and Metabolic Disorders*, 21(1), 881–888. DOI: 10.1007/s40200-021-00913-0 PMID: 35673436

Hensing, W., Santa-Maria, C. A., Peterson, L. L., & Sheng, J. Y. (2020, October). Landmark trials in the medical oncology management of early stage breast cancer. [). WB Saunders.]. *Seminars in Oncology*, 47(5), 278–292. DOI: 10.1053/j.seminoncol.2020.08.001 PMID: 32933761

Ho, D., Quake, S. R., McCabe, E. R., Chng, W. J., Chow, E. K., Ding, X., Gelb, B. D., Ginsburg, G. S., Hassenstab, J., Ho, C.-M., Mobley, W. C., Nolan, G. P., Rosen, S. T., Tan, P., Yen, Y., & Zarrinpar, A. (2020). Enabling technologies for personalized and precision medicine. *Trends in Biotechnology*, 38(5), 497–518. DOI: 10.1016/j.tibtech.2019.12.021 PMID: 31980301

Hobbs, B. P., Pestana, R. C., Zabor, E. C., Kaizer, A. M., & Hong, D. S. (2022). Basket trials: Review of current practice and innovations for future trials. *Journal of Clinical Oncology*, 40(30), 3520–3528. DOI: 10.1200/JCO.21.02285 PMID: 35537102

Horgan, D., Ciliberto, G., Conte, P., Curigliano, G., Seijo, L., Montuenga, L. M., Garassino, M., Penault-Llorca, F., Galli, F., Ray-Coquard, I., Querleu, D., Riegman, P., Kerr, K., Van Poppel, H., Bjartell, A., Codacci-Pisanelli, G., Koeva-Balabanova, J., Paradiso, A., Maravic, Z., & Capoluongo, E. (2021). Bringing onco-innovation to Europe's healthcare systems: The potential of biomarker testing, real world evidence, tumour agnostic therapies to empower personalised medicine. *Cancers (Basel)*, 13(3), 583. DOI: 10.3390/cancers13030583 PMID: 33540773

Hossain, M. A. (2023). Precision Medicine: An emerging approach for patient care. *Bangladesh Journal of Medicine*, 201-201.

Jin, B., Tao, Q., Peng, J., Soo, H. M., Wu, W., Ying, J., Fields, C. R., Delmas, A. L., Liu, X., Qiu, J., & Robertson, K. D. (2008). DNA methyltransferase 3B (DNMT3B) mutations in ICF syndrome lead to altered epigenetic modifications and aberrant expression of genes regulating development, neurogenesis and immune function. *Human Molecular Genetics*, 17(5), 690–709. DOI: 10.1093/hmg/ddm341 PMID: 18029387

Johnson, K. B., Wei, W. Q., Weeraratne, D., Frisse, M. E., Misulis, K., Rhee, K., Zhao, J., & Snowdon, J. L. (2021). Precision medicine, AI, and the future of personalized health care. *Clinical and Translational Science*, 14(1), 86–93. DOI: 10.1111/cts.12884 PMID: 32961010

Kantae, V., Krekels, E. H., Esdonk, M. J. V., Lindenburg, P., Harms, A. C., Knibbe, C. A., Van der Graaf, P. H., & Hankemeier, T. (2017). Integration of pharmacometabolomics with pharmacokinetics and pharmacodynamics: Towards personalized drug therapy. *Metabolomics*, 13(1), 1–11. DOI: 10.1007/s11306-016-1143-1 PMID: 28058041

Karouzakis, E., Gay, R. E., Michel, B. A., Gay, S., & Neidhart, M. (2009). DNA hypomethylation in rheumatoid arthritis synovial fibroblasts. *Arthritis and Rheumatism*, 60(12), 3613–3622. DOI: 10.1002/art.25018 PMID: 19950268

Kato, S., Subbiah, V., & Kurzrock, R. (2017). Counterpoint: successes in the pursuit of precision medicine: biomarkers take credit. *Journal of the National Comprehensive Cancer Network : JNCCN*, 15(7), 863–866. DOI: 10.6004/jnccn.2017.0127 PMID: 28687573

Kenner, B., Chari, S. T., Kelsen, D., Klimstra, D. S., Pandol, S. J., Rosenthal, M., Rustgi, A. K., Taylor, J. A., Yala, A., Abul-Husn, N., Andersen, D. K., Bernstein, D., Brunak, S., Canto, M. I., Eldar, Y. C., Fishman, E. K., Fleshman, J., Go, V. L. W., Holt, J. M., & Wolpin, B. (2021). Artificial intelligence and early detection of pancreatic cancer: 2020 summative review. *Pancreas*, 50(3), 251–279. DOI: 10.1097/MPA.0000000000001762 PMID: 33835956

Kojima, T., Kawai, K., Miyazaki, J., & Nishiyama, H. (2017). Biomarkers for precision medicine in bladder cancer. *International Journal of Clinical Oncology*, 22(2), 207–213. DOI: 10.1007/s10147-016-1068-8 PMID: 27896485

König, I. R., Fuchs, O., Hansen, G., von Mutius, E., & Kopp, M. V. (2017). What is precision medicine? *The European Respiratory Journal*, 50(4), 1700391. DOI: 10.1183/13993003.00391-2017 PMID: 29051268

Liu, X., Luo, X., Jiang, C., & Zhao, H. (2019). Difficulties and challenges in the development of precision medicine. *Clinical Genetics*, 95(5), 569–574. DOI: 10.1111/cge.13511 PMID: 30653655

Liu, Y., Aryee, M. J., Padyukov, L., Fallin, M. D., Hesselberg, E., Runarsson, A., Reinius, L., Acevedo, N., Taub, M., Ronninger, M., Shchetynsky, K., Scheynius, A., Kere, J., Alfredsson, L., Klareskog, L., Ekström, T. J., & Feinberg, A. P. (2013). Epigenome-wide association data implicate DNA methylation as an intermediary of genetic risk in rheumatoid arthritis. *Nature Biotechnology*, 31(2), 142–147. DOI: 10.1038/nbt.2487 PMID: 23334450

Liu, Y., Song, F., Li, Z., Chen, L., Xu, Y., Sun, H., & Chang, Y. (2023). A comprehensive tool for tumor precision medicine with pharmaco-omics data analysis. *Frontiers in Pharmacology*, 14, 1085765. DOI: 10.3389/fphar.2023.1085765 PMID: 36713829

Love-Koh, J., Peel, A., Rejon-Parrilla, J. C., Ennis, K., Lovett, R., Manca, A., Chalkidou, A., Wood, H., & Taylor, M. (2018). The future of precision medicine: Potential impacts for health technology assessment. *PharmacoEconomics*, 36(12), 1439–1451. DOI: 10.1007/s40273-018-0686-6 PMID: 30003435

Mahler, M., Martinez-Prat, L., Sparks, J. A., & Deane, K. D. (2020). Precision medicine in the care of rheumatoid arthritis: Focus on prediction and prevention of future clinically-apparent disease. *Autoimmunity Reviews*, 19(5), 102506. DOI: 10.1016/j.autrev.2020.102506 PMID: 32173516

Makarov, D. V., Loeb, S., Getzenberg, R. H., & Partin, A. W. (2009). Biomarkers for prostate cancer. *Annual Review of Medicine*, 60(1), 139–151. DOI: 10.1146/annurev.med.60.042307.110714 PMID: 18947298

Mavaddat, N., Michailidou, K., Dennis, J., Lush, M., Fachal, L., Lee, A., & MacInnis, R. J. (2019). Polygenic risk scores for prediction of breast cancer and breast cancer subtypes. *American Journal of Human Genetics*, 104(1), 21–34. DOI: 10.1016/j.ajhg.2018.11.002 PMID: 30554720

Mezinska, S., Gallagher, L., Verbrugge, M., & Bunnik, E. M. (2021). Ethical issues in genomics research on neurodevelopmental disorders: A critical interpretive review. *Human Genomics*, 15(1), 1–14. DOI: 10.1186/s40246-021-00317-4 PMID: 33712057

Miao, F., Smith, D. D., Zhang, L., Min, A., Feng, W., & Natarajan, R. (2008). Lymphocytes from patients with type 1 diabetes display a distinct profile of chromatin histone H3 lysine 9 dimethylation: An epigenetic study in diabetes. *Diabetes*, 57(12), 3189–3198. DOI: 10.2337/db08-0645 PMID: 18776137

Mitri, Z., Constantine, T., & O'Regan, R. (2012). The HER2 receptor in breast cancer: Pathophysiology, clinical use, and new advances in therapy. *Chemotherapy Research and Practice*, 2012(1), 743193. DOI: 10.1155/2012/743193 PMID: 23320171

Muteeb, G., Rehman, M. T., Shahwan, M., & Aatif, M. (2023). Origin of antibiotics and antibiotic resistance, and their impacts on drug development: A narrative review. *Pharmaceuticals (Basel, Switzerland)*, 16(11), 1615. DOI: 10.3390/ph16111615 PMID: 38004480

Naithani, N., Atal, A. T., Tilak, T. V. S. V. G. K., Vasudevan, B., Misra, P., & Sinha, S. (2021). Precision medicine: uses and challenges. *medical journal armed forces india, 77*(3), 258-265.

Naithani, N., Sinha, S., Misra, P., Vasudevan, B., & Sahu, R. (2021). Precision medicine: Concept and tools. *medical journal armed forces india, 77*(3), 249-257.

National Academies Press (US) (2017 Jul 10). Case Studies in Precision Drug Development. National Academies of Sciences, Engineering, and Medicine; Health and Medicine Division; Board on Health Sciences Policy; Roundtable on Genomics and Precision Health; Forum on Drug Discovery, Development, and Translation. Enabling Precision Medicine: The Role of Genetics in Clinical Drug Development: Proceedings of a Workshop. Washington (DC): Available from: https://www.ncbi.nlm.nih.gov/sites/books/NBK458878/

NHS England. (n.d.). Personalised medicine. www.england.nhs.uk/ourwork/qual-clin-lead/personalisedmedicine/ [Accessed 23 July 2024].

Nisar, H., & Srivastava, R. (2018). Fundamental concept of psychosomatic disorders: a review. *International Journal of contemporary Medicine surgery and radiology, 3*(1), 12-18.

Nordberg, G. F. (2010). Biomarkers of exposure, effects and susceptibility in humans and their application in studies of interactions among metals in China. *Toxicology Letters*, 192(1), 45–49. DOI: 10.1016/j.toxlet.2009.06.859 PMID: 19540908

Ota, M., & Fujio, K. (2021). Multi-omics approach to precision medicine for immune-mediated diseases. *Inflammation and Regeneration*, 41(1), 23. DOI: 10.1186/s41232-021-00173-8 PMID: 34332645

Parker, L. S. (2024). Ethical considerations in precision medicine. In *The New Era of Precision Medicine* (pp. 143–172). Academic Press. DOI: 10.1016/B978-0-443-13963-5.00002-9

Peterson, B. S. (2020). Biomarkers in precision medicine for mental illnesses. *Journal of Child Psychology and Psychiatry, and Allied Disciplines*, 61(12), 1279–1281. DOI: 10.1111/jcpp.13357 PMID: 33252151

Pritchard, D. E., Moeckel, F., Villa, M. S., Housman, L. T., McCarty, C. A., & McLeod, H. L. (2017). Strategies for integrating personalized medicine into healthcare practice. *Personalized Medicine*, 14(2), 141–152. DOI: 10.2217/pme-2016-0064 PMID: 29754553

Proper, S. P., Azouz, N. P., & Mersha, T. B. (2021). Achieving precision medicine in allergic disease: Progress and challenges. *Frontiers in Immunology*, 12, 720746. DOI: 10.3389/fimmu.2021.720746 PMID: 34484229

Ramaswami, R., Bayer, R., & Galea, S. (2018). Precision medicine from a public health perspective. *Annual Review of Public Health*, 39(1), 153–168. DOI: 10.1146/annurev-publhealth-040617-014158 PMID: 29166244

Schomaker, S., Ramaiah, S., Khan, N., & Burkhardt, J. (2019). Safety biomarker applications in drug development. *The Journal of Toxicological Sciences*, 44(4), 225–235. DOI: 10.2131/jts.44.225 PMID: 30944276

Singla, A. (2024). Precision Medicine: Tailoring Treatment to Individual Genetic Profiles. *Shodh Sagar Journal for Medical Research Advancement*, 1(1), 27–37. DOI: 10.36676/ssjmra.v1.i1.04

Sisodiya, S. M. (2021). Precision medicine and therapies of the future. *Epilepsia*, 62(S2), S90–S105. DOI: 10.1111/epi.16539 PMID: 32776321

Sitapati, A., Kim, H., Berkovich, B., Marmor, R., Singh, S., El-Kareh, R., Clay, B., & Ohno-Machado, L. (2017). Integrated precision medicine: The role of electronic health records in delivering personalized treatment. *Wiley Interdisciplinary Reviews. Systems Biology and Medicine*, 9(3), e1378. DOI: 10.1002/wsbm.1378 PMID: 28207198

Slikker, W.Jr. (2018). Biomarkers and their impact on precision medicine. *Experimental Biology and Medicine*, 243(3), 211–212. DOI: 10.1177/1535370217733426 PMID: 28927290

Stolfi, F., Abreu, H., Sinella, R., Nembrini, S., Centonze, S., Landra, V., Brasso, C., Cappellano, G., Rocca, P., & Chiocchetti, A. (2024). Omics approaches open new horizons in major depressive disorder: From biomarkers to precision medicine. *Frontiers in Psychiatry*, 15, 1422939. DOI: 10.3389/fpsyt.2024.1422939 PMID: 38938457

Subramanian, M., Wojtusciszyn, A., Favre, L., Boughorbel, S., Shan, J., Letaief, K. B., Pitteloud, N., & Chouchane, L. (2020). Precision medicine in the era of artificial intelligence: Implications in chronic disease management. *Journal of Translational Medicine*, 18(1), 1–12. DOI: 10.1186/s12967-020-02658-5 PMID: 33298113

Udegbe, F. C., Ebulue, O. R., Ebulue, C. C., & Ekesiobi, C. S.Francisca Chibugo UdegbeOgochukwu Roseline EbulueCharles Chukwudalu EbulueChukwunonso Sylvester Ekesiobi. (2024). Precision Medicine and Genomics: A comprehensive review of IT-enabled approaches. *International Medical Science Research Journal*, 4(4), 509–520. DOI: 10.51594/imsrj.v4i4.1053

Urdinguio, R. G., Sanchez-Mut, J. V., & Esteller, M. (2009). Epigenetic mechanisms in neurological diseases: Genes, syndromes, and therapies. *Lancet Neurology*, 8(11), 1056–1072. DOI: 10.1016/S1474-4422(09)70262-5 PMID: 19833297

Vargas, A. J., & Harris, C. C. (2016). Biomarker development in the precision medicine era: Lung cancer as a case study. *Nature Reviews. Cancer*, 16(8), 525–537. DOI: 10.1038/nrc.2016.56 PMID: 27388699

Visvikis-Siest, S., Theodoridou, D., Kontoe, M. S., Kumar, S., & Marschler, M. (2020). Milestones in personalized medicine: From the ancient time to nowadays—the provocation of COVID-19. *Frontiers in Genetics*, 11, 569175. DOI: 10.3389/fgene.2020.569175 PMID: 33424917

Vogelstein, B., Papadopoulos, N., Velculescu, V. E., Zhou, S., Diaz, L. A.Jr, & Kinzler, K. W. (2013). Cancer genome landscapes. *Science*, 339(6127), 1546–1558. DOI: 10.1126/science.1235122 PMID: 23539594

Wang, E., Cho, W. C., Wong, S. C., & Liu, S. (2017). Disease biomarkers for precision medicine: Challenges and future opportunities. *Genomics, Proteomics & Bioinformatics*, 15(2), 57–58. DOI: 10.1016/j.gpb.2017.04.001 PMID: 28392478

Weinstein, D., Leininger, J., Hamby, C., & Safai, B. (2014). Diagnostic and prognostic biomarkers in melanoma. *The Journal of Clinical and Aesthetic Dermatology*, 7(6), 13. PMID: 25013535

Woodcock, J., & LaVange, L. M. (2017). Master protocols to study multiple therapies, multiple diseases, or both. *The New England Journal of Medicine*, 377(1), 62–70. DOI: 10.1056/NEJMra1510062 PMID: 28679092

Xi, X., Li, T., Huang, Y., Sun, J., Zhu, Y., Yang, Y., & Lu, Z. J. (2017). RNA biomarkers: Frontier of precision medicine for cancer. *Non-Coding RNA*, 3(1), 9. DOI: 10.3390/ncrna3010009 PMID: 29657281

Yogeshappa, V. G. (2024). AI-driven precision medicine: Revolutionizing personalized treatment plans. [IJCET]. *International Journal of Computer Engineering and Technology*, 15(5), 455–474.

Yuan, B. (2022). What personalized medicine humans need and way to it——also on the practical significance and Scientific Limitations of Precision Medicine. *Pharmacogenomics and Personalized Medicine*, 15, 927–942. DOI: 10.2147/PGPM.S380767 PMID: 36353711

Chapter 8

Revolutionizing Pharmaceutical Sciences: A Comprehensive Review of Artificial Intelligence's Impact on Drug Discovery and Development

Ch. Hima Bindu
http://orcid.org/0000-0002-1268-7802
QIS College of Engineering and Technology, India

Raveendra Babu
http://orcid.org/0000-0002-2219-8551
QIS College of Pharmacy, India

Sasidhar Bhimana
https://orcid.org/0000-0002-4813-7773
QIS College of Pharmacy, India

M. Kishore Babu
https://orcid.org/0009-0005-1837-4099
QIS College of Pharmacy, India

Ch. Devadas
http://orcid.org/0000-0002-2985-9762
QIS College of Pharmacy, India

M. Jalaiah
https://orcid.org/0000-0002-2118-3008
QIS College of Pharmacy, India

ABSTRACT

Artificial Intelligence (AI) is revolutionizing the pharmaceutical sciences by significantly impacting drug discovery and development, which traditionally is a lengthy, difficult, and costly process often resulting in unrewarded investments. AI has given a way in accelerating the drug discovery cycle, reducing costs, and integrating the 3R principles (Replacement, Reduction, and Refinement). It has been instrumental in predicting drug-target interactions (DTIs) and understanding mechanisms of action, as demonstrated by the AI-DTI model which successfully rediscovered DTIs for COVID-19 treatments. AI's role extends to predicting drug toxicity, bioactivity, and physicochemical properties, complementing conventional

DOI: 10.4018/979-8-3693-6597-7.ch008

experiments and accelerating the drug discovery process. This is further supported by machine learning (ML) and deep learning (DL), used for computer facilitated drug discovery, in addressing molecular design and reaction prediction. AI tools and methodologies enhance decision making, efficiency in drug development to improve the human health outcomes.

1. INTRODUCTION

The conventional method of drug detection and expansion process is fraught with numerous challenges, making it a long, complex, and costly endeavor. In order to identify possible molecular targets for therapeutic intervention, the first step of target identification requires a deeper comprehension of the basic mechanisms underlying diseases (Lu et al., 2023). Following this, lead compound discovery involves screening thousands of compounds for easy identification with the potential and to interact effectively with the target. Preclinical testing, which includes laboratory and animal studies, is essential to evaluate the safety and efficacy of the lead compounds (Raza et al., 2022). If successful, the drug's safety, efficiency, and dosage in humans can be tested on candidate progresses to clinical trials with multiple phases of tests. This entire process can take more than a decade and expensive, making it a high-risk venture for pharmaceutical companies. The integration of AI into this process makes easy in the drug discovery cycle, reducing costs, and integrating the 3R principles (Replacement, Reduction, and Refinement) (Dashpute et al., 2023). AI-based models, such as deep generative models, have enabled the exploration of unknown chemical spaces, although the challenge remains in prioritizing molecular structures for subsequent development.

AI techniques have also been applied to predict drug permeability and human intestinal absorption, enhancing early-stage drug screening. Furthermore, AI has been instrumental in predicting DTIs and understanding mechanisms of action, as demonstrated by the AI-DTI model which successfully rediscovered DTIs for COVID-19 treatments. The AI uses in reaction yield prediction has also shown to reduce the number of unsuccessful synthesis attempts, thereby saving resources and time. AI's role extends to predicting drug toxicity, bioactivity, and physicochemical properties, complementing conventional experiments and accelerating the drug discovery process (Sonawane & Gaikwad, 2022). In clinical pharmacology integrating with AI (AI/CP) requires a comprehensive ecosystem with digitized data collection and standardized processes. AI's potential in drug discovery is further supported by advancements in ML and DL, which have been extensively used for computer-facilitated drug discovery, addressing issues such as molecular design and reaction prediction. Despite the challenges, AI continues to offer promising tools and methodologies that enhance decision-making and efficiency in drug development, ultimately transforming the pharmaceutical industry and improving human health outcomes (Vora et al., 2023).

The initiation of AI in pharmaceutical sciences marks a significant evolution in the drug discovery and development landscape. Traditionally, this has been laborious, involving extensive trial and error, which often leads to high costs and long development times. However, with the integration of AI, this paradigm is rapidly changing. AI techniques, particularly ML and DL, have become pivotal in the process of drug research and development (R&D), from target discovery to post-market surveillance. AI's capability to examine vast and complex datasets has revolutionized in the process of drug identification and the design of latest drugs. For instance, the AlphaFold2 system demonstrated AI's potential by precisely predicting the 3-dimensional structures of proteins, a task that is crucial for understanding biological processes and designing effective drugs. Similarly, AI models like graph neural networks (GNNs) combined with transformer models have excelled in predicting molecular properties, showcasing AI's capability to handle high-dimensional biomedical data. In the pharmaceutical industry, AI has been adopted to reduce the process time and cost. State-of-the-art AI models are now integral to various pipelines, facilitating the discovery of lead compounds and optimizing drug synthesis processes (Kolluri et al., 2022). For example, automated AI-based chemical synthesis technologies have significantly reduced the manual workload for researchers, enabling high-throughput experiments and improving reaction throughput. Moreover, AI has shown promise in predicting drug permeability, human intestinal absorption, and drug-target interactions (DTIs), which are critical for early-stage drug screening and understanding mechanisms of action. The AI-DTI model, for instance, successfully rediscovered DTIs for COVID-19 treatments, highlighting AI's potential in addressing urgent medical needs. Despite these advancements, challenges remain in fully integrating AI into the traditional pharmaceutical system. Some of the obstacles that must be overcome are the volume and quality of medical data, the requirement for standardized procedures, and the complexity of biological systems. Nevertheless, the continuous evolution of AI technologies promises to further transform pharmaceutical sciences, making drug discovery and development more efficient and effective (Sultana et al., 2023).

1.1. Objectives and Scope of the Review

The primary objective of this review is to systematically explore and elucidate the primary role of AI in the pharma studies. This section address the AI technologies, particularly ML and DL, are being integrated into various stages of pharma R&D for the challenges of traditional methods. By examining the applications of AI in target discovery, drug design, preclinical research, clinical trials, and post-market surveillance, this review seeks to highlight the significant advancements and efficiencies introduced by AI in these areas. Furthermore, this review will delve into the specific AI models and algorithms that have shown promise in revolutionizing drug discovery (Sampene & Nyirenda, 2024). For instance, importance of DL models to explore unknown chemical spaces and the application of convolutional neural networks (CNNs) for predicting drug permeability and human intestinal absorption will be discussed in detail. The review will also cover the function of AI in drug-target interactions (DTIs) and understanding mechanisms of action, as demonstrated by models like AI-DTI, which have successfully rediscovered DTIs for COVID-19 treatments. In addition to highlighting the successes, this review will discuss the limitations and challenges for the use of AI in drug R&D. Issues such as the need for larger quality data, the complexity of biological systems, and the requirement for standardized processes will be examined. Potential solutions and future directions for overcoming these challenges will also be proposed, emphasizing the need for continuous research and development in this field (Rehman et al., 2024). Overall, this review aims to provide a detailed and systematic analysis of AI in pharmaceutical sciences, offering valuable insights for researchers, practitioners, and stakeholders in the industry.

2. METHODOLOGY

2.1. Literature Search Strategy

The literature strategy for this systematic review was carefully designed to capture a detailed range of study in AI applications in pharma R&D. The search was conducted across multiple reputed and standard databases to ensure a broad and inclusive collection of literature. Keywords and search terms were carefully selected to cover various aspects of AI and its integration into pharmaceutical sciences (Paul et al., 2021). To refine the search, Boolean operators (AND, OR, NOT) were used to combine keywords and exclude irrelevant studies. The search restricted to peer-reviewed articles published from 2000 to 2023 to capture the most recent advancements and trends in the field. Additionally, the identified selected articles screened further relevant studies. The initial search yielded a large number of articles related to rigorous screening process. Titles and abstracts were reviewed to assess review of potentially eligible studies (Yadav et al., 2024). The extraction of data was performed to capture key information, including study objectives, AI models used, application areas, outcomes, and limitations. This systematic approach ensured a comprehensive and unbiased review of the literature, providing a solid foundation for understanding AI importance in pharma R&D.

2.2. Study Selection

2.2.1. Criteria for Selecting Studies

The selection of studies for this systematic review was based on several stringent criteria to ensure the inclusion of high-quality and relevant research. Studies were considered if they specifically addressed the AI applications in the process of drug discovery and development, including target discovery, drug design, preclinical research, medical trials, and post-market observation. The types of AI models and algorithms used, such as ML, DL, generative models, and variational autoencoders (VAEs), were also key factors in the selection process. Studies that did not directly relate to AI applications in pharmaceutical sciences or those that were purely theoretical without practical implementation or validation were excluded (Liebman, 2022).

2.2.2. Process for Screening and Selecting Studies

During this phase, key information such as study objectives, AI models used, application areas, outcomes, and limitations were extracted using a standardized form. This data extraction process was designed to capture comprehensive details about each study, facilitating a thorough analysis of the role of AI in pharmaceutical sciences (Dhudum et al., 2024). This systematic approach ensured a comprehensive and unbiased review of the literature, providing a solid foundation for understanding the role of AI.

2.2.3. Data Extraction and Synthesis

2.2.3.1. Procedures for Extracting Data from Selected Studies

The data extraction process was meticulously designed to ensure the accurate and comprehensive set data from the selected studies. AI models and algorithms used application areas within drug discovery and development, outcomes, and limitations. This form was piloted on a subset of studies to ensure its effectiveness and was subsequently refined based on feedback. Data extraction was performed independently by two reviewers to minimize bias and errors (Vora et al., 2023). This rigorous approach ensured that all relevant data were systematically and accurately captured from each study.

2.2.4. Methods for Synthesizing and Analyzing the Extracted Data

The synthesis and analysis of the extracted data involved both qualitative and quantitative methods. Qualitative synthesis was conducted to identify common themes, trends, and gaps in the application of AI across different stages of drug discovery and development. This involved thematic analysis to categorize the various AI models and their specific applications, such as: target discovery; drug design; preclinical research; clinical trials, and post-market surveillance. Quantitative synthesis included the use of meta-analysis techniques where applicable, to aggregate data on the performance and outcomes of AI models. Statistical methods were employed to compare the effectiveness of different AI approaches and to identify factors influencing their success. Additionally, network meta-analysis was used to explore the relationships between various AI models and their applications in pharma (Alanazi, 2024). The combined qualitative and quantitative synthesis provided a comprehensive understanding of the current state and potential of AI in revolutionizing pharmaceutical sciences.

3. THE ROLE OF AI IN DRUG DISCOVERY AND DEVELOPMENT

3.1. Accelerating the Drug Discovery Cycle

AIhas significantly accelerated the drug discovery cycle, transforming the traditionally lengthy and costly process into a effective and expensive process. One notable case study is the application of AI in predicting drug-target interactions (DTIs). The AI-DTI model, for instance, successfully rediscovered DTIs for COVID-19 treatments, demonstrating AI's potential in rapidly identifying promising drug candidates during a global health crisis ((Blanco-González et al., 2023). By using DL to link the macroscopic experimental findings with the microscopic characteristics of molecules, this model contributes to our understanding of the biology of immunoglobulin and facilitates the productive production of novel antibodies.

Figure 1. Diagram for accelerating the drug discovery cycle

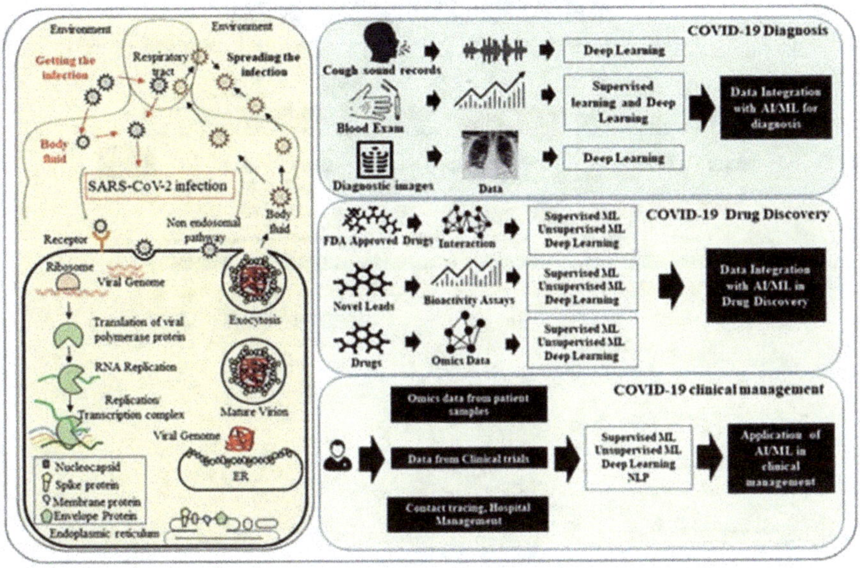

Another significant impact of AI is observed in the prediction of reaction yields, which has reduced the number of unsuccessful synthesis attempts. This not only saves resources and time but also accelerates the overall drug development process. AI-based automated chemical synthesis technologies, such as the Chemputer system, have further streamlined the synthesis of drug compounds (Malkawi, 2024). Automated chemical synthesis in laboratories is being widely adopted thanks to the Chemputer system, which synthesises chemicals using a modular robotic platform. The system records the synthesis process and generates digital code that can be shared across platforms.

Figure 2. The drug discovery and development pipeline

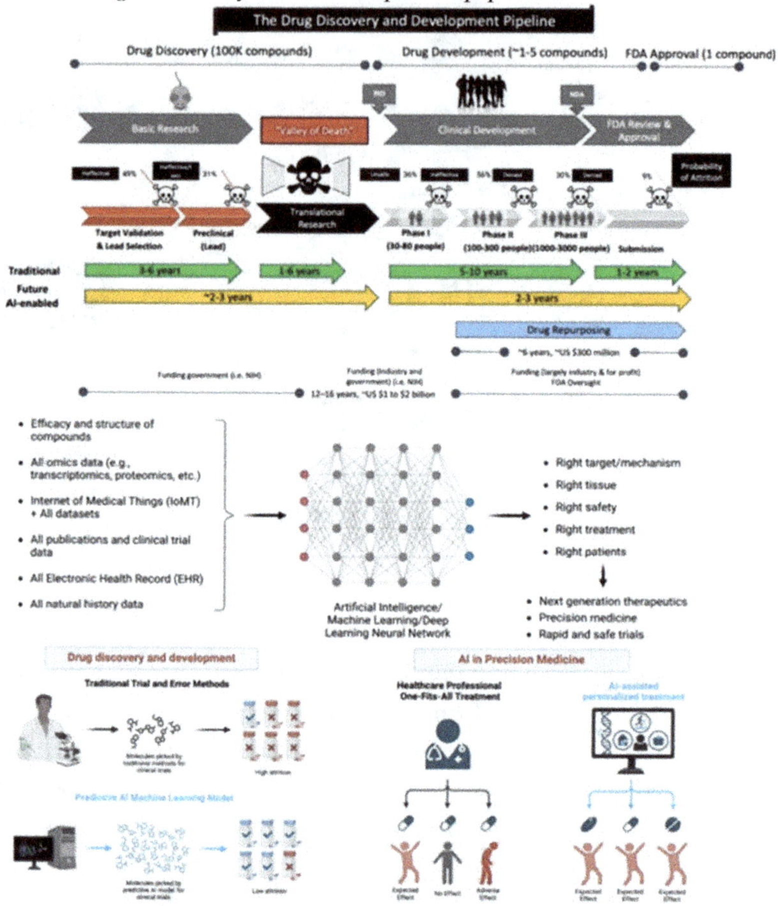

Moreover, AI has been instrumental in predicting drug permeability and human intestinal absorption, enhancing early-stage drug screening. This capability allows for the rapid identification of drug candidates with favorable absorption properties, thereby accelerating the transition from preclinical to clinical stages. AI's role extends to predicting drug toxicity, bioactivity, and physicochemical properties, complementing conventional experiments and further speeding up the drug discovery process.

Figure 3. Diagram representation for learning of artificial intelligence

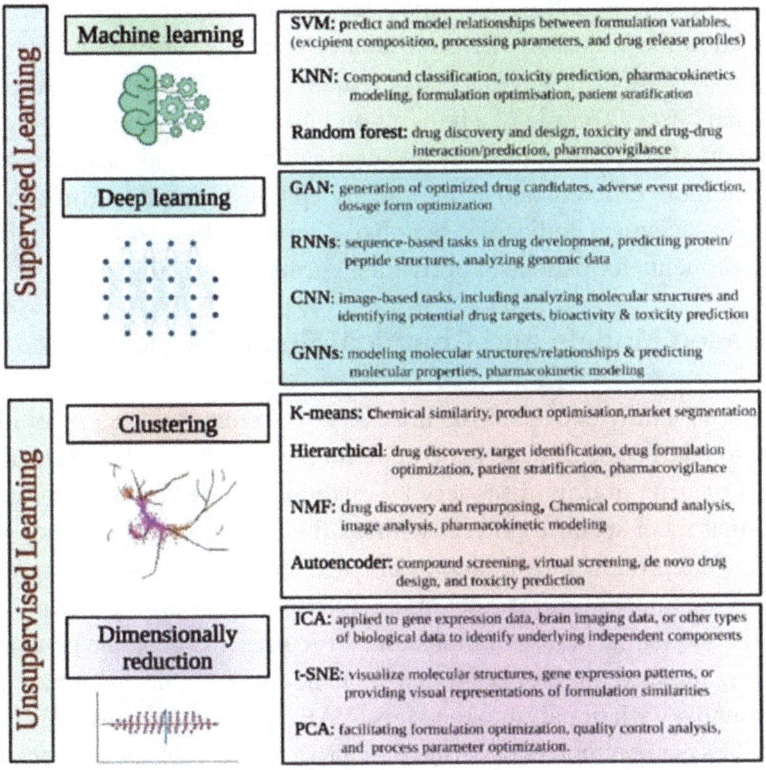

In summary, AI has demonstrated its ability to significantly shorten the drug discovery cycle through various applications, including DTI prediction, reaction yield prediction, automated synthesis, and early-stage screening. These advancements highlight AI's transformative impact on pharmaceutical sciences, ultimately leading to more efficient drug development and improved human health outcomes (Tiwari et al., n.d.).

3.2. Discussion on Cost Reduction and Resource Savings

AI has revolutionized the pharmaceutical industry by considerably reduce the costs and resources associated with drug discovery and development. Traditional drug discovery processes are notoriously long and expensive, often resulting in unrewarded investments due to high failure rates and lengthy timelines. AI addresses these challenges by using large data and novel algorithms to update various stages of drug development, from target classification to clinical trials. In clinical trials, AI assists in designing more efficient and rational trial protocols, which can lead to faster and more cost-effective studies. AI's capable to analyze huge amounts of clinical data helps in identifying potential issues early, thereby reducing the likelihood of costly trial failures. Post-market surveillance and long-term regulatory work are also enhanced by AI, ensuring that drugs remain safe and effective while minimizing additional costs. Overall, AI's integration into the pharmaceutical sciences has led to substantial cost reductions and resource savings, making drug discovery and development more efficient and economically viable. This transformation not only benefits pharmaceutical companies but also accelerates the availability of new and effective treatments for patients worldwide (Verma & Awasthi, 2024).

3.3. AI-Based Models and Chemical Space Exploration

AIhas significantly expanded the horizons of chemical space exploration, particularly through the use of deep generative models. These models have developed the field of pharma in drug discovery by enabling the design of novel molecular structures that meet specific criteria without the need for explicit prior chemical knowledge. One prominent example of deep generative models in drug discovery is the use of Recurrent Neural Networks (RNNs). RNN-based generative models can learn the sequential nature of molecular structures and generate new molecules by predicting the next atom or bond in a sequence, thus exploring a vast chemical space efficiently. Variational Autoencoders (VAEs) are another powerful tool in this domain. VAEs encode molecular structures into a continuous latent space and then decode them back into molecular structures, allowing for the generation of novel compounds that are structurally diverse and potentially bioactive. Adversarial Auto encoders (AAEs) and Generative Adversarial Networks (GANs) have also been employed in de novo drug design. These models generate new molecular structures by learning the distribution of existing molecules and then creating new samples that follow the same distribution. This approach leads the inventions of entirely new classes of compounds that were previously unexplored. The stochastic nature of these generative models ensures a high degree of variability in the generated molecules, which is crucial for exploring the vast chemical space and identifying novel drug candidates. Reinforcement learning algorithms further enhance the capabilities of generative models by fine-tuning the generated molecules to meet specific drug-like properties (Shah, 2023). This targeted optimization ensures that the generated compounds not only explore new chemical spaces but also possess desirable pharmacological properties. The fusion of these advanced AI models into the pharma pipeline has the potential to bypass the limitations of traditional empirical-based drug design, making the process faster, more cost-effective, and more innovative. In summary, deep generative models such as RNNs, VAEs, AAEs, and GANs, along with reinforcement learning, have significantly advanced the exploration of chemical space in drug discovery. These AI-based models enable the generation of novel molecular structures with specific properties, thereby accelerating the discovery of new drugs and expanding the boundaries of pharmaceutical sciences.

3.4. Challenges in Prioritizing Molecular Structures for Development

The addition of AI in drug finding has opened new avenues for exploring chemical space, yet it also presents significant challenges in prioritizing molecular structures for development. One of the primary challenges is the "activity cliff" phenomenon, where small changes in molecular structure can lead to significant changes in biological activity, making it difficult to predict which modifications will yield the most promising drug candidates. This unpredictability necessitates the use of sophisticated AI algorithms that can accurately model and predict these changes. The computational cost associated with running advanced AI models is another significant challenge. Deep learning models, for instance, require substantial computational resources and time, which can be a restricted access in the drug detection process. Moreover, the interpretability of AI models remains a critical issue. Deep learning models in particular often function as "black boxes," it is challenging for researchers to comprehend the reasoning behind their predictions.

This lack of transparency can hinder the acceptance and trust of AI-generated results in the scientific community. Despite these challenges, ongoing advancements in AI and computational power are gradually addressing these issues. Researchers are developing more interpretable models and leveraging transfer learning to make better use of limited data. As AI continues to evolve, it holds the potential to overcome these challenges, thereby enhancing the competence and usefulness of drug discovery and development.

4. EARLY-STAGE DRUG SCREENING

4.1. Applications of AI in Predicting Drug Permeability and Human Intestinal Absorption

AI has become an indispensable tool in early-stage drug screening, particularly in predicting drug permeability and human intestinal absorption. These properties are crucial for determining the bioavailability of drug candidates, which directly impacts their efficacy and safety profiles. AI models, especially those based on ML and DL, have shown remarkable capabilities in this domain. One of the primary applications of AI in this context is the prediction of Absorption, Distribution,

Metabolism, Excretion, and Toxicity (ADMET) properties. CNNs and RNNs are frequently employed to predict these properties due to their ability to handle complex and high-dimensional data. These models can analyze molecular structures and predict their permeability across biological membranes, which is a critical factor in drug absorption. Moreover, AI models can integrate various types of data, including physicochemical properties, molecular descriptors, and biological assay results, to provide comprehensive predictions. For instance, graph-based AI algorithms are particularly effective in modeling the complex interactions between drug molecules and biological systems, thereby enhancing the accuracy of permeability and absorption predictions (Bai et al., 2024). This multi-faceted approach allows for a more holistic about a drug candidate behavior and changes in the human body. In addition to predicting permeability, AI can also assist in optimizing drug formulations to improve intestinal absorption. By analyzing the interaction of drug excipients with biomolecules, AI models can suggest modifications to enhance solubility and stability, thereby improving overall bioavailability. This capability is particularly valuable in the development of oral drugs, where intestinal absorption is a major determinant of therapeutic success. Despite these advancements, challenges remain in the application of AI for predicting drug permeability and absorption. The quality and diversity of training data are critical for model accuracy, and there is a need for more extensive datasets that cover a wide range of molecular structures and biological conditions. In addition, the importance of AI models is an ongoing concern, as understanding the rationale behind predictions is essential for gaining regulatory approval and clinical acceptance. In summary, AI has significantly advanced the early-stage screening of drug candidates by accurately predicting drug permeability and human intestinal absorption. During the use of sophisticated ML and DL models, AI can analyze complex datasets and provide valuable insights that enhance the drug development process. However, ongoing efforts are needed to address data quality and model interpretability challenges to fully realize the potential of AI in this field.

4.2. Contributions to Enhanced Early-Stage Drug Screening

AI has significantly contributed to enhancing early-stage drug screening by improving the accuracy and efficiency of identifying potential drug candidates. One of the primary applications of AI in this phase is the prediction of ADMET properties, which are critical for determining the viability of drug candidates before they proceed to more costly and time-consuming stages of development. AI models, particularly those based on ML and DL, can analyze vast datasets to predict these properties with high accuracy, decreasing the possibility of late-stage failure in the process. AI also has a significant impact in predicting drug permeability and human intestinal absorption, which are essential for assessing the bioavailability of oral drugs. By leveraging large-scale biological and chemical data, AI algorithms can model the complex interactions between drug molecules and biological membranes, providing insights into how well a drug will be absorbed in the human intestine. This capability is particularly valuable in the early stages of drug development, where quick and accurate predictions can save significant time and resources. Moreover, AI-driven approaches can integrate various types of data, including molecular structures, physicochemical properties, and biological assay results, to provide a comprehensive assessment of drug candidates (Tade et al., 2024). This multi-faceted analysis allows for a more holistic understanding of a drug's potential efficacy and safety, thereby enhancing the overall screening process. Additionally, AI can assist in optimizing drug formulations by analyzing the interaction of drug excipients with biomolecules, which can improve solubility and stability, further enhancing bioavailability. Despite these advancements, challenges remain, such as the need for high-quality, diverse datasets and the interpretability of AI models. However, ongoing research and development efforts are addressing these issues, making AI an increasingly indispensable tool in early-stage drug screening.

4.3. Predicting Drug-Target Interactions (DTIs)

AI has revolutionized the prediction of Drug-Target Interactions (DTIs), a critical step in drug discovery and development. Various AI techniques, combining ML & DL, have been employed to enhance the accuracy and efficiency of DTI prediction. These techniques can be broadly categorized into ligand, structure, and chemogenomic methods, each with its own set of advantages and limitations. Ligand-based methods utilize the chemical properties of ligands to predict their interactions with target proteins. These methods often rely on molecular fingerprints and descriptors to represent the ligands, which are then used as inputs for ML models such as SVMs and RFs. Structure-based methods, on the other hand, focus on the three-dimensional structures of target proteins and ligands. Techniques like molecular docking and virtual screening are commonly used in this approach, leveraging the spatial compatibility between the drug and its target. Chemogenomic methods integrate drug-related and target-related information from multiple sources, including public databases like DrugBank, ChEMBL, and BindingDB, to predict DTIs. In order to improve prediction accuracy, these techniques frequently use ML & DL algorithms to absorb pertinent functional information and decrease noise from heterogeneous data.

Knowledge graphs (KGs) have also emerged as powerful tools in DTI prediction, integrating diverse biomedical data to provide a comprehensive understanding of drug-target relationships. KG embedding (KGE) models represent entities and relations as low-rank vectors, which can be used for link prediction to discover new

DTIs. Recent advancements in DL have further enhanced DTI prediction (Pareek et al., 2023). Models like DeepDTA and DeepTrans utilize sequence information and transcriptome data, respectively, to predict binding affinities and potential interactions. Network-based methods, such as Network-based Inference (NBI) and Network-based Random Walk with Restart on the Heterogeneous network (NRWRH), leverage the topological properties of drug-target networks to predict interactions. Hybrid methods combine multiple approaches to harness the strengths of each, providing more robust and accurate predictions. In summary, AI techniques have significantly advanced the field of DTI prediction, offering powerful tools to identify potential drug candidates and their targets. By integrating diverse data sources and employing sophisticated algorithms, these techniques have improved the efficiency and accuracy of early-stage drug discovery, paving the way for the development of new and effective therapeutics.

4.4. Case Studies, Including the AI-DTI Model and Its Applications, Such as COVID-19

AI has been instrumental in predicting Drug-Target Interactions (DTIs), significantly advancing drug discovery and development. One notable case study is the use of AI in the rapid identification of potential drug candidates for COVID-19. During the pandemic, AI models were employed to screen existing drugs and novel compounds for their potential to interact with SARS-CoV-2 targets, such as the main protease (Mpro) and the spike protein. These models utilized various AI techniques, including DL and machine learning (ML), to analyze vast datasets of molecular structures and biological activities, thereby accelerating the drug repurposing process. For instance, the DeepDTA model, was adapted to identify potential inhibitors of SARS-CoV-2. This model demonstrated high accuracy in predicting interactions, thereby facilitating the rapid screening of large compound libraries. Another example is the use of knowledge graphs (KGs) to integrate heterogeneous data from multiple sources, such as DrugBank, ChEMBL, and BindingDB, to construct comprehensive models like BioKG and PharmKG. Moreover, AI-driven chemogenomic approaches have been pivotal in predicting DTIs by connecting information to multiple sources of biomedical data ((Prajapati et al., 2016). These methods have been particularly effective in identifying new targets for existing drugs, thereby facilitating drug repositioning efforts during the COVID-19 crisis. The integration of omics data with biomedical data networks has further enhanced the accuracy of DTI predictions, enabling the identification of promising drug candidates with high precision. In summary, AI models have played a crucial role in predicting DTIs, particularly in the context of urgent health crises like COVID-19. By leveraging advanced AI techniques and integrating diverse data sources, these models have significantly accelerated the drug discovery process, highlighting the transformative potential of AI in pharmaceutical sciences.

5. AI IN REACTION YIELD PREDICTION AND SYNTHESIS

5.1. AI's Role in Reaction Yield Prediction Techniques and Models Used for Predicting Reaction Yields

AI has significantly advanced the field of reaction yield prediction, a crucial aspect of chemical synthesis. Various AI techniques and models have been developed to predict reaction yields with high accuracy, thereby optimizing the synthesis process and reducing experimental costs. One of the earliest AI applications in this domain was the computer-aided synthetic design (CASP) proposed by Corey and Wipke in the 1960s, which has seen a resurgence with the advent of modern ML methods. Retrosynthetic planning is a key technique in AI-driven reaction yield prediction. This method consists of deconstruction of target molecules into simpler, commercially available starting materials. Advanced models, such as those combining Monte Carlo tree search (MCTS) with neural networks, have been developed to predict the best retrosynthetic routes, significantly speeding up the process compared to conventional algorithms. Additionally, AI models like the one proposed by Gao et al. use neural networks to predict suitable response situation, including temperature and solvents, based on extensive datasets from sources like Reaxys. Forward reaction prediction is another critical technique, which verifies the feasibility of designed synthetic routes by predicting the products of chemical reactions (Nwosu, 2024). For instance, Coley et al. developed a neural network model trained on 15,000 reaction examples from the United States Patent and Trademark Office (USPTO) literature, achieving an accuracy of 71.8% in predicting reaction outcomes. This model uses an edit-based representation of candidate reactions, allowing for the ranking of generated compounds to select the most promising ones. HTEs have also been integrated with AI to enhance reaction yield prediction. HTEs can perform dozens to thousands of reactions simultaneously, providing large datasets for training AI models. For example, Ahneman et al. utilized a random forest (RF) algorithm. In conclusion, AI techniques and models have revolutionized reaction yield prediction by integrating advanced ML and DL algorithms with extensive chemical datasets. These innovations have not only improved the accuracy and efficiency of predicting reaction yields but also facilitated the exploration of new chemical spaces, ultimately accelerating the drug discovery and development process.

5.2. Impact on Reducing Unsuccessful Synthesis Attempts and Resource Utilization

AI has revolutionized the field of reaction yield prediction, significantly impacting the efficiency of chemical synthesis by reducing unsuccessful synthesis attempts and optimizing resource utilization. Traditional methods of reaction yield prediction often involve extensive trial and error, which is both time-consuming and resource-intensive. AI models, however, can analyze vast datasets to predict reaction outcomes with high accuracy, thereby minimizing the need for repetitive experiments (Pareek et al., 2023). One of the key techniques in AI-driven reaction yield prediction is the use of neural networks trained on large datasets of chemical reactions. This model demonstrated a high success rate in predicting reaction conditions, thereby reducing the time and resources spent on screening reaction conditions manually. High-throughput experiments (HTEs) have also been integrated with AI to enhance reaction yield prediction. HTEs can perform dozens to thousands of reactions simultaneously, providing large datasets for training AI models. For example, Ahneman et al. utilized a random forest (RF) algorithm trained on high-throughput datasets to predict the tolerance of palladium catalysts in C-N bond formation reactions, outperforming traditional linear regression methods. This integration of HTEs with AI not only improves the accuracy of reaction yield predictions but also accelerates the discovery of optimal reaction conditions. Moreover, AI models like the one proposed by Coley et al. use neural networks to predict reaction outcomes based on extensive datasets from the USPTO literature. These models can rank generated compounds to select the most promising ones, thereby reducing the number of unsuccessful synthesis attempts and optimizing the use of resources. In conclusion, AI has significantly impacted reaction yield prediction by reducing unsuccessful synthesis attempts and optimizing resource utilization. By leveraging advanced AI techniques and integrating diverse data sources, these models have enhanced the efficiency and accuracy of chemical synthesis, ultimately accelerating the drug discovery and development process.

6. PREDICTING DRUG TOXICITY, BIOACTIVITY, AND PHYSICOCHEMICAL PROPERTIES

6.1. AI Applications in Toxicity Prediction

6.1.1. AI Models for Predicting Drug Toxicity

Artificial intelligence (AI) has been applied to anticipate medication toxicity, greatly improving the speed and precision of the drug development process. To forecast the toxicological characteristics of medication candidates and lower the risk of side effects while enhancing patient safety, several artificial intelligence models have been created. Large datasets and sophisticated ML and DL algorithms are utilized by these models to detect possible toxicity at an early stage of the medication development process. ADMET lab 2.0, a well-known AI model for toxicity prediction, is built on the multi-task graph attention (MGA) architecture. This model includes 88 relevant characteristics, including as 23 ADME attributes, 27 toxicity endpoints, and 8 toxicophore rules, and it can predict numerous ADMET properties of medicines. ADMET lab 2.0 is an all-inclusive tool for medicinal chemists because the data used to train it came from toxicity prediction software, relevant literature, and bioactivity data found in open-access databases.

In addition to ADMET lab 2.0, CNNs and RNNs have shown outstanding results in predicting the ADMET properties of molecules (Keerthana et al., 2022). These models can analyze complex relationships within the data, such as drug-drug interactions and protein-protein interactions, to provide accurate toxicity predictions. For example, CNNs are particularly effective in extracting features from high-dimensional data, while RNNs are adept at handling sequential data, making them suitable for time-series toxicity predictions. Furthermore, AI has been employed in post-market surveillance to monitor and evaluate drug safety after approval. Electronic health record (EHR) mining, using models like the self-control case series (SCCS) and convolutional SCCS (ConvSCCS), has been utilized to detect adverse drug reactions (ADRs) in patient cohorts. These models can process both structured and unstructured data, enabling the detection of undocumented toxic effects and improving the overall safety profile of drugs. In conclusion, AI models have revolutionized the prediction of drug toxicity by leveraging advanced algorithms and extensive datasets. These models not only enhance the accuracy of toxicity predictions but also reduce the time and resources required for drug development, ultimately leading to safer and more effective pharmaceuticals.

6.2. Integration with Conventional Experimental Approaches

The integration of AI with conventional experimental approaches has significantly enhanced the prediction of drug toxicity, providing a more comprehensive and accurate assessment of potential adverse effects. Traditional methods of toxicity testing, such as in vitro and in vivo assays, are often time-consuming, costly, and ethically challenging. AI models, on the other hand, can analyze large datasets to predict toxicological outcomes, thereby complementing and refining these conventional methods. AI models, such as CNNs and RNNs, have been employed to predict the ADMET properties of drug candidates. These models can process complex datasets, including high-throughput screening data and electronic health records (EHRs), to identify potential toxicities early in the drug development process. For instance, the use of AI in EHR mining has enabled the detection of ADRs through models like the SCCS and ConvSCCS, which process both structured and unstructured data to improve the accuracy of toxicity predictions. Moreover, AI-driven approaches have been integrated with HTEs to enhance the prediction of drug toxicity. HTEs can generate large datasets of chemical reactions, which are then used to train AI models for predicting toxicological outcomes. In addition, AI models have been utilized to predict the physicochemical properties of drug candidates, which are crucial for assessing their pharmacokinetic and toxicological profiles. By analyzing data from various sources, including chemical databases and literature, AI models can predict properties such as solubility, stability, and bioavailability, thereby informing the design of safer and more effective drugs. In conclusion, the integration of AI with conventional experimental approaches has revolutionized the prediction of drug toxicity. By leveraging advanced algorithms and extensive datasets, AI models provide a more accurate and efficient assessment of toxicological risks, ultimately enhancing the safety and efficacy of new pharmaceuticals (Sayam et al., 2023).

7. BIOACTIVITY AND PHYSICOCHEMICAL PROPERTIES

7.1. Techniques for AI-driven Prediction of Bioactivity and Physicochemical Attributes

AI-driven techniques have revolutionized the prediction of bioactivity and physicochemical properties of drug candidates, significantly enhancing the drug discovery and development. These techniques leverage advanced ML & DL algorithms to analyze large and complex datasets, providing insights that are often beyond the reach of traditional methods. One of the primary techniques used in AI-driven prediction is Quantitative Structure-Activity Relationship (QSAR) modeling. QSAR models, which can be based on various ML algorithms such as SVM and RF, predict the bioactivity of compounds by correlating their chemical structure with biological activity data. These models have been successfully applied in drug-protein affinity prediction, enabling researchers to identify potential drug candidates more efficiently. Deep learning methods, particularly CNNs and RNNs, have also shown great promise in predicting bioactivity and physicochemical properties. CNNs are adept at extracting features from high-dimensional data, making them suitable for tasks such as protein-folding prediction and molecular property prediction. RNNs, with their ability to handle sequential data, are particularly useful for time-series predictions, such as those required in pharmacokinetics and toxicology studies. Generative models, including VAEs and GANs, have been employed in de novo drug design. These models can generate novel compounds that meet specific bioactivity and physicochemical criteria, bypassing the limitations of traditional empirical-based drug design paradigms. By learning the underlying molecular features and drug targets, these models can produce compounds with desired properties at a faster and less costly rate. GNNs have emerged as powerful tools for molecular feature learning and representation. GNNs can model the complex relationships between atoms in a molecule, providing a detailed representation that can be used for predicting various physicochemical properties. For instance, GNNs combined with MPNNs have been used to predict properties such as solubility and lipophilicity, achieving state-of-the-art performance on multiple datasets. In conclusion, AI-driven techniques for predicting bioactivity and physicochemical properties have significantly advanced the field of drug discovery. By leveraging sophisticated algorithms and extensive datasets, these techniques provide accurate and efficient predictions, ultimately accelerating the development of safer and more effective pharmaceuticals (B V et al., 2024).

7.2. Influence on the Overall Drug Discovery Process

The prediction of bioactivity and physicochemical properties using AI has a profound influence on the overall drug discovery process. AI-driven techniques enable the efficient identification of lead compounds by predicting their bioactivity against specific targets, which is crucial for the initial stages of drug discovery. These techniques utilize various ML and DL models to analyze large datasets, thereby accelerating the identification of promising drug candidates. One significant advantage of AI in predicting bioactivity is its ability to handle high-dimensional data, which is often generated in drug discovery research. This integration helps in identifying potential drug targets and understanding the mechanisms of action, which are critical for the development of effective therapeutics. In addition to bioactivity, the prediction of physicochemical properties, such as solubility, stability, and lipophilicity, is essential for assessing the pharmacokinetic and toxicological profiles of drug candidates. AI models, including GNNs and MPNNs, have been employed to predict these properties with high accuracy. Accurate prediction of physicochemical properties facilitates the optimization of drug candidates, ensuring they meet the desired criteria for efficacy and safety. Moreover, AI-driven predictions can significantly reduce the time and cost associated with traditional experimental methods. By providing early insights into the bioactivity and physicochemical properties of compounds, AI models help in prioritizing the most promising candidates for further development, thereby streamlining the drug discovery pipeline. This efficiency is particularly beneficial in the preclinical stages, where early evaluation of drug properties can decrease the failure rate and loss of clinical studies. In conclusion, the integration of AI in predicting bioactivity and physicochemical properties has revolutionized the drug discovery process. By leveraging advanced algorithms and extensive datasets, AI provides accurate and efficient predictions, ultimately enhancing the development of safer and more effective pharmaceuticals.

8. THE FUTURE OF AI IN CLINICAL PHARMACOLOGY

8.1. Integration of AI in Clinical Pharmacology (AI/CP)

8.1.1. Requirements for a Comprehensive AI-Driven Ecosystem

The integration of AI in clinical pharmacology (AI/CP) necessitates a comprehensive ecosystem that can support the multifaceted requirements of modern drug development and patient care. A robust AI-driven ecosystem must first ensure the availability and accessibility of high-quality, structured, and unstructured data. This includes extensive pharmaceutical databases that provide detailed information on drug targets, molecular structures, pharmacokinetics, and clinical outcomes. The quality and quantity of this data are critical, as they form the foundation upon which AI models are trained and validated. These algorithms must be capable of handling diverse data types and structures, from genomic sequences to EHRs. Techniques such as DL, CNNs, and GNNs are particularly valuable for their ability to model complex biological interactions and predict clinical outcomes with high accuracy. Furthermore, the ecosystem must support the seamless integration of AI tools into existing clinical workflows. This involves the development of user-friendly interfaces and decision support systems that can assist clinicians in making informed decisions based on AI-driven insights.

Moreover, regulatory structures need to change to reflect the distinct challenges posed by AI in clinical pharmacology. This includes establishing guidelines for the validation and approval of AI-driven tools and ensuring that these tools meet stringent standards for accuracy, reliability, and ethical use. Post-market surveillance systems must also be enhanced to monitor the long-term safety and efficacy of AI-driven interventions, leveraging AI for the detection and analysis of ADRs. In conclusion, this process requires a comprehensive ecosystem that encompasses high-quality data, advanced algorithms, seamless integration into clinical workflows, and robust regulatory frameworks. This ecosystem will enable the full potential of AI to be realized, ultimately improving drug development processes and patient outcomes (Dominic et al., 2024).

8.2. Role of Digitized Data Collection and Standardized Processes

The AI in clinical pharmacology (AI/CP) hinges significantly on the role of digitized data collection and standardized processes. The high-quality, structured, and unstructured data is paramount for the effective application of AI in this field. These applications are greatly aided by the availability of comprehensive and organized large data from pharmaceutical databases. For AI-driven analysis, these databases contain data on targets, sequences, 3D structures, ligand-binding characteristics, pertinent enzymes, and related pharmacological information.

Digitized data collection facilitates the seamless integration of various data types, such as EHRs, genomic data, and clinical trial results, into AI models. EHR mining, for instance, where structured data simplifies the process of data pre-processing. The use of standardized processes ensures that data collected from different sources is consistent and reliable, which is critical for training accurate and robust AI models. Techniques such as DL, CNNs, and GNNs are particularly valuable for modeling complex biological interactions and predicting clinical outcomes. These algorithms can analyze multi-patient and multi-factor data to improve the accuracy of prognosis predictions, as demonstrated in various cancer prognoses. In conclusion, the role of digitized data collection and standardized processes is fundamental to the successful integration of AI in clinical pharmacology. By ensuring the availability of high-quality data and the use of advanced AI algorithms, the AI/CP ecosystem can enhance drug development processes and improve patient outcomes.

9. DISCUSSION

9.1. Summary of Key Findings

The integration of AI in drug R&D has considerably transformed the pharmaceutical industry by enhancing efficiency and reducing costs. Effective development of therapeutic antibodies has been made easier by AI algorithms, especially ML and DL, which have demonstrated promising outcomes in identification and discovery. These algorithms have also offered novel approaches to antibody design. In order to find possible treatment targets and comprehend the mechanisms underlying disease, artificial intelligence (AI) has been used to analyze omics data.

For example, with unsupervised ML algorithms: analyze molecular datasets, leading to the identification of key genes associated with diseases such as lung adenocarcinoma (LUAD). Additionally, AI approaches have been employed to integrate multi-omics data, revealing composite biological systems and syndrome mechanisms, which is crucial for discovering candidate therapeutic targets. AI has also been applied in predicting the ADMET properties of drug candidates, which is essential for ensuring their safety and efficacy. Various ML & DL have been used to predict these properties, demonstrating satisfactory performance and reducing the need for extensive animal studies.

Furthermore, AI-assisted designing clinical trials and monitoring after the sale have enhanced the rationality and safety of clinical trials, ultimately providing more realistic responses to clinical trial results. In conclusion, AI is becoming a crucial tool in the medication research and development process, offering significant advantages in terms of efficiency, cost reduction, and the ability to handle complex biological data.

9.2. Implications for Future Research

The findings from the reviewed studies, yet they also highlight several areas that warrant further investigation. The integration of AI in drug R&D has demonstrated significant improvements in efficiency and cost reduction, particularly in target discovery, drug design, preclinical research, and automated drug synthesis. However, the current applications of AI have not fully dismantled the traditional pharmaceutical system, indicating that there is substantial room for optimization and expansion.

9.3. Advancements in AI Algorithms and Models

Future research should focus on developing more sophisticated AI algorithms and models that can handle the complexity and high-dimensionality of biomedical data more effectively. Existing limitations in drug discovery technologies are likely to be solved; easing the development of novel medications, as AI algorithms get more complicated and high-quality data becomes more available. For example, DL techniques are being used more and more in the study and creation of therapeutic antibodies, offering fresh possibilities for the pipeline of antibody design in the future. Further exploration into DL and other advanced AI models could yield even more efficient and accurate drug discovery processes.

9.4. Integration of Multi-Omics Data

In order to identify potential treatment targets, it is essential to understand complicated biological systems and disease causes, which can be accomplished through the integration of multi-omics data utilizing AI techniques. Future research should continue to explore and refine these integrative approaches, potentially incorporating additional omics layers such as metabolomics and proteomics. This could lead to a more comprehensive knowledge of disease mechanisms and the discovery of new therapeutic targets.

9.5. AI in Preclinical and Clinical Research

AI has shown promise in predicting the ADMET properties of drug candidates, which is essential for ensuring their safety and efficacy. Future research should aim to enhance these predictive models, making them more accurate and reliable. Additionally, AI-assisted clinical trial design has the potential to significantly reduce malfunction of clinical trials by improving patient recruitment and monitoring. Further research is needed to refine these AI-based approaches, ensuring they can be effectively implemented in real-world clinical settings.

9.6. Automated Drug Synthesis

The automation of drug synthesis using AI has already begun to free researchers from repetitive tasks and improve reaction throughput. Future research should focus on expanding the types of reactions that can be automated and improving the accuracy and efficiency of these automated systems. This could involve the development of new AI algorithms for reaction prediction and the integration of these algorithms into existing automated synthesis platforms.

9.7. Post-Market Surveillance and Prognosis Prediction

AI applications in post-market surveillance and prognosis prediction have shown potential in enhancing drug safety and predicting disease outcomes. Future research should aim to develop more robust AI models for these applications, potentially incorporating real-time data from electronic health records (EHRs) and wearable devices. This could lead to more accurate and timely identification of adverse drug reactions and better disease prognosis predictions.

9.8. Addressing Limitations and Ethical Considerations

Despite the promising advancements, there are still limitations and challenges in these applications. The issues associated to data quality, model interpretability, and the need for extensive validation. Future research should address these limitations, potentially through the development of new AI techniques and the establishment of standardized validation protocols. Additionally, ethical considerations related to the use of AI in pharma, such as data privacy and algorithmic bias, should be carefully examined and addressed. In conclusion, while AI has already made significant strides in transforming drug discovery and development, there is still substantial potential for future research to further enhance its applications. By addressing the current limitations and exploring new avenues for AI integration, researchers can continue to drive innovation in the pharmaceutical industry and improve the efficiency and effectiveness of drug R&D.

9.9. Limitations

The current systematic review on drug discovery and development with the application of AI has limitations these can be discussed in future studies. One of the primary limitations is the dependency on high-quality and accurate data, which is crucial for the development of effective AI algorithms. Despite the availability of several excellent publicly accessible databases such as TTD, ChEMBL, DrugBank, CMAP, and PRIDE, the lack data to support more complex research endeavors. The acquisition and processing of high-quality data are costly and time-consuming, and the variability in data acquisition methods, times, and locations further complicates the standardization of data. Another significant limitation is the interpretability of DL methods. Due to their lack of decision-making transparency, DL approaches are viewed as a "black box" in contrast to standard ML methods.

The challenge task for researchers is to understand and trust the results, especially when new findings contradict existing knowledge. In this field, where understanding the mechanisms of action, metabolic pathways, and regulatory mechanisms is crucial, this limitation is particularly problematic. Furthermore, the current AI models the conventional pharmaceutical system has not yet reached its maximum potential. Optimization of AI, and the application areas of AI in drug development stages require further expansion. For instance, while AI has been used to study the interaction of drug excipients with biomolecules, more in-depth research is needed to fully leverage AI's potential in pharmaceutical preparations. The huge fault rate of clinical trials is another area where AI has shown promise but still faces challenges. Approximately 90% of drug candidates are eliminated in clinical trials, with each failed trial costing between 0.8 billion to 1.4 billion USD. AI-based approaches for improving patient recruitment and monitoring have been developed, but further refinement is needed to ensure their effectiveness in real-world settings. Lastly, the ethical considerations related to the use of AI in drug discovery, such as data privacy and algorithmic bias, have not been thoroughly examined. Future research should address these ethical concerns to ensure the responsible use of AI in the pharmaceutical industry. In conclusion, while AI has made significant strides in transforming drug discovery and development, several limitations need to be addressed to fully realize its potential. Future research should focus on improving data quality, enhancing the interpretability of AI models, optimizing traditional pharmaceutical processes, refining AI-based clinical trial approaches, and addressing ethical considerations. By tackling these challenges, researchers can continue to drive innovation in the pharmaceutical industry and improve the efficiency and effectiveness of drug R&D.

10. CONCLUSION

AI became pivotal technology in transforming drug discovery and development, offering the potential to significantly improve human health outcomes. The integration of AI into various stages of drug R&D has demonstrated it's capability to enhance efficiency, reduce costs, and accelerate the overall process. AI's capability to analyze large datasets of high-dimensional biomedical information has been particularly beneficial, enabling the classification of novel drug targets and the design of novel therapeutic compounds more effectively than traditional methods. One of the most promising applications of AI in drug discovery is its role in virtual screening (VS), both structure-based (SBVS) and ligand-based (LBVS). AI algorithms have shown remarkable success in identifying potential lead compounds, thereby addressing the challenges associated with the activity cliff and facilitating de novo drug design. Additionally, AI has revolutionized the synthesis of new drugs by automating the exploration of reaction spaces and optimizing synthetic routes, which has significantly shortened the drug development cycle and reduced the rate of synthetic failures. AI's impact extends beyond the preclinical stages of drug development. In clinical trials, AI has been instrumental in improving patient recruitment, monitoring, and overall trial design. By leveraging AI to analyze patient biomarkers and predict treatment responses, researchers can improve the success rates of clinical trials and reduce associated costs. Furthermore, AI has importance in post-market surveillance and prognosis prediction, ensuring the safety and efficacy of new drugs through the analysis of EHRs and other data sources. Despite these advancements, several challenges remain. The need on high-quality data, the comfort of AI models, and the need for further optimization of traditional pharmaceutical processes are significant hurdles that need to be addressed. Moreover, to guarantee the appropriate application of AI in drug research, ethical issues pertaining to algorithmic bias and data privacy must be carefully investigated. In conclusion, artificial intelligence (AI) has unquestionably emerged as a crucial tool for drug research and development, with the ability to revolutionize the field at all levels of the process.

While there are still challenges to overcome, the continued advancement and integration of AI technologies hold the promise of unraveling the complexities of biological systems and revolutionizing the pharmaceutical industry. By addressing the current limitations, enhancing the quality, improving model interpretability, and ensuring ethical considerations are met, AI can pave the way for more capable and useful drug discovery processes, ultimately leading to improved human health outcomes.

REFERENCES

Alanazi, R. J. (2024). Role of Artificial Intelligence in Pharmacy Practice: A Systematic Review. *Archives of Pharmacy Practice (Mumbai)*, 15(2), 34–42. DOI: 10.51847/FMHsMFwgTP

Bai1, , FLi, , SLi, , H. (2024). AI enhances drug discovery and development. *National Science Review*, 11(303), 1–3.

Blanco-González, A., Cabezón, A., Seco-González, A., Daniel, C.-T. P. A.-R., Piñeiro, Á., & Garcia-Fandino, R. (2023). The Role of AI in Drug Discovery: Challenges, Opportunities, and Strategies. *Pharmaceuticals (Basel, Switzerland)*, 16(6), 891. DOI: 10.3390/ph16060891 PMID: 37375838

Dhudum, R., Ganeshpurkar, A., & Pawar, A. (2024). Revolutionizing Drug Discovery: A Comprehensive Review of AI Applications. *Drugs and Drug Candidates*, 3(1), 148–171. DOI: 10.3390/ddc3010009

Dominic, F., Ekpan, M., Ori, M. O., Samuel, H. S., & Egwuatu, O. P. (2024). The Synergy of AI and Drug Delivery: A Revolution in Healthcare. *International Journal of Advanced Biological and Biomedical Research*, 12(1), 44–66.

Keerthana, V., Mohaideen, S. S., Vigneshwaran, L. V., & Kumar, M. S. (2022). Role of artificial intelligence in drug development. *International Journal of Research in Pharmaceutical Sciences and Technology, 3*(1), 09-14.

Kolluri, S., Lin, J., Liu, R., Zhang, Y., & Zhang, W. (2022). Machine learning and artificial intelligence in pharmaceutical research and development: A review. *The AAPS Journal*, 24, 1–10.

Lalitkumar, K. (2023). Artificial Intelligence in Pharmaceutical Technology and Drug Delivery Design. *Pharmaceutics*, 15(7), 1916. PMID: 37514102

Liebman, I. (2022). The Role of Artificial Intelligence in Drug Discovery and Development. *Chemistry International*, 44(1), 16–19. DOI: 10.1515/ci-2022-0105

Lu, M., Yin, J., Zhu, Q., Lin, G., Mou, M., Liu, F., Pan, Z., You, N., Lian, X., Li, F., Zhang, H., Zheng, L., Zhang, W., Zhang, H., Shen, Z., Gu, Z., Li, H., & Zhu, F. (2023). Smart Process Manufacturing toward Carbon Neutrality-Review Artificial Intelligence in Pharmaceutical Sciences. *Engineering (Beijing)*, 27, 37–69. DOI: 10.1016/j.eng.2023.01.014

Malkawi, R. (2024). Revolutionizing drug delivery innovation: Leveraging AI-driven Chabot's for enhanced efficiency. *International Journal of Applied Pharmaceutics*, 16(2), 52–56. DOI: 10.22159/ijap.2024v16i2.50182

Nwosu, S. N. (2024). Revolutionizing Drug Discovery: The Impact of Artificial Intelligence on Efficiency and Accuracy. *Journal of Advances in Medical and Pharmaceutical Sciences*, 26(6), 56–63. DOI: 10.9734/jamps/2024/v26i6695

Pareek, V., Tuteja, L., Sharma, L., Kumar, S., & Verma, N. (2023). Revolutionizing Drug Design with Artificial Intelligence: A Comprehensive Review of Techniques, Applications, and Case Studies. *Journal of Pharmacy Research*, 22(3), 103–112.

Paul, D., Sanap, G., Shenoy, S., Kalyane, D., Kalia, K., & Tekade, R. K. (2021). Artificial intelligence in drug discovery and development. *Drug Discovery Today*, 26(1), 80–93. DOI: 10.1016/j.drudis.2020.10.010 PMID: 33099022

Pranay, O. V., Prajapati, A. P., Patel, B. S., Narkhede, S. B., & Luhar, S. (2016). A review on integrating artificial intelligence into drug development: Revolutionizing the pharmaceutical landscape. *I J R D.*, 8(16), 205–210.

Raza, M. A., Aziz, S., Noreen, M., Saeed, A., Anjum, I., Ahme, M., & Raza, S. M. (2022). AIin Pharmacy: An Overview of Innovations. *Pharmacy Practice and Practice-Based Research.*, 13(2), 1–13.

Rehman, A. U., Li, M., Wu, B., Ali, Y., Rasheed, S., Shaheen, S., Liu, X., Luo, R., & Zhang, J. (2024, May). Role of Artificial Intelligence in Revolutionizing Drug Discovery. *Fundamental Research (Beijing).* Advance online publication. DOI: 10.1016/j.fmre.2024.04.021

Sakshi, V. (2023). Artificial Intelligence and Machine Learning in the Pharmaceutical Industry. *IJPPR*, 28(2), 111–131.

Sampene, A. K., & Nyirenda, F. (2024). Evaluating the effect of artificial intelligence on pharmaceutical product and drug discovery in China. *Future Journal of Pharmaceutical Sciences*, 10(58), 1–11. DOI: 10.1186/s43094-024-00632-2

Sanjay, S. T., & Vishal, G. (2022). Artificial Intelligence in the Pharmaceutical Industry-An Overview of Innovations. *International Journal of Research Publication and Reviews.*, 3(7), 822–834.

Sayam, A. G., Pradhan, M., & Choudhury, A. K. (2023). Artificial Intelligence the Futuristic Technology in the Drug Discovery Process: A Review. *Journal of Young Pharmacists : JYP*, 15(3), 390–396. DOI: 10.5530/jyp.2023.15.54

Shah, B. (2023). Revolutionizing Drug Discovery on the role of Artificial Intelligence. *International Journal of Scientific Research*, 12(12), 1848–1852.

Sultana, A., Maseera, R., Rahamanulla, A., & Misiriya, A. (2023). Emerging of artificial intelligence and technology in pharmaceuticals [review]. *Future Journal of Pharmaceutical Sciences*, 9(65), 1–9. DOI: 10.1186/s43094-023-00517-w

Swapna, B. V., Shetty, S., Shetty, M., & Shetty, S. S. (2024). Shibani Shetty, Manjunath Shetty, Smitha Sammith Shetty, Smart science: How artificial intelligence is revolutionizing pharmaceutical medicine. *Acta Marisiensis Seria Medica*, 70(1), 8–15. DOI: 10.2478/amma-2024-0002

Tade, R. S., Jain, S. N., Satyavijay, J. T., Shah, P. N., Bari, T. D., Patil, T. M., & Shah, R. P. (2024). Artificial Intelligence in the Paradigm Shift of Pharmaceutical Sciences: A Review. *Nano Biomedicine and Engineering*, 16(1), 64–77. DOI: 10.26599/NBE.2023.9290043

Tiwari, P. C., Pal, R., Chaudhary, M. J., & Nath, R. (n.d.). Artificial intelligence revolutionizing drug development: Exploring opportunities and challenges, https://doi.org/DOI: 10.1002/ddr.22115

Verma, A., & Awasthi, A. (2024). Revolutionizing Drug Discovery: The Role of Artificial Intelligence and Machine Learning. *Current Pharmaceutical Design*, 30(1), 807–810. DOI: 10.2174/0113816128298691240222054120 PMID: 38409722

Vora, L. K., Gholap, A. D., Jetha, K., Raghu, R. S. T., & Hetvi, K. (1916). Artificial Intelligence in Pharmaceutical Technology and Drug Delivery Design. *Pharmaceutics*, 2023(15), 1–46.

Yadav, S., Singh, A., Singhal, R., & Yadav, J. P. (2024). Revolutionizing drug discovery: The impact of artificial intelligence on advancements in pharmacology and the pharmaceutical industry. *Intelligent Pharmacy.*, 2(3), 367–380. DOI: 10.1016/j.ipha.2024.02.009

Chapter 9
Trends of Pharmacogenomics in Personalized Medicine

John J. Georrge
https://orcid.org/0000-0002-9858-6112
University of North Bengal, India

Saurav Kumar Mishra
University of North Bengal, India

Sneha Roy
University of North Bengal, India

Tabsum Chhetri
University of North Bengal, India

Kusum Gurung
University of North Bengal, India

ABSTRACT

Personalized medicine is a medical approach in which treatment focuses on individual characteristics or genetic profiles. Integrating advanced technology and Pharmacogenomics emerged as a new way that can precisely analyze based on the genetic information of an individual's reaction to a drug. It also acts as a bridge between pharmacokinetics and pharmacodynamics in optimizing strategies for a better and safer understanding of treatment options at an individual level. Though pharmacogenomics has some established roles in personalized medications, it is not free from challenges and ethical considerations. The prospects of pharmacogenomics

DOI: 10.4018/979-8-3693-6597-7.ch009

are very optimistic; it is potent enough to revolutionize treatments and open new ways of direction in personalized medicine, and the genome-wide approaches and omics integration related to pharmacogenomics provide a comprehensive scope of drug reactions and many others. So, this study aims to provide a detailed understanding of pharmacogenomics and its role in personalized medicine, followed by the technologies, advancement, and future directions.

1.0 INTRODUCTION

Pharmacogenomics is a fast-developing area with huge potential for individualizing therapy. It discovers the effect an individual's genetic composition has on their response to drugs, so pharmacological intervention might be tailored because of every patient's genetic profile. Pharmacogenomics seeks improved therapeutic outcomes through studies of Genetic variability affecting drug effectiveness and safety. The new vision of medicine enabled by the Human Genome Project and the HapMap Project would be the use of molecular data for the better classification of diseases, development of targeted therapies, and more accurate determination of disease predisposition and drug response. Pharmacogenomics is perceived to be important in achieving this aim by handling genetically determined variations in people's response to medications (Pirmohamed, 2023; Maaike van der Lee et al., 2020). There is huge potential for the impact of pharmacogenomics on personalized treatment. The genomic high-throughput biotechnologies could be exploited to identify new pharmacological targets and therapies by incorporating genetic information into medication development. In clinical practice, knowledge of the individual's genetic makeup may help the clinician tailor treatment options to shift emphasis away from symptomatic illness care and empirical regimes for drug prescription. Pharmacogenomics plays a key role in personalized medicine in optimizing pharmacological efficacy (Sadee et al., 2023; Singh, 2020). Traditional research into medication uses a one-size-fits-all approach and can have wide-ranging responses among many people. Some patients may show a good response to the treatment, while others will show a bad or no response. Pharmacogenomics is the research into individual variances in response to medications in consideration of genomic information. Genetic information may guide appropriate changes in clinical care when a patient comes with a gene variation associated with some specified pharmacologic response, such as dosage adjustment or alternative treatment. Scientists can think about gene variants that influence a person's response to medication just as they think about gene variants related to diseases that are discovered by first identifying genetic loci associated with known medication responses and then searching for people whose responses are unknown. Two modern methods of doing this multigene analysis and whole-genome single nucleotide polymorphism profiles are only just now being applied in clinical trials for drug discovery (Arafah et al., 2021; Cai & Wang, 2020). It can predict medicine reactions by studying the genetic factors that influence drug metabolism and ensure that patients get the most effective drugs with the lowest possibilities of side effects. For example, variations in the family of enzymes responsible for metabolizing many medicines, known as the CYP450 enzyme family, may very significantly affect how a patient responds to treatment. Genotyping these enzymes will guide dosage adjustments and drug choice for optimal treatment outcomes (Mostafa et al., 2022; Taylor et al., 2020). More importantly, one of the critical aspects of pharmacogenomics is the minimization of Adverse Drug Reactions (ADR), which remain one of the leading causes of morbidity and mortality. In this respect, pharmacogenomics will study the genetic basis for these reactions that may have very wide variations between persons resulting from genetic differences. In that case, ADRs are the primary cause of morbidity and mortality, so prevention becomes one of the cardinal goals of personalized medicine. These variations may change the way drugs get metabolized, how they work, and also the safety profile, making the drug have some undesired effects on some individuals and making everything work perfectly for other people who might have been prescribed the same drug (Cacabelos et al., 2021; Wang et al., 2022). Pharmacogenomics is one component of personalized medicine that studies gene involvement in modifying the response to drugs of organisms. It allows health providers to make decisions about the best medicine and individualized therapy to suit the unique requirements of patients by considering an individual's genetic makeup (Hassan et al., 2021). It helps detect the optimal drug dosage concerning individual metabolic rates and genetic variances. This prevents overmedication or undermedication. Moreover, identifying patients with a high risk for adverse drug responses would help ensure the safety of patients and also facilitate the adoption of proactive management. The role of this discipline in medication research is also immense, as the discovery time of the medicine targeted towards the genes gets reduced by studying the genetic variables that influence drug response. Apart from optimizing treatments, pharmacogenomics also prevents diseases by identifying individuals with predispositions to certain conditions and facilitating genetic testing for early interventions and changes in lifestyle. Finally, pharmacogenomics is revolutionizing health care by taking a data-driven approach to medical decision-making and ensuring the proper and effective medication is given to the patient (Argueta et al., 2022; Balogun et al., 2024; Vakhariya et al., 2023) (**Figure 1**).

Figure 1. Roles of pharmacogenomics in personalized medicine.

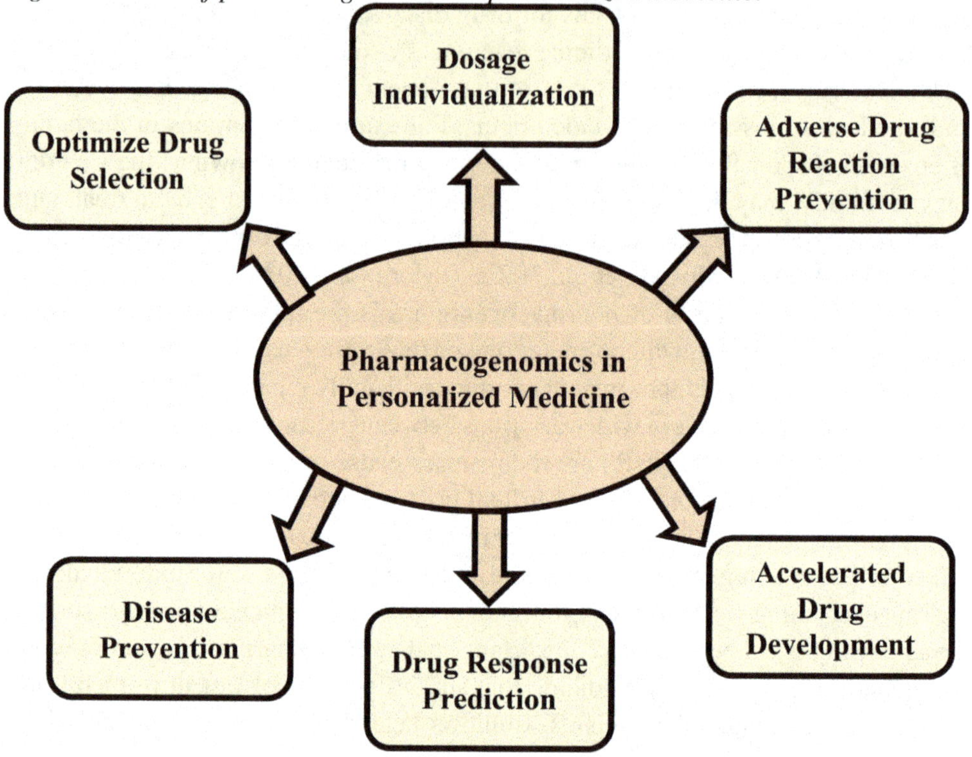

Although genetic testing to identify particular mutations within a tumor is used in oncology, allowing the selection of targeted therapies that would be more likely effective in personalized medicine, pharmacogenomics in oncology has been An essential player in the development of targeted treatments. On the other hand, in the last two decades, several studies have proved that the treatment of hypertension, diabetes, and depression may mandatorily require lifelong pharmaceutical management, with highly variable individual patient responses to these drugs. Genetic testing can assist in discovering the optimum drugs for patients as individuals, thus better managing sickness and quality of life (Miteva-Marcheva et al., 2020; Reizine & O'Donnell, 2022). When considering medication action in humans, researchers consider two issues: the amount of a drug necessary to get into its target in the body, and how much the target cells respond to the medicine, such as heart tissue or neurons. Pharmacokinetics (PK) and Pharmacodynamics (PD) are the scientific expressions for these two factors, both of which are major issues in pharmacogenomics. PK and PD are the two major subfields of pharmacogenomics within personalized medicine (**Figure 2**).

Figure 2. Role of pharmacokinetics and pharmacogenomics in pharmacogenomics.

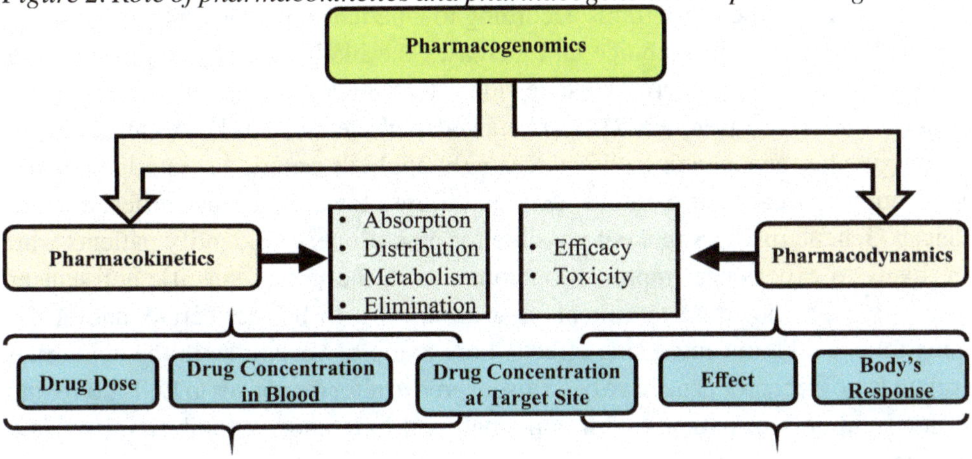

Understanding how an individual's genotype influences the ADME of medications and how the administered drug affects a person allows practitioners to adjust therapies to achieve the best potential therapeutic outcomes, reducing the associated side effects. More importantly, pharmacokinetics deals with drug absorption in the body, and genetic variants can significantly influence this. Genetics has been studied mostly concerning its influence on drug-metabolizing enzymes. These types of enzymes metabolize the vast majority of medications. Genetic variation in the genes encoding these enzymes results in various metabolizer statuses, including poor, intermediate, extensive, and ultra-rapid. Poor metabolizers may accumulate the drug and develop drug toxicity, while ultrarapid metabolizers may require higher doses to achieve a therapeutic effect. Genotyping of the enzymes before prescription allows a health expert to adjust the dosage for better efficacy and safety (Eke, 2022; Singh et al., 2022; Turner et al., 2022). In contrast, pharmacodynamics is concerned with the mechanism by which a medicine affects the body, involving its mode of action, therapeutic effects, and side effects. Genetic variation could change medication targets, such as receptors, enzymes, or ion channels, reducing the drug's effectiveness or enhancing the risk of resulting side effects. PK-PD interplay in pharmacogenomics impacts many therapeutic areas; some of the most prominent ones include oncology, cardiology, psychiatry, and infectious diseases (Michael et al., 2021; Yehya, 2022).

Pharmacogenomics drives personalized medicine by optimizing medicinal regimes by individual tailoring, maximizing therapeutic success, and reducing side effects. One such example it is known for is Type 2 Diabetes Mellitus (T2DM) treatment. A great deal of genetically determined heterogeneity occurs in the response to T2DM medicines among patients. For example, pharmacogenomic investigations of gene variants associated with glucose metabolism and insulin sensitivity have allowed personalized treatment programs according to a patient's genetic makeup. Such an approach optimizes drug choice and increases the likelihood of compliance with therapy by reducing potential adverse side effects and assuring better therapeutic responses (Guo & Priefer, 2021; Venkatachalapathy et al., 2021). Additionally, in oncology, pharmacogenomics is increasingly applied; genetic testing dictates the application of treatment therapies that are tailored toward patients suffering from cancer. Genetic markers make it possible for practitioners to identify patients who are likely to experience improved outcomes with the use of specific anti-cancer medications, avoid those treatments that are known to be ineffective, and at the same time, avoid unwanted side effects. For example, Variations in the epidermal growth factor receptor gene have been discovered to alter responses to EGFR tyrosine kinase inhibitors in patients with non-small cell lung cancer. Inhibition by a few such targeted drugs is more effective in a small subset of patients with particular EGFR mutations. Another modality where pharmacogenomics has made giant headway in managing cardiovascular disorders (Astras et al., 2020; Saulsberry & Olopade, 2021; Sethi et al., 2023; Zhu et al., 2020). Genetic differences in the genes that code for certain metabolizing enzymes, such as cytochrome P450 enzymes, may change the pharmacokinetics and pharmacodynamics of cardiovascular drugs. This allows clinicians to identify genetic polymorphisms and initiate appropriate dosage regimes to minimize the risk of adverse drug reactions. Pharmacogenomics has already enabled the development of some personalized medicines from natural products. Some natural products have already demonstrated encouraging anti-cancer potential, and pharmacogenomic approaches could identify the genetic factors influencing their activity and tolerance. This information may be used to better select appropriate natural products for individual patients, moving further into the area of personalized treatment in oncology (Singh, 2020; Zhou et al., 2024). Despite this success of pharmacogenomics in entering clinical practice, huge barriers still exist for large-scale prospective studies to validate the clinical utility of pharmacogenomic tests, the creation of cost-effective and user-friendly testing platforms, and the integration of pharmacogenomic data into electronic health records to facilitate seamless clinical decision support (Singh, 2020). However, modern technologies are used within a span to identify the genetic variants controlling the efficacy and safety of drugs in pharmacogenomics research. Pharmacogenomics mainly comes under next-generation sequencing, genotyping arrays, polymerase chain reaction, bioinformatics tools, and various molecular biology techniques. Next-generation sequencing has revolutionized pharmacogenomics by enabling fast sequencing of whole genomes or specific regions of interest at low costs. Techniques like Whole Genome Sequencing (WGS) and WES Whole Exome Sequencing (WES) generate high-accuracy information on genetic variants, while DNA microarrays analyze thousands of variants simultaneously (S. M. Caspar et al., 2020; Dietmar Enko et al., 2023). GWAS identifies common genetic variations associated with drug response. Polymerase chain reaction (PCR) and its derivatives amplify and quantify DNA sequences (Qahwaji et al., 2024; Maaike van der Lee et al., 2020). Other molecular biology techniques include RNA sequencing, epigenetic studies, and gene editing technology used in pharmacogenomics in personalized medicine (M. van der Lee et al., 2020; Zampatti et al., 2021). Apart from personalized medicine, the integration of computational approaches evolved and enhanced the therapeutic design in terms of the vaccine and potential inhibitors (Kachhadiya et al., 2024; Mishra & Georrge; Mishra et al., 2023; Vaghasia et al., 2024; Vakhariya Sakina et al., 2023; Vinjoda et al., 2024). Therefore, this chapter aims to provide information about pharmacogenomics and its application in personalized medicine including its technologies and clinical practices.

2.0 TECHNOLOGIES USED IN PHARMACOGENOMICS

Technologies used in pharmacogenomics are pivotal in advancing personalized medicine by tailoring drug therapies to individual genetic profiles. These cutting-edge tools and technologies enable researchers and clinicians to learn how genetic variants affect drug response, metabolism, and toxicity (**Figure 3**). From sequencing technologies to bioinformatics, integrating these technologies into pharmacogenomics has revolutionized the field, offering the potential for more effective and safer drug treatments (Maaike van der Lee et al., 2020). Furthermore, Single nucleotide variant (SNV) panels are The most widely used technique in pharmacogenomics practice, whether through commercially accessible microarray systems or custom-designed assays. These panels include a predetermined set of genetic variations associated with drug response, making them suitable for clinical use (Maaike van der Lee et al., 2020). SNV panels offer several advantages, including a short turnaround time, straightforward interpretation, and cost-effectiveness. The limited number of pre-selected variants in SNV panels simplifies the interpretation of results compared to sequencing approaches. Additionally, SNV panels are generally more cost-effective than sequencing methods for targeted pharmacogenomics testing. However, SNV panels also have limitations. They are restricted to the pre-selected variants and may miss rare variants or structural variations that could be important for drug response. SNV panels often do not provide information about the phase of variants, which can be crucial for understanding haplotypes and their functional consequences. The selection of variants in SNV panels may not always include all the variants potentially associated with drug response, as the evidence underlying the selected variants can vary (Medwid & Kim, 2022; Tafazoli et al., 2021; Wang et al., 2024). To address some of these limitations, researchers have developed customized SNV panels that focus on specific genes or variants of interest. These custom panels can be a cost-effective and accurate approach to clinical pharmacogenomics testing (Tafazoli et al., 2020). However, Next-generation sequencing (NGS) technologies have emerged as an innovator in pharmacogenomics, allowing for a more thorough understanding of how genetic variants influence individual pharmaceutical reactions. NGS refers to a set of high-throughput sequencing methods that allow for the simultaneous study of numerous genes, offering a complete perspective of genetic variations such as SNVs, insertions, deletions, and structural variants across the genome. This capability is particularly valuable in pharmacogenomics, where identifying genetic factors influencing drug metabolism, efficacy, and toxicity is crucial for personalized medicine (Mishra et al., 2024; Russell et al., 2021; M. van der Lee et al., 2020). One of the primary benefits of NGS in pharmacogenomics is its capacity to gather large amounts of data quickly and cost-effectively. Unlike traditional genotyping methods, which often focus on a limited number of known variants, NGS can uncover both common and rare genetic variants across a broad spectrum of pharmacogenes. Studies have shown that a significant proportion of pharmacogenetic variability is attributed to rare variants, which may not be detected by standard genotyping approaches (D. Enko et al., 2023; Kim et al., 2020). NGS can be applied in various forms, including whole-genome sequencing (WGS) and whole-exome sequencing (WES). WGS gives a complete image of an individual's genome, whereas WES focuses on the coding sections, or exons, that are more likely to harbor functionally relevant variants. Both approaches have demonstrated high concordance rates with traditional genotyping methods (Sylvan M Caspar et al., 2020; Caspar et al., 2021).This high level of accuracy makes NGS a reliable tool for pharmacogenomic testing, although challenges remain in translating research findings into routine clinical practice. Despite its potential, the clinical implementation of NGS in pharmacogenomics faces several hurdles. These include the complexity of data interpretation, the need for robust bioinformatics pipelines, and the integration of NGS results into clinical workflows. Due to these challenges, many clinical laboratories still rely on targeted genotyping approaches, which can limit the full utility of NGS technologies (Bienfait et al., 2022; Zhou et al., 2022). Recent advancements in NGS technology, such as developing targeted NGS panels specifically designed for pharmacogenomic applications, aim to streamline the testing process. These panels focus on clinically actionable variants associated with drug response, allowing for a more efficient and cost-effective approach to pharmacogenomics testing. By selecting genes with strong evidence linking them to drug metabolism and efficacy, these targeted panels enhance the relevance of genetic testing in clinical decision-making (Arbitrio et al., 2021; Zhou & Lauschke, 2024). Additionally, long-read sequencing technologies are increasingly recognized for their potential in pharmacogenomics, providing a deeper understanding of complex genetic variations that influence drug responses. Unlike traditional short-read sequencing, which is limited in its ability to resolve intricate genomic regions, long-read sequencing can capture longer DNA fragments, allowing for the detection of structural variants, haplotype phasing, and rare variants that are often missed by other methods. This capability is particularly important in pharmacogenomics, where genes involved in drug metabolism, such as CYP450 enzymes, exhibit a high degree of genetic variability, including complex haplotypes and structural variations that can impact therapeutic outcomes (Graansma et al., 2023; Lauschke et al., 2024; van der Lee et al., 2022). One of the primary benefits of long-read sequencing is the capacity to make extremely accurate variant calls across complex pharmacogenomic regions. Long-read sequencing has been shown to reach variant calling accuracies in single nucleotide variations (SNVs), as well as insertions and deletions. This high level of accuracy is crucial for pharmacogenomic applications, as even minor discrepancies in variant identification can lead to significant differences in drug response predictions. Furthermore, long-read sequencing facilitates the phasing of variants, allowing for a more comprehensive understanding of how multiple variants interact within a single gene or across linked genes. Research has shown that long-read sequencing can phase target regions and significantly outperform short-read technologies, which typically phase (van der Lee et al., 2022; Wohlers et al., 2022). The integration of long-read sequencing into clinical pharmacogenomics is still evolving, but its potential is evident. Moreover, resolving large repeat regions and structural variants enhances our understanding of gene-drug interactions, ultimately leading to more personalized and effective therapeutic strategies. As technology advances, long-read sequencing holds promise for improving the resolution of pharmacogenomic loci, thereby enhancing the ability to tailor drug therapy based on a person's distinct genetic profile. While challenges related to cost and turnaround time remain, ongoing innovations in long-read sequencing platforms are expected to facilitate broader adoption in clinical settings, paving the way for more precise and individualized approaches to medication management in the future (Zhou & Lauschke, 2023; Zhou et al., 2022). Furthermore, the most commonly used for amplifying and measuring a given DNA sequence in pharmacogenomics is the polymerase chain reaction and its derivatives, including real-time PCR and digital PCR. They are extremely sensitive and specific, thus constituting excellent tools for identifying known variations that link with medication responses (Attapong et al., 2024). RT-PCR in pharmacogenomics is useful in finding SNPs related to metabolism enzymes of drugs, such as the genes for CYP450, that play a very important role in the pharmacokinetics of many drugs. For example, by utilizing TaqMan SNP Genotyping Assays, laboratories can quickly genotype common markers for drug metabolism, such as CYP2D6 and CYP2C19. These two markers are key to identifying the optimum dosage of drugs and minimizing adverse drug reactions (de Andrés et al., 2021; Khedr et al., 2020). Moreover, RT-PCR has a very short turn-around time of only a few hours, allowing one to make timely clinical decisions of paramount importance in the optimization of regimes of therapy depending on the genetic profile of every patient. The broad use of RT-PCR in pharmacogenomic testing enhances patients' safety through the reduced possibility of adverse medication reactions and contributes to the cumulative effectiveness of techniques in personalized medicine (Li & Qiu, 2022; Mroz et al., 2021). In addition, microarray technology has played a key role in pharmacogenomics, which allows for the simultaneous analysis of hundreds of genetic variants related to pharmaceutical response. High-throughput technologies allow researchers and physicians to investigate complicated interactions between genetic variants and pharmaceutical efficacy or toxicity (M. van der Lee et al., 2020). Microarrays can interrogate hundreds of thousands of loci across the human genome, making them an invaluable tool for large-scale genetic studies to identify actionable pharmacogenomic markers (Cacabelos et al., 2021). Therefore, the power of microarray technology in delivering full perspective gene expression and genotypic data in understanding the therapeutic response at an individual level is one of its greatest strengths. Microarrays have helped generate extensive datasets that feed into important pharmacogenomic databases like PharmGKB and dbSNP, expanding knowledge in drug-gene interactions (Lenz et al., 2024). One of the important advantages of microarrays is their cost-effectiveness for large populations under test, as they allow for analyzing hundreds or thousands of genetic markers concurrently without extensive sequencing efforts. However, applying microarray technology in clinical pharmacogenomics requires detailed validation to ensure its accuracy and reliability. This level of performance is important in clinical practice, whereby accurate genotype calls may be applied to medication choices and dosages that truly improve patient outcomes and reduce adverse drug reactions (Ji & Shaaban, 2024; Scionti et al., 2022).

Figure 3. The technologies used in pharmacogenomics.

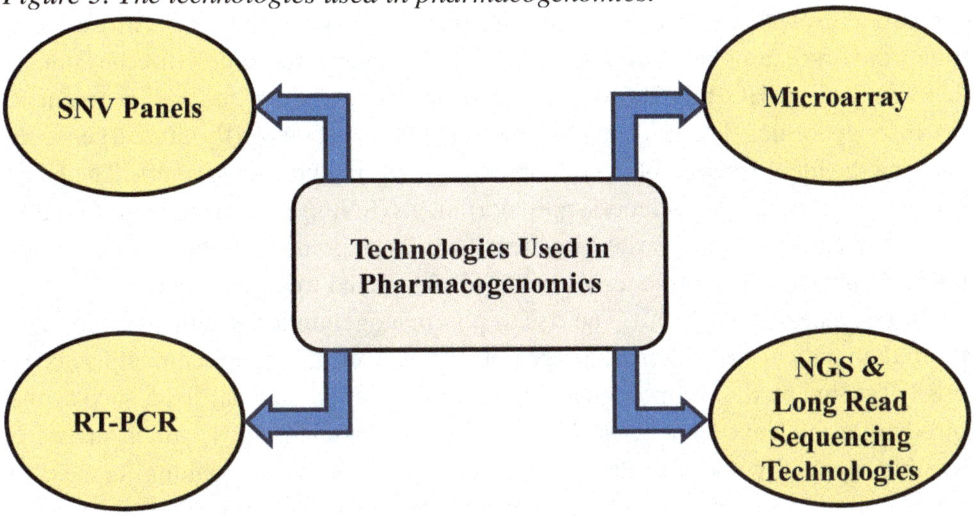

3.0 PHARMACOGENOMICS IN PERSONALIZED DRUG DEVELOPMENT

The process of development of drugs is crucial wherein various clinical and genetic factor comes into play. The drugs need a positive response in the human body to improve drug efficacy. However, it has been noticed that various drugs have different reactions in different individuals. This difference is caused mainly because of the individual's genetic makeup. In the current age where the healthcare industry is booming, personalized medicine has emerged as a cornerstone wherein the needs of an individual are curated based on their genetic makeup. However, it is essential to have information on a person's genetic makeup and pharmacogenomics helps us to do the same. Pharmacogenomics is the study of the response of drugs to different individual's genetic makeup. It also focuses on an individual's inherited or acquired genetic traits and how human genetic variants affect drug metabolism, absorption, and distribution (Mirsadeghi & Larijani, 2017). Important variables include genetic factors such as Single Nucleotide Polymorphisms (SNPs), gene expression and regulation, and copy number variations (CNVs). Gene expression influences protein levels, SNPs impact drug metabolism, and CNVs impact gene dosage and medication responsiveness. For the effective treatment of the drugs, understanding these factors is essential (Jarvis et al., 2019; Kim et al., 2015). The concept that a person's genetic composition may influence how they respond to medicine was first proposed in the 1950s. Arno Motulsky's discovery that some genetic characteristics influenced how patients responded to medications like the antimalarial medicine primaquine is among the first examples (Auwerx et al., 2022). Furthermore, completing the Human Genome Project (HGP) in 2003 has hastened the development and advancement of pharmacogenomics. The HGP produced a comprehensive map of the human genome, enabling the identification of genetic variants, affecting how drugs work. The HGP found millions of single nucleotide polymorphisms (SNPs) and other genetic variants to comprehend individual variability in medication response. This has made it easier to design pharmacogenomic testing and individualized treatment plans (Collins et al., 2003; Nebert et al., 2008). The rise of pharmacogenomics is significant, and its use in the medical field is incomprehensible; however, this advancement has been largely favored by the technological advancements in the medical field, such as the introduction of Next Generation Sequencing (NGS) technology, which makes it possible to quickly and affordably sequence a whole genome, making it easier to find uncommon genetic variants that have a big influence on how medicine works. Similarly, Genome-Wide Association Studies (GWAS) have facilitated the study of various genes and their responses to drugs and therapeutics. This study has highly favored the identification of genetic polymorphism related to drug response characteristics (Demkow, 2016). Further, extensive databases like PharmGKB, which compile genetic data related to medication response, have been built to support the study and integrate pharmacogenomic research into clinical practice (Whirl-Carrillo et al., 2012). The creation of clinical guidelines by international consortia, such as the Dutch Pharmacogenetics Working Group (DPWG) and the Clinical Pharmacogenetics Implementation Consortium (CPIC), is one of the major developments in pharmacogenomics. With the help of these guidelines, which offer evidence-based suggestions for drug selection and dosage based on particular genetic profiles, pharmacogenomic knowledge can be more easily applied in clinical settings (Cecchin & Stocco, 2020). Identifying patient populations most likely to respond to new therapies is made possible by pharmacogenomics, speeding up the drug discovery and development process. A remarkable development that has garnered attention is the identification of unique genetic markers that explain the variance observed in reactions to novel medications. Genetic variations previously disregarded, including rare polymorphisms, are now being studied since they may significantly impact medication metabolism. Understanding these new signals will enable clinicians to see individual differences in drug reactions more thoroughly (Cecchin & Stocco, 2020; Ross et al., 2005). Clinical practice has included pharmacogenomic biomarkers in a variety of treatment domains. Pharmacogenomic data is a feature of many FDA-approved medications. It covers various therapeutic areas, including cardiology, psychiatry, and oncology (Kim et al., 2021). Patients likely to respond to specific treatments or at risk for adverse drug reactions can be identified using these crucial biomarkers. Drug metabolization in a patient can be greatly impacted by changes in genes encoding drug-metabolizing enzymes, for example, which can affect a drug's efficacy and safety (Lauschke et al., 2018). Another field of ongoing study is the integration of pharmacogenomics with other omics technologies, such as proteomics and metabolomics. This all-encompassing strategy aims to develop a more thorough understanding of the patient's health, enabling the discovery of new pharmacological targets and more accurate therapeutic approaches (Cecchin & Stocco, 2020).

Although the pharmacogenomic has brought significant changes, its limitations include genetic diversity, ethical and legal issues, and its clinical implementation. Most research has been conducted on a particular population group, limiting the global implementation of personalized medicine. Therefore, expanding the research to a different population group is essential. Another key determining factor is the concern regarding privacy and data security. The use of genetic data raises ethical questions regarding possible discrimination. Privacy and ethical issues must be adequately addressed before utilizing an individual's genetic information. There are currently few clinical applications for pharmacogenomic biomarkers. Robust guidelines, integration into regular procedures, and clinician education are among the challenges. Multi-omics techniques, machine learning, and artificial intelligence will play a significant role in pharmacogenomics in the future. This area can potentially enable safer, more effective treatments through personalized medicine (**Figure 4**). However, overcoming obstacles and moral dilemmas is essential for integration to be successful.

Figure 4. An overview of pharmacogenomics in personalized medicine.

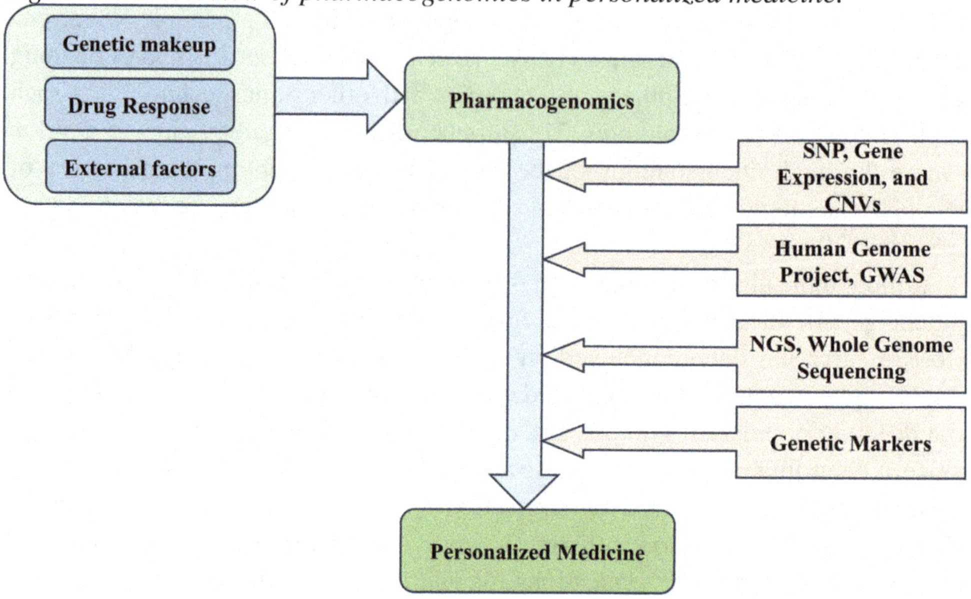

4.0 INTEGRATION OF PHARMACOGENOMICS IN CLINICAL PRACTICE

The study of genetic variety in drug response phenotypes, which can result in the finding of unfavorable drugs and can help tailor medications to individual needs, is known as pharmacogenomics. It can help lessen the drug-drug interaction and increase the treatment efficacy, wherein the drug can be tailored to a patient's needs (Sauver et al., 2016). The strategy of customizing and tailoring medicine to one's needs is becoming increasingly significant, particularly in fields like cardiology and primary care. Pharmacogenomics can enhance patient outcomes by minimizing adverse drug reactions (ADRs), increasing medication efficacy, and lowering medical expenses. Identifying those who are most likely to respond to a certain medication helps avoid unneeded treatment and adverse effects and reduces healthcare costs by identifying people who are at risk. Pharmacogenomics' potential to enhance pharmacological therapy results is becoming more widely acknowledged. Several procedures involve clinical implementation, such as genotyping patients, analyzing genetic data, and using the results to advise medication choice and dosage. Technologies such as polymerase chain reaction (PCR) and next-generation sequencing (NGS) are making genetic testing more accessible and affordable (Giannopoulou et al., 2019; Hippman & Nislow, 2019). With the use of these technologies, genetic variations influencing medication metabolism, effectiveness, and toxicity can be identified. To apply Pharmacogenomics in clinical practice, it is essential to evaluate the available data, examine medications, identify relevant drug-gene combinations, and identify variant alleles that are associated with them. Drug regulatory agencies like the FDA and EMA and expert groups like the Pharmacogenomics Knowledge Base and the Clinical Pharmacogenetics Implementation Consortium (CPIC) Can provide data for this purpose. Making a genotyping strategy decision and distinguishing reactive point-of-care testing from preventive testing is crucial. It is essential to determine the most effective means of interpreting and reporting data from Pharmacogenomics tests, possibly integrating them with Clinical Data Systems (CDS) and Electronic Health Records (EHRs). The implementation strategy should engage all relevant stakeholders, such as hospital administration, pharmacy and therapeutics committees, hospital laboratories, IT departments, and healthcare professionals (HCPs) (Kabbani et al., 2023). When pharmacogenomic data is integrated with the Clinical Decision Support System (CDSS) and Electronic Health Records (EHR), practitioners can get real-time assistance choosing the right drugs. Pharmacogenomics must be successfully integrated, and this requires the education of healthcare professionals (Weinshilboum & Wang, 2017). Pharmacists are crucial in integrating pharmacogenomics testing into clinical practice, enhancing patient outcomes, and minimizing adverse drug reactions. They recommend testing, evaluating the data, and developing personalized treatment programs. They also assist medical teams in developing decision support systems in electronic health records, conducting research, and advocating for appropriate compensation to assure access to these services (Balogun et al., 2024). Academic medical institutes and health systems are putting funds into networks to apply pharmacogenomics since genetic testing costs are decreasing and the technology is becoming more trustworthy. However, the integration of pharmacogenomics has proven difficult to integrate into clinical practice due to issues with the accessibility of testing centers, evidence-based guidelines, and electronic health records integration. As reporting and interpreting genetic information can be complicated, physicians involved in the clinical diagnosis must have an idea and better understanding of integrating genetic data into everyday care activities (Unertl et al., 2015). Another major obstacle is that Pharmacogenomics tests are not covered by insurance. Subsequently, to integrate clinical Pharmacogenomics, reviewing the available data, talking about pertinent drugs, and concentrating on significant genetic variants and drug-gene pairings that can be affected are crucial steps to be considered (Wayne T Nicholson et al., 2021). Pharmacogenomics (**Figure 5**) has the potential to revolutionize and enhance the field of medication therapy, personalized medicine, and patient response to medication; nevertheless, it is essential to address the obstacles before applying it widely.

Additionally, research and development are required to understand and create user-friendly tools for professional use. It is also essential to make the technique cost-efficient and handle ethical and legal concerns to develop and enhance personalized medicine. The data privacy, consent, and genetic discrimination guidelines must be designed and adhered to (W. T. Nicholson et al., 2021). High-quality evidence may be produced more quickly when government, business, and academic institutions collaborate. Raising public knowledge and comprehension of pharmacogenomics has the potential to improve acceptability and involvement in genetic testing.

Figure 5. Advantages of integrating pharmacogenomics in clinical practices.

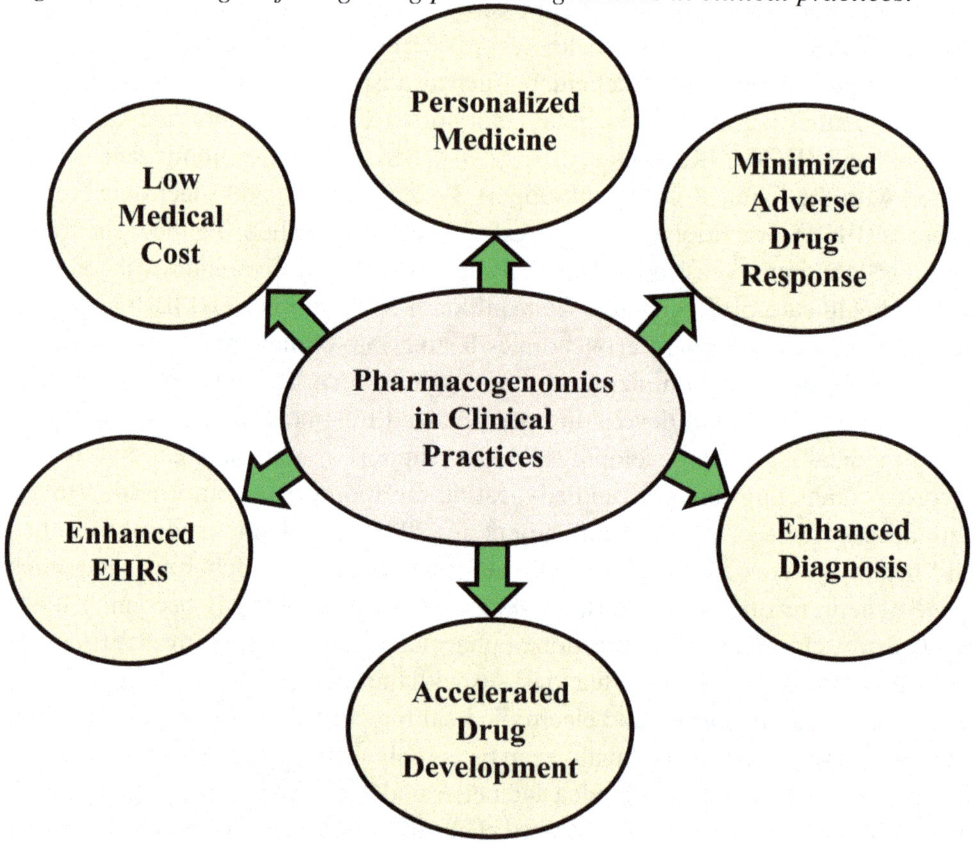

5.0 ADVANCEMENT AND CHALLENGES

Pharmacogenomics is a newly emerged method that utilizes or analyses the genetic information of the patients to tailor medication therapy accordingly, which enhances therapeutic efficacy and safety, providing more precise, effective, and personalized treatments. A significant advantage is enhanced drug efficacy; with pharmacogenomics study, the overall genetic makeup of the patient can be understood, which allows the healthcare professionals to provide the most effective or tailored form of the medications and treatments accordingly. This minimizes the risks and maximizes the chances of successful outcomes in the patients' health. This method guarantees appropriate and effective treatments and medications for the individual, eliminating the trial-and-error techniques. As a result, it dramatically reduces the time required to identify suitable treatments, and early treatment can be initiated. Pharmacogenomics helps understand the patient's genetic makeup and can predict the harmful reactions and the side effects of the prescribed drugs and treatments; therefore, the drugs and treatments can be tailored accordingly to minimize the risks, increase the benefits, and reduce the adverse reactions of the medications, also promotes patient trust and satisfaction with their treatment plan. Pharmacogenomics enhances personalized medicine by providing optimized drug dosage, i.e., adjusting the medication dosage according to the individual's genetic composition, reducing risk and side effects, and improving the drug's efficacy. Above all, the significant advantage of pharmacogenomics is to provide personalized treatment or personalized medicine, where the medication and treatments are designed and altered according to the individual's genetic composition; these patient-specific treatments offer much better outcomes in healthcare. The pharmacogenomic method eliminates trial and error methods as it minimizes the medication response, increases drug efficacy, and specific correct treatment can be started from the very early stage of the illness; as a result, it reduces and eliminates subsequent and follow-up sessions, which results in a significant cost saving of the treatments and hospital bills and also overall improves the patient's treatment quality with a higher chance of success. Chronic illness management is one of the essential advantages of Pharmacogenomics in personalized medicine; patients suffering from different chronic diseases can suffer a lot during their treatments with many risk factors, adverse side effects, and trials of treatments; personalized medicine can be very beneficial as it studies and understands the overall genetic makeup of the patients and plans treatments and medication accordingly which results in more precise and effective interventions and better illness control; additionally, it also reduces the side effects, slows down disease progression and a higher chance of long term well-being. Pharmacogenomics provides the overall information of the genetic composition. It can also predict the response to drugs and treatments, giving the patient an idea of the outcome. It empowers them by understanding the process and making an intelligent healthcare decision for their health. Furthermore, it gives patients confidence and control over their healthcare journey, increasing positive health outcomes. The complex treatment plans contain multiple high-dosage drugs and treatments which can adversely affect health; understanding the genetic profile of the individual personalized medicine can be very effective in reducing polypharmacy, which is a very significant benefit as it reduces the risk of drug-drug interaction and reducing unneeded drugs ultimately increasing patient outcome and quality of life. Pharmacogenomics is a developing field in healthcare that aims to provide personalized treatments to all demographic groups, thereby reducing health inequities. This method, which tailors interventions to individual genetic profiles, can deliver more precise and effective therapies, especially for underserved populations. This approach fosters equitable access to advanced medicines, closing the healthcare outcome gap and ultimately improving health outcomes on a larger scale.

Pharmacogenomics greatly impacts personalized medicine, adapting therapies to each individual's unique genetic composition, which decreases risks and improves health outcomes. However, in addition to the benefits, other obstacles must be addressed, one of which is interpreting the genetic variance of individuals accurately since different genetic makeup will show different drug responses, toxicity, and efficacy, so it is very crucial to be very accurate during identifying and studying the genetic makeup additionally, not only the genetic variation but the surrounded environmental factors can affect gene expression and drug metabolism and overall the response in the body and these can complicate the genetic data and predicted response. The studying of genetic profiles of individuals can be problematic since it raises the risk of invading the privacy and security of the person, the genetic data can reveal susceptible information about health and other personal matters. Everyone has a unique genetic composition and should be protected, but giving access to it can be risky. There are high chances that personalized treatments and medication built according to the individual's genetic profile can be very effective but the pharmacogenomics method for the genetic testing to identify the genetic variation is very costly. The technique consists of multiple steps like testing, sequencing, analysis, research expenses, interpretation, and other hospital implementation costs. This makes the method very costly for the low-income population and may also create economic discrimination since the low-income population will be restricted from getting health facilities. Giving access to one's genetic data can raise serious concerns about genetic discrimination in different fields; employers can use genetic information to make career decisions, which may not be fair and unethical. Also, insurers may utilize the information in decision-making or reject insurance based on the genetic makeup. Patient consent should be taken before any decision, and the impact and other concerns should be clarified. They should have control over their genetic data. They should have privacy and security to make sure that the data is not used negatively or exploited and that the data is only used the way the patient has explicitly agreed to; public awareness about genetic discrimination concerns and protections should be boosted to protect and avoid the utilization of the patient's genetic data in an unethical manner.

6.0 FUTURE PROSPECTS

With the development of pharmacogenomics, the future of personalized medicine seems bright with the introduction of numerous trends and directions that could revolutionize therapies and improve healthcare. To understand the genetic construction of the patients, more advanced methods and tools can be utilized to enhance the accuracy of the data to predict adverse effects and drug efficacy, resulting in a more precise treatment. Genetic testing allows healthcare professionals to analyze the composition to make treatment decisions accordingly, and there will be greater chances to positively affect patient's health outcomes. Pharmacogenomics has increasing impacts on drug research, their development, and drug response, and also focuses on treatment response direct target selection. Incorporating pharmacogenomics data into clinical trial design increases trial precision by stratifying individuals based on genetic markers. Studying genetic profiles of different individuals for treatment can be very helpful, but with it comes privacy and security concerns; therefore, ethical considerations, rights, and principles should be framed to safeguard the data to avoid misuse against the patient's will. Global applicability is critical for creating pharmacogenomics techniques that are applicable and effective across different genetic variations all over the place. Introduction to modern technologies, such as artificial intelligence and machine learning in the field to analyze complex genetic profiles and to predict drug responses can be helpful additionally, the blockchain technique can be utilized for maintaining the data's privacy and security. The future of pharmacogenomics in personalized medicine is bright; with the application of advanced technologies and development, the patient's health outcome can be positively enhanced worldwide.

7.0 CONCLUSION

The genetic information is precisely analyzed to interpret the available data for the proper medication based on personalized medication. To improve and enhance the treatments based on personalized medicine, pharmacogenomics, along with their attributes and application, emerged as a new hope to enhance accuracy. This study specifically discussed the current knowledge inter os the trend of pharmacogenomics, its application, and their involvement movements in medication. Enhancing these precise steps requires proper advancement to gain the actual accuracy for more precise results.

8.0 FUNDING

None

9.0 DATA AVAILABILITY STATEMENT

Not applicable

10.0 DECLARATION OF COMPETING INTEREST

The authors declare no known competing interests.

REFERENCES

Arafah, A., Ali, S., Majid, S., Rashid, S., Rasool, S., Wani, H. A., Rasool, I., & Rehman, M. U. (2021). Single Nucleotide Polymorphisms and Pharmacogenomics. *Genetic Polymorphism and cancer susceptibility*, 23-52.

Arbitrio, M., Scionti, F., Di Martino, M. T., Caracciolo, D., Pensabene, L., Tassone, P., & Tagliaferri, P. (2021). Pharmacogenomics Biomarker Discovery and Validation for Translation in Clinical Practice. *Clinical and Translational Science*, 14(1), 113–119. DOI: 10.1111/cts.12869 PMID: 33089968

Argueta, N., Notari, E., & Szigeti, K. (2022). Role of pharmacogenomics in individualizing treatment for Alzheimer's disease. *CNS Drugs*, 36(4), 365–376. DOI: 10.1007/s40263-022-00915-3 PMID: 35352296

Astras, G., Papagiannopoulos, C. I., Kyritsis, K. A., Markitani, C., & Vizirianakis, I. S. (2020). Pharmacogenomic Testing to Guide Personalized Cancer Medicine Decisions in Private Oncology Practice: A Case Study. *Frontiers in Oncology*, 10, 521. DOI: 10.3389/fonc.2020.00521 PMID: 32411592

Attapong, J., Kaset, C., Nakkam, N., Tassaneeyakul, W., Wichukchinda, N., & Chomean, S. (2024). Dual Approaches in Pharmacogenetics: Developing PCR-SSP and RT-PCR Methods for HLA-B* 13: 01 Screening to Prevent Dapsone and Co-Trimoxazole SCARs. *Heliyon*, 10(15), e34977. DOI: 10.1016/j.heliyon.2024.e34977 PMID: 39144953

Auwerx, C., Sadler, M. C., Reymond, A., & Kutalik, Z. (2022). From pharmacogenetics to pharmaco-omics: Milestones and future directions. *HGG Advances*, 3(2), 100100. DOI: 10.1016/j.xhgg.2022.100100 PMID: 35373152

Balogun, O. D., Ayo-Farai, O., Ogundairo, O., Maduka, C. P., Okongwu, C. C., Babarinde, A. O., & Sodamade, O. T. (2024). The role of pharmacists in personalised medicine: A review of integrating pharmacogenomics into clinical practice. *International Medical Science Research Journal*, 4(1), 19–36. DOI: 10.51594/imsrj.v4i1.697

Bienfait, K., Chhibber, A., Marshall, J. C., Armstrong, M., Cox, C., Shaw, P. M., & Paulding, C. (2022). Current challenges and opportunities for pharmacogenomics: Perspective of the Industry Pharmacogenomics Working Group (I-PWG). *Human Genetics*, 141(6), 1165–1173. DOI: 10.1007/s00439-021-02282-3 PMID: 34081195

Cacabelos, R., Naidoo, V., Corzo, L., Cacabelos, N., & Carril, J. C. (2021). Genophenotypic Factors and Pharmacogenomics in Adverse Drug Reactions. *International Journal of Molecular Sciences*, 22(24), 13302. Advance online publication. DOI: 10.3390/ijms222413302 PMID: 34948113

Cai, W., & Wang, Z. (2020). Introduction and Principles of Pharmacogenomics in Precision Medicine. *Pharmacogenomics in Precision Medicine: From a Perspective of Ethnic Differences*, 1-19.

Caspar, S. M., Schneider, T., Meienberg, J., & Matyas, G. (2020). Added value of clinical sequencing: WGS-based profiling of pharmacogenes. *International Journal of Molecular Sciences*, 21(7), 2308. DOI: 10.3390/ijms21072308 PMID: 32225115

Caspar, S. M., Schneider, T., Meienberg, J., & Matyas, G. (2020). Added Value of Clinical Sequencing: WGS-Based Profiling of Pharmacogenes. *International Journal of Molecular Sciences*, 21(7), 2308. Advance online publication. DOI: 10.3390/ijms21072308 PMID: 32225115

Caspar, S. M., Schneider, T., Stoll, P., Meienberg, J., & Matyas, G. (2021). Potential of whole-genome sequencing-based pharmacogenetic profiling. *Pharmacogenomics*, 22(3), 177–190. DOI: 10.2217/pgs-2020-0155 PMID: 33517770

Collins, F. S., Morgan, M., & Patrinos, A. (2003). The Human Genome Project: Lessons from large-scale biology. *Science*, 300(5617), 286–290. DOI: 10.1126/science.1084564 PMID: 12690187

de Andrés, F., Altamirano-Tinoco, C., Ramírez-Roa, R., Montes-Mondragón, C. F., Dorado, P., Peñas-Lledó, E. M., & LLerena, A. (2021). Relationships between CYP1A2, CYP2C9, CYP2C19, CYP2D6 and CYP3A4 metabolic phenotypes and genotypes in a Nicaraguan Mestizo population. *The Pharmacogenomics Journal*, 21(2), 140–151. DOI: 10.1038/s41397-020-00190-9 PMID: 33024249

Demkow, U. (2016). Next generation sequencing in pharmacogenomics. In *Clinical Applications for Next-Generation Sequencing* (pp. 217–240). Elsevier. DOI: 10.1016/B978-0-12-801739-5.00011-8

Eke, A. C. (2022). An update on the physiologic changes during pregnancy and their impact on drug pharmacokinetics and pharmacogenomics. *Journal of Basic and Clinical Physiology and Pharmacology*, 33(5), 581–598. DOI: 10.1515/jbcpp-2021-0312 PMID: 34881531

Enko, D., Michaelis, S., Schneider, C., Schaflinger, E., Baranyi, A., Schnedl, W. J., & Muller, D. J. (2023). The Use of Next-Generation Sequencing in Pharmacogenomics. *Clinical Laboratory*, 69(8). Advance online publication. DOI: 10.7754/Clin.Lab.2023.230103 PMID: 37560847

Enko, D., Michaelis, S., Schneider, C., Schaflinger, E., Baranyi, A., Schnedl, W. J., & Muller, D. J. (2023). The Use of Next-Generation Sequencing in Pharmacogenomics. *Clinical Laboratory*, 69(8). Advance online publication. DOI: 10.7754/Clin.Lab.2023.230103 PMID: 37560847

Giannopoulou, E., Katsila, T., Mitropoulou, C., Tsermpini, E.-E., & Patrinos, G. P. (2019). Integrating next-generation sequencing in the clinical pharmacogenomics workflow. *Frontiers in Pharmacology*, 10, 384. DOI: 10.3389/fphar.2019.00384 PMID: 31024324

Graansma, L. J., Zhai, Q., Busscher, L., Menafra, R., van den Berg, R. R., Kloet, S. L., & van der Lee, M. (2023). From gene to dose: Long-read sequencing and *-allele tools to refine phenotype predictions of CYP2C19. *Frontiers in Pharmacology*, 14, 1076574. DOI: 10.3389/fphar.2023.1076574 PMID: 36937863

Guo, Z., & Priefer, R. (2021). Current progress in pharmacogenomics of Type 2 diabetes: A systemic overview. *Diabetes & Metabolic Syndrome*, 15(5), 102239. DOI: 10.1016/j.dsx.2021.102239 PMID: 34371302

Hassan, R., Allali, I., Agamah, F. E., Elsheikh, S. S. M., Thomford, N. E., Dandara, C., & Chimusa, E. R. (2021). Drug response in association with pharmacogenomics and pharmacomicrobiomics: Towards a better personalized medicine. *Briefings in Bioinformatics*, 22(4), bbaa292. Advance online publication. DOI: 10.1093/bib/bbaa292 PMID: 33253350

Hippman, C., & Nislow, C. (2019). Pharmacogenomic Testing: Clinical Evidence and Implementation Challenges. *Journal of Personalized Medicine*, 9(3), 40. Advance online publication. DOI: 10.3390/jpm9030040 PMID: 31394823

Jarvis, J. P., Peter, A. P., & Shaman, J. A. (2019). Consequences of CYP2D6 copy-number variation for pharmacogenomics in psychiatry. *Frontiers in Psychiatry*, 10, 432. DOI: 10.3389/fpsyt.2019.00432 PMID: 31281270

Ji, Y., & Shaaban, S. (2024). Interrogating Pharmacogenetics Using Next-Generation Sequencing. *The Journal of Applied Laboratory Medicine*, 9(1), 50–60. DOI: 10.1093/jalm/jfad097 PMID: 38167765

Kabbani, D., Akika, R., Wahid, A., Daly, A. K., Cascorbi, I., & Zgheib, N. K. (2023). Pharmacogenomics in practice: A review and implementation guide. *Frontiers in Pharmacology*, 14, 1189976. DOI: 10.3389/fphar.2023.1189976 PMID: 37274118

Kachhadiya, D. K., Pooja, K., Mishra, S. K., & George, J. J. (2024). In Silico Based Identification Of Novel Inhibitors For Selected MDR Protein From Shigella Species: A Validation Through Molecular Docking Analysis. *Educational Administration: Theory And Practice, 30*(6 (S)), 309-316.

Khedr, S., Donia, T., Salim, E., & Hessien, M. (2020). CYP2D6 expression is cumulatively up regulated in multidrug treated hepatoma cells: A predective pharmacogenetics in vitro model. *The Egyptian Journal of Biochemistry and Molecular Biology*, 38(1), 39–55. DOI: 10.21608/ejb.2020.136526

Kim, I. W., Han, N., Kim, M. G., Kim, T., & Oh, J. M. (2015). Copy number variability analysis of pharmacogenes in patients with lymphoma, leukemia, hepatocellular, and lung carcinoma using The Cancer Genome Atlas data. *Pharmacogenetics and Genomics*, 25(1), 1–7. DOI: 10.1097/FPC.0000000000000097 PMID: 25379720

Kim, J. A., Ceccarelli, R., & Lu, C. Y. (2021). Pharmacogenomic Biomarkers in US FDA-Approved Drug Labels (2000-2020). *Journal of Personalized Medicine*, 11(3), 179. Advance online publication. DOI: 10.3390/jpm11030179 PMID: 33806453

Kim, W. Y., Kim, H. S., Oh, M., & Shin, J. G. (2020). Survey of physicians' views on the clinical implementation of pharmacogenomics-based personalized therapy. *Translational and Clinical Pharmacology*, 28(1), 34–42. DOI: 10.12793/tcp.2020.28.e6 PMID: 32274379

Lauschke, V. M., Milani, L., & Ingelman-Sundberg, M. (2018). Pharmacogenomic biomarkers for improved drug therapy—Recent progress and future developments. *The AAPS Journal*, 20(1), 1–16. DOI: 10.1208/s12248-017-0161-x PMID: 29181807

Lauschke, V. M., Zhou, Y., & Ingelman-Sundberg, M. (2024). Pharmacogenomics Beyond Single Common Genetic Variants: The Way Forward. *Annual Review of Pharmacology and Toxicology*, 64(1), 33–51. DOI: 10.1146/annurev-pharmtox-051921-091209 PMID: 37506333

Lenz, C., Narang, A., & Bousman, C. A. (2024). Pharmacogenomic allele coverage of genome-wide genotyping arrays: A comparative analysis. *Pharmacogenetics and Genomics*, 34(4), 130–134. DOI: 10.1097/FPC.0000000000000523 PMID: 38359167

Li, J., & Qiu, C. (2022). Recent advances in pharmacogenomics research of anti-asthmatic drugs: A narrative review. *Annals of Translational Medicine*, 10(6), 369. DOI: 10.21037/atm-22-291 PMID: 35434004

Medwid, S., & Kim, R. B. (2022). Implementation of pharmacogenomics: Where are we now? *British Journal of Clinical Pharmacology*. PMID: 36366858

Michael, M., Liauw, W., McLachlan, S.-A., Link, E., Matera, A., Thompson, M., Jefford, M., Hicks, R. J., Cullinane, C., Hatzimihalis, A., Campbell, I. G., Rowley, S., Beale, P. J., Karapetis, C. S., Price, T., & Burge, M. E. (2021). Pharmacogenomics and functional imaging to predict irinotecan pharmacokinetics and pharmacodynamics: The predict IR study. *Cancer Chemotherapy and Pharmacology*, 88(1), 39–52. DOI: 10.1007/s00280-021-04264-8 PMID: 33755789

Mirsadeghi, S., & Larijani, B. (2017). Personalized Medicine: Pharmacogenomics and Drug Development. *Acta Medica Iranica*, 55(3), 150–165. https://www.ncbi.nlm.nih.gov/pubmed/28282716 PMID: 28282716

Mishra, S. K., & Georrge, J. J. In Silico investigation of Epigallocatechin-3-Gallate and its derivatives against the VEGF-A cancerous protein through molecular docking and MD simulation studies.

Mishra, S. K., Jeba Praba, J., & Georrge, J. J. (2024). An emerging trends of bioinformatics and big data analytics in healthcare. *Digital Transformation in Healthcare 5.0: Volume 2: Metaverse, Nanorobots and Machine Learning*, 159.

Mishra, S. K., Priya, P., Rai, G. P., Haque, R., & Shanker, A. (2023). Coevolution based immunoinformatics approach considering variability of epitopes to combat different strains: A case study using spike protein of SARS-CoV-2. *Computers in Biology and Medicine*, 163, 107233. DOI: 10.1016/j.compbiomed.2023.107233 PMID: 37422941

Miteva-Marcheva, N. N., Ivanov, H. Y., Dimitrov, D. K., & Stoyanova, V. K. (2020). Application of pharmacogenetics in oncology. *Biomarker Research*, 8(1), 32. DOI: 10.1186/s40364-020-00213-4 PMID: 32821392

Mostafa, S., Polasek, T. M., Bousman, C. A., Müeller, D. J., Sheffield, L. J., Rembach, J., & Kirkpatrick, C. M. (2022). Pharmacogenomics in psychiatry–the challenge of cytochrome P450 enzyme phenoconversion and solutions to assist precision dosing. *Pharmacogenomics*, 23(15), 857–867. DOI: 10.2217/pgs-2022-0104 PMID: 36169629

Mroz, P., Michel, S., Allen, J. D., Meyer, T., McGonagle, E. J., Carpentier, R., Vecchia, A., Schlichte, A., Bishop, J. R., Dunnenberger, H. M., Yohe, S., Thyagarajan, B., Jacobson, P. A., & Johnson, S. G. (2021). Development and Implementation of In-House Pharmacogenomic Testing Program at a Major Academic Health System. *Frontiers in Genetics*, 12, 712602. DOI: 10.3389/fgene.2021.712602 PMID: 34745204

Nebert, D. W., Zhang, G., & Vesell, E. S. (2008). From human genetics and genomics to pharmacogenetics and pharmacogenomics: Past lessons, future directions. *Drug Metabolism Reviews*, 40(2), 187–224. DOI: 10.1080/03602530801952864 PMID: 18464043

Nicholson, W. T., Formea, C. M., Matey, E. T., Wright, J. A., Giri, J., & Moyer, A. M. (2021). Considerations When Applying Pharmacogenomics to Your Practice. *Mayo Clinic Proceedings*, 96(1), 218–230. DOI: 10.1016/j.mayocp.2020.03.011 PMID: 33308868

Nicholson, W. T., Formea, C. M., Matey, E. T., Wright, J. A., Giri, J., & Moyer, A. M. (2021). Considerations when applying pharmacogenomics to your practice. Mayo Clinic Proceedings, Pirmohamed, M. (2023). Pharmacogenomics: Current status and future perspectives. *Nature Reviews. Genetics*, 24(6), 350–362.

Qahwaji, R., Ashankyty, I., Sannan, N. S., Hazzazi, M. S., Basabrain, A. A., & Mobashir, M. (2024). Pharmacogenomics: A Genetic Approach to Drug Development and Therapy. *Pharmaceuticals (Basel, Switzerland)*, 17(7), 940. Advance online publication. DOI: 10.3390/ph17070940 PMID: 39065790

Reizine, N. M., & O'Donnell, P. H. (2022). Modern developments in germline pharmacogenomics for oncology prescribing. *CA: a Cancer Journal for Clinicians*, 72(4), 315–332. DOI: 10.3322/caac.21722 PMID: 35302652

Ross, J. S., Symmans, W. F., Pusztai, L., & Hortobagyi, G. N. (2005). Pharmacogenomics and clinical biomarkers in drug discovery and development. *Pathology Patterns Reviews*, 124(suppl_1), S29–S41. DOI: 10.1309/XYQAFANAPYNC6X59 PMID: 16468416

Russell, L. E., Zhou, Y., Almousa, A. A., Sodhi, J. K., Nwabufo, C. K., & Lauschke, V. M. (2021). Pharmacogenomics in the era of next generation sequencing - from byte to bedside. *Drug Metabolism Reviews*, 53(2), 253–278. DOI: 10.1080/03602532.2021.1909613 PMID: 33820459

Sadee, W., Wang, D., Hartmann, K., & Toland, A. E. (2023). Pharmacogenomics: Driving personalized medicine. *Pharmacological Reviews*, 75(4), 789–814. DOI: 10.1124/pharmrev.122.000810 PMID: 36927888

Saulsberry, L., & Olopade, O. I. (2021). Precision oncology: Directing genomics and pharmacogenomics toward reducing cancer inequities. *Cancer Cell*, 39(6), 730–733. DOI: 10.1016/j.ccell.2021.04.013 PMID: 34019805

Sauver, J. L. S., Bielinski, S. J., Olson, J. E., Bell, E. J., Mc Gree, M. E., Jacobson, D. J., McCormick, J. B., Caraballo, P. J., Takahashi, P. Y., & Roger, V. L. (2016). Integrating pharmacogenomics into clinical practice: promise vs reality. *The American journal of medicine, 129*(10), 1093-1099. e1091.

Scionti, F., Di Martino, M. T., Caracciolo, D., Pensabene, L., Tagliaferri, P., & Arbitrio, M. (2022). Tools in Pharmacogenomics Biomarker Identification for Cancer Patients. *Methods in Molecular Biology (Clifton, N.J.)*, 2401, 1–12. DOI: 10.1007/978-1-0716-1839-4_1 PMID: 34902118

Sethi, Y., Patel, N., Kaka, N., Kaiwan, O., Kar, J., Moinuddin, A., Goel, A., Chopra, H., & Cavalu, S. (2023). Precision Medicine and the future of Cardiovascular Diseases: A Clinically Oriented Comprehensive Review. *Journal of Clinical Medicine*, 12(5), 1799. Advance online publication. DOI: 10.3390/jcm12051799 PMID: 36902588

Singh, D. B. (2020). The impact of pharmacogenomics in personalized medicine. *Current Applications of Pharmaceutical Biotechnology*, 369-394.

Singh, S., Singh, D. B., Gautam, B., Singh, A., & Yadav, N. (2022). Pharmacokinetics and pharmacodynamics analysis of drug candidates. In *Bioinformatics* (pp. 305–316). Elsevier. DOI: 10.1016/B978-0-323-89775-4.00001-8

Tafazoli, A., Guchelaar, H.-J., Miltyk, W., Kretowski, A. J., & Swen, J. J. (2021). Applying next-generation sequencing platforms for pharmacogenomic testing in clinical practice. *Frontiers in Pharmacology*, 12, 693453. DOI: 10.3389/fphar.2021.693453 PMID: 34512329

Tafazoli, A., Wawrusiewicz-Kurylonek, N., Posmyk, R., & Miltyk, W. (2020). Pharmacogenomics, how to deal with different types of variants in next generation sequencing data in the personalized medicine area. *Journal of Clinical Medicine*, 10(1), 34. DOI: 10.3390/jcm10010034 PMID: 33374421

Taylor, C., Crosby, I., Yip, V., Maguire, P., Pirmohamed, M., & Turner, R. M. (2020). A Review of the Important Role of CYP2D6 in Pharmacogenomics. *Genes*, 11(11), 1295. DOI: 10.3390/genes11111295 PMID: 33143137

Turner, R. M., Magavern, E. F., & Pirmohamed, M. (2022). Pharmacogenomics: Relevance and opportunities for clinical pharmacology. *British Journal of Clinical Pharmacology*, 88(9), 3943–3946. DOI: 10.1111/bcp.15329 PMID: 35347744

Unertl, K. M., Jaffa, H., Field, J. R., Price, L., & Peterson, J. F. (2015). Clinician perspectives on using pharmacogenomics in clinical practice. *Personalized Medicine*, 12(4), 339–347. DOI: 10.2217/PME.15.10 PMID: 26635887

Vaghasia, V. V., Sharma, K., Mishra, S. K., & Georrge, J. J. (2024). In silico identification of natural product inhibitor for multidrug resistance proteins from selected gram-positive bacteria. In *Nanotechnology and In Silico Tools* (pp. 309–317). Elsevier. DOI: 10.1016/B978-0-443-15457-7.00015-0

Vakhariya, S., Kumar Mishra, S., & Georrge, J. J. (2023). Identification of Novel Curcumin Analogues to Inhibit Mitogen-Activated Protein Kinase 1 Using in Silico Combinatorial Library Design and Molecular Docking Approach. *Available atSSRN* 4649193.

Vakhariya Sakina, S., Mishra, S. K., Sharma, K., & Georrge, J. J. (2023). Designing of a novel curcumin analogue to inhibit mitogen-activated protein kinase: A cheminformatics approach. *Journal of Phytonanotechnology and Pharmaceutical Sciences*, 3(1), 37–47.

van der Lee, M., Kriek, M., Guchelaar, H.-J., & Swen, J. J. (2020). Technologies for pharmacogenomics: A review. *Genes*, 11(12), 1456. DOI: 10.3390/genes11121456 PMID: 33291630

van der Lee, M., Kriek, M., Guchelaar, H. J., & Swen, J. J. (2020). Technologies for Pharmacogenomics: A Review. *Genes*, 11(12), 1456. Advance online publication. DOI: 10.3390/genes11121456 PMID: 33291630

van der Lee, M., Rowell, W. J., Menafra, R., Guchelaar, H. J., Swen, J. J., & Anvar, S. Y. (2022). Application of long-read sequencing to elucidate complex pharmacogenomic regions: A proof of principle. *The Pharmacogenomics Journal*, 22(1), 75–81. DOI: 10.1038/s41397-021-00259-z PMID: 34741133

Venkatachalapathy, P., Padhilahouse, S., Sellappan, M., Subramanian, T., Kurian, S. J., Miraj, S. S., Rao, M., Raut, A. A., Kanwar, R. K., Singh, J., Khadanga, S., Mondithoka, S., & Munisamy, M. (2021). Pharmacogenomics and Personalized Medicine in Type 2 Diabetes Mellitus: Potential Implications for Clinical Practice. *Pharmacogenomics and Personalized Medicine*, 14, 1441–1455. DOI: 10.2147/PGPM.S329787 PMID: 34803393

Vinjoda, P., Mishra, S. K., Sharma, K., & Georrge, J. J. (2024). In silico identification of novel drug target and its natural product inhibitors for herpes simplex virus. In *Nanotechnology and In Silico Tools* (pp. 377–383). Elsevier. DOI: 10.1016/B978-0-443-15457-7.00007-1

Wang, C. W., Preclaro, I. A. C., Lin, W. H., & Chung, W. H. (2022). An Updated Review of Genetic Associations With Severe Adverse Drug Reactions: Translation and Implementation of Pharmacogenomic Testing in Clinical Practice. *Frontiers in Pharmacology*, 13, 886377. DOI: 10.3389/fphar.2022.886377 PMID: 35548363

Wang, D., Bolleddula, J., Coenen-Stass, A., Grombacher, T., Dong, J. Q., Scheuenpflug, J., Locatelli, G., & Feng, Z. (2024). Implementation of whole-exome sequencing for pharmacogenomics profiling and exploring its potential clinical utilities. *Pharmacogenomics*, 25(4), 197–206. DOI: 10.2217/pgs-2023-0243 PMID: 38511470

Weinshilboum, R. M., & Wang, L. (2017). Pharmacogenomics: Precision medicine and drug response. Mayo Clinic Proceedings, Whirl-Carrillo, M., McDonagh, E. M., Hebert, J. M., Gong, L., Sangkuhl, K., Thorn, C. F., Altman, R. B., & Klein, T. E. (2012). Pharmacogenomics knowledge for personalized medicine. *Clinical Pharmacology and Therapeutics*, 92(4), 414–417. DOI: 10.1038/clpt.2012.96

Wohlers, I., Garg, S., & Hehir-Kwa, J. Y. (2022). Editorial: Long-read sequencing-Pitfalls, benefits and success stories. *Frontiers in Genetics*, 13, 1114542. DOI: 10.3389/fgene.2022.1114542 PMID: 36685894

Yehya, A. (2022). Interplay between Pharmacokinetics and Pharmacogenomics. In *Dosage Forms-Innovation and Future Perspectives*. IntechOpen.

Zampatti, S., Fabrizio, C., Ragazzo, M., Campoli, G., Caputo, V., Strafella, C., Pellicano, C., Cascella, R., Spalletta, G., Petrosini, L., Caltagirone, C., Termine, A., & Giardina, E. (2021). Precision Medicine into Clinical Practice: A Web-Based Tool Enables Real-Time Pharmacogenetic Assessment of Tailored Treatments in Psychiatric Disorders. *Journal of Personalized Medicine*, 11(9), 851. Advance online publication. DOI: 10.3390/jpm11090851 PMID: 34575628

Zhou, Y., & Lauschke, V. M. (2023). Challenges Related to the Use of Next-Generation Sequencing for the Optimization of Drug Therapy. *Handbook of Experimental Pharmacology*, 280, 237–260. DOI: 10.1007/164_2022_596 PMID: 35792943

Zhou, Y., & Lauschke, V. M. (2024). Next-generation sequencing in pharmacogenomics - fit for clinical decision support? *Expert Review of Clinical Pharmacology*, 17(3), 213–223. DOI: 10.1080/17512433.2024.2307418 PMID: 38247431

Zhou, Y., Peng, S., Wang, H., Cai, X., & Wang, Q. (2024). Review of Personalized Medicine and Pharmacogenomics of Anti-Cancer Compounds and Natural Products. *Genes*, 15(4), 468. Advance online publication. DOI: 10.3390/genes15040468 PMID: 38674402

Zhou, Y., Tremmel, R., Schaeffeler, E., Schwab, M., & Lauschke, V. M. (2022). Challenges and opportunities associated with rare-variant pharmacogenomics. *Trends in Pharmacological Sciences*, 43(10), 852–865. DOI: 10.1016/j.tips.2022.07.002 PMID: 36008164

Zhu, Y., Swanson, K. M., Rojas, R. L., Wang, Z., St Sauver, J. L., Visscher, S. L., Prokop, L. J., Bielinski, S. J., Wang, L., Weinshilboum, R., & Borah, B. J. (2020). Systematic review of the evidence on the cost-effectiveness of pharmacogenomics-guided treatment for cardiovascular diseases. *Genetics in Medicine*, 22(3), 475–486. DOI: 10.1038/s41436-019-0667-y PMID: 31591509

Conclusion

A. PHARMACOGENOMICS: CHANGING THE THERAPEUTIC LANDSCAPE

In this context, pharmacogenomics has appeared as a strong enabling tool in drug discovery and development. It offers new approaches toward the goal of personalized medicine. The interaction of an individual's genetic profile with his or her response to a certain medication- that is, the basis of pharmacogenomics- has opened new dimensions toward developing more effective and safer medicines for well-defined patient groups. This has phenomenally accelerated new drug discovery using a genomics-driven approach, improved clinical trial outcomes, and enhanced our understanding of disease mechanisms at a molecular level. Integrating genomic information within drug discovery and development has brought about a paradigm shift in how we think about creating new therapeutics. In more conventional, traditional drug discovery, the model has often been 'one-size-fits-all,' where treatments worked for some but proved inefficient or even harmful for others. Pharmacogenomics has shifted this landscape and enabled researchers to trace genetic variations in people's response to drugs and the metabolism of drugs and show side effects (Konda et al., 2023).

B. KEY ADVANCES AND APPLICATIONS

Through this book, we have covered several ways through which genomics and pharmacogenomics have impacted the process of new drug discovery:

1. Target Identification and Validation

The genomic approach allowed the identification of new targets and better validation of the existing targets, thus resulting in targeted therapies, especially in areas such as oncology and rare genetic disorders (Heerlie et al., 2024). It was possible to understand the disease's genetic basis and pinpoint specific molecular pathways and proteins that play a key role in disease progression. For example, in oncology, genomic profiling of tumors has revealed many potential therapeutic targets, such as determining driver mutations in genes like EGFR, ALK, and BRAF. These findings have translated into targeted therapies, including erlotinib, crizotinib, and vemurafenib, significantly improving outcomes in patients harboring these genetic profiles. Genomic approaches in rare diseases have been instrumental in elucidating the genetic etiology of many disorders, thus improving diagnosis and opening new avenues for treatment. For instance, the discovery of the genetic basis of cystic fibrosis led to the development of ivacaftor, a drug targeting specific CFTR gene mutations (Senft et al., 2017).

2. Biomarker Discovery

These facilitate pharmacogenomic studies to identify genetic biomarkers predictive of drug response, efficacy, and possible adverse effects. Many such biomarkers have also begun to be employed in stratifying patients within clinical trials and guiding treatment decisions within clinical practice. Biomarker discovery has had a greater impact on oncology. For example, the identification of HER2 overexpression in breast cancer allowed the use of the targeted therapy trastuzumab, which substantially improved outcomes in those with HER2-positive disease (Margaretha et al., 2023; Piccart-Gebhart et al., 2005; Ramanto et al., 2019). Similarly, the discovery of the presence of the KRAS mutation to predict the poor response of colon cancers to inhibitors of EGFR allows better therapy choices associated with an improvement in outcome. Outside of oncology, a number of pharmacogenomic biomarkers have been applied across many therapeutic areas. HLA-B*57:01 testing is routine in predicting hypersensitivity to abacavir in HIV treatment, and CYP2C19 genotyping guides antiplatelet therapy in cardiovascular disease (Ventola, 2013).

3. Optimizing Drug Development

Today, incorporating pharmacogenomic information early in the drug development process can allow pharmaceutical firms to construct more efficient clinical trials with lower attrition rates, bringing new drugs to market faster and at lower cost (Blanco-Gonzalez et al., 2022; Blaschke et al., 2020). Pharmacogenomics is making a big difference in optimizing drug development by improving patient stratification in clinical trials. Identification of genetic markers of drug response or toxicity allows for the design of trials that focus on the patients most likely to benefit from a treatment, hence increasing the probability of success and reducing the time and cost of development. Besides, pharmacogenomic approaches may save drugs that proved futile in larger populations by defining genetic subgroups of those showing positive responses (Parikesit, 2018; Parikesit et al., 2020). Further, this "precision medicine" approach rescues potentially useful treatments but, most importantly, ensures that drugs reach those in dire need (Surendiran et al., 2008).

4. Personalized Medicine

When incorporated into clinical decisions, pharmacogenomic information promises true personalization of treatment regimens through optimization of drug selection and dosing based on the patient's genetic profile. Personalized medicine is more than selecting the right drug for the right patient; it also involves optimizing dosing regimens. Many genetic variations exist in these drug-metabolizing enzymes, such as CYP2D6 and CYP2C19, which may affect individual drug processing. Thus, being aware of patient genetics will allow the clinician to tailor dosing for optimal efficacy with minimum side effects. Recently, psychiatric pharmacogenomic tests inform choices related to antidepressants and antipsychotics, therefore allowing a reduction in trial-and-error approaches to finding appropriate drugs. Similarly, in the pain management field, a patient's genetic background could provide relevant information about his response to opioid medications, thus avoiding or at least diminishing dependency or other adverse effects (Oates & Lopez, 2018).

C. CHALLENGES AND FUTURE DIRECTIONS

Whereas immense progress has been made in the area of genomics-driven drug discovery, a few challenges persist, including:

1. Data Integration and Interpretation

Since genomic data sets are much larger and increasingly complex, we have to find new, sophisticated bioinformatics tools and AI algorithms that will allow us to analyze and interpret these data effectively. It has a two-sided effect: On one hand, large datasets allow greater statistical power to find rare genetic variants and complex gene-drug interactions; on the other hand, such data processing and analysis require significant computational resources and bioinformatic expertise.

The problems being encountered increasingly utilize machine learning and artificial intelligence algorithms in an attempt to improve. Such tools can uncover patterns and relationships in genomic data not directly observable by traditional analytic methods. However, one major challenge still remains: how to make AI-driven insights reliable and reproducible. Integrating genomic data with other biological and clinical data levels, such as proteomics, metabolomics, and electronic health records, will also be necessary for more in-depth insights into drug response and disease mechanisms. Such a multi-omics approach poses further challenges to data standardization, integration, and interpretation (Ahmed et al., 2024).

2. Ethical and Regulatory Considerations

Genetic information applied to drug development and clinical practice raises serious ethical questions and regulatory challenges. Genetic data is the most sensitive when it comes to privacy. There is a very thin line between keeping the confidentiality and security of patient's genetic information while enabling research and drug development based on such information. The line ranges from regulatory frameworks that include but are not limited to, Europe's GDPR and the United States HIPAA, which are only partly updated in time with the rapid pace at which genomic technologies change (Alberto et al., 2023; Hongladarom, 2006). Genetic discrimination represents yet another ethical issue. Though partial protection through legislation such as the Genetic Information Nondiscrimination Act exists in the United States, many questions persist about how genetic information could be used in employment and insurance (NIH, 2024). The regulatory landscape for pharmacogenomic tests and genomics-based therapies is also in flux. Regulatory agencies, like the FDA, are still developing frameworks for the clinical validity and utility assessment of pharmacogenomic tests and biomarkers. However, due to the field's complexity and rapid evolution, this usually remains a challenge to regulators (Braverman et al., 2018).

3. Diversity and Inclusion

Pharmacogenomic studies must include a diverse population to avoid health disparities and ensure that all patient groups will benefit from genomics-driven therapies. Historically, most genomic studies have been skewed towards populations of European ancestry. Inadequate diversity in genomic databases may result in an incomplete and inaccurate understanding of genetic variations in other populations, possibly leading to ineffective or harmful treatment recommendations for underrepresented groups. Consequently, some efforts are being made to change this situation; for instance, the “All of Us Research Program” of NIH aims to establish a genetically and health-diverse database (https://allofus.nih.gov/) . However, a long way remains to be traveled if genomic research and translation into drug development are to reflect global genetic diversity (Jordan et al., 2023).

4. Clinical Implementation

Despite its promise, the wide application of pharmacogenomics in clinical practice has been limited. Some of the barriers to be surmounted before this branch of science reaches its full potential include physician education, reimbursement issues, and a need for standardized testing protocols. One of the major challenges to clinical implementation is the need for more standardization regarding how pharmacogenomic testing and results are reported. Laboratories might be using different testing platforms or interpretation algorithms, so potential inconsistencies in results or recommendations should be highlighted. Another critical component is education for healthcare providers. Many clinicians might need more confidence in interpreting and acting on pharmacogenomic test results. Moreover, comprehensive educational programs and decision-support tools will also play an essential role in broad clinical adoption. In most health systems, reimbursement presents a major barrier to the performance of pharmacogenomic testing. Only a few tests enjoy insurance coverage, while many still need to. Access to such potentially useful tools has, therefore, been limited. Robust health economic studies will be important in demonstrating the cost-effectiveness of pharmacogenomic testing and may lead to an improvement in reimbursement policies (Pearce et al., 2022).

D. THE ROAD AHEAD

Looking ahead, several intriguing trends are foreseen that may shape the continued evolution of genomics-driven drug discovery:

1. Multi-Omics Integration

Integrating pharmacogenomics with other 'omics' technologies, such as but not limited to transcriptomics, proteomics, and metabolomics, will provide an integrated view of drug response and disease mechanisms. Although very useful, genomics provides much-needed knowledge on the individual's genetic composition; it is just a part of explaining gene expression and interactions on an environmental level. Integrated with other omics results, genomic data provides a complete picture of biological processes and drug responses. For example, integrating genomic data with transcriptomic data will provide insights into how genetic variations affect gene expression and, therefore, are likely to point out new drug targets or biomarkers. Similarly, integration with metabolomic data will indicate how genetic variation affects drug metabolism and efficacy. This is particularly promising in complex diseases such as cancer, where the interplay between genetics, gene expression, and metabolism can greatly impact treatment response. As technologies continue to evolve and improve for both the generation and analysis of multi-omics data, holistic and precise strategies toward drug discovery and personalized medicine could be more realized (Resurreccion & Fong, 2022).

2. Advanced Gene Editing Technologies

That is where rapid improvement in the gene editing tool, including but not limited to CRISPR-Cas9, will provide more precise manipulation of genetic targets and facilitate drug discovery and gene therapy development. CRISPR-Cas9 and other gene-editing methodologies have revolutionized genetic research by affording unparalleled precision in generating disease models and validating pharmaceutical targets. As these technologies evolve, applications along the drug discovery pipeline will similarly evolve. Gene editing technologies are opening up new therapeutic possibilities. Gene therapies, which seek to treat diseases by changing genes in a patient, are increasingly feasible. For example, CRISPR-based therapies are already in clinical trials for conditions such as sickle cell disease and some types of blindness. Coupled with pharmacogenomics, gene editing technologies have the potential to develop "genetic medicines" tailored to a certain genetic profile, thus making treatments very personalized to the actual genetic causes of the disease (Aljabali et al., 2024; Ansori et al., 2023).

3. Real-World Data and Digital Health

The increasing availability of real-world genomic and clinical data and digital health technologies will further improve our ability to carry out large-scale pharmacogenomic studies and implement personalized medicine approaches. This growth in consumer genetic testing, electronic health records, and wearable health devices further gives rise to large volumes of real-world data. Provided these are analyzed and integrated adequately, this will yield good insight into drug efficacy and safety among different populations. Digital health technologies, such as those delivered via mobile phone applications or wearable devices, will continue to enable real-time data collection(Parikesit, Agustriawan, et al., 2022). This should provide a complete picture of how individuals respond to treatments over time and shed light on pharmacogenomic signals that may have otherwise not been visible in traditional clinical trials. Integrating real-world data and digital health technologies could lead to more adaptive and personalized treatments. For example, a patient's genetic profile might be combined with real-time health data to constantly optimize the dosage of a certain drug or predict potential side effects (Abul-Husn & Kenny, 2019).

4. New Drug Modalities

Genomics-driven approaches will continue enabling the development of new therapeutic modalities, including RNA-based therapies, cell therapies, and gene therapies. The recent few years have witnessed heightened interest in RNA-based therapies comprised of siRNA and mRNA therapeutics following the COVID-19 successes with mRNA vaccines (Parikesit, Ansori, et al., 2022; Parikesit et al., 2018; Parikesit & Nurdiansyah, 2018, 2020). Such therapies have a huge potential to specifically target the disease-causing gene in highly specific manners, while design and manufacturing can also be accomplished much faster (Barik, 2010; Carthew & Sontheimer, 2009; Chang & Hu, 2006; Grimm, 2009; Khaitov et al., 2021; Koli et al., 2014). Another frontier in genomics-driven drug discovery is cell therapies, including CAR-T cell therapy for cancer (Anurogo et al., 2024; Ikrar & Anurogo, 2021). Such a treatment involves genetically modifying a patient's cells to fight disease. With our continually improving knowledge in genomics and cellular biology, even more sophisticated and effective forms of cell therapies are likely to emerge. Gene therapies, which involve inserting functional genes into patients' cells to treat disease, are also moving at a rapid pace. With the enhancement of gene editing technologies such as CRISPR, the applications of gene therapy are expanding to a wider range of diseases (Agustriawan et al., 2019; Parikesit, 2021; Sparmann & Vogel, 2023).

5. Artificial Intelligence and Machine Learning

Applying AI and machine learning algorithms to genomic and pharmacogenomic data will accelerate target discovery, lead optimization, and patient stratification strategies (Parikesit et al., 2020). AI and machine learning have applications in analyzing complex genomic data, identifying potential drug targets, and predicting drug response. As the technologies continue to evolve, so does the complexity of the applications we have begun seeing in drug discovery and development. Such predictive modeling will employ AI algorithms to predict structure-function changes in proteins arising from genetic variations, a possibility that might reveal new drug targets. Machine learning models can be used to predict, from genetic profiles, the interactions of drug combination treatments as a means to optimize combination therapies. In clinical trials, AI may leverage genetic and other biomarker data to identify the optimal patient subgroups benefiting most from particular treatments.

This offers the opportunity for more lean and successful clinical trials, which may hasten drug development (Taherdoost & Ghofrani, 2024).

E. CONCLUDING REMARK

Pharmacogenomics-driven drug discovery ushers in a new era of precision medicine through genomics, affording unparalleled opportunities to develop more effective and safer therapies. As we understand more about the intricacies of the human genome and its impact on drug response, we can only expect much more personalized health care. The future of drug discovery lies in harnessing the power of genomics and translating this knowledge into tangible benefits for patients worldwide. Meeting these challenges and seizing such opportunities in this rapidly evolving field allows one to look forward with the promise that treatments will be truly individualized, maximizing efficacy with minimum adverse effects. However, the integration of genomics into drug discovery is not only changing how we think about developing new treatments but also changing our understanding of health and disease. Going forward, we will be moving away from symptomatic treatments to the genetic root causes of a disease. This paradigm can potentially revolutionize healthcare toward a future in which diseases are prevented or cured, not managed. Thus, the effects of genomics-driven drug discovery will not stop at medicine. The revolution will touch public health policy, healthcare economics, and even our social conception of health and disease. As genetic testing becomes more common and treatments more personalized, new ethical, legal, and social issues are coming into focus.

The journey of genomics-driven drug discovery is anything but over. With recent scientific breakthroughs and newer technological innovations on the horizon, opportunities for a sea change in changing the face of patient care and improving health outcomes across the globe have never been better. Indeed, it is an exciting time for all: the researcher, the clinician, and the patient are together, hand in glove, working to unlock all potential that genomics holds vis-à-vis drug discovery and development. We dream that one day, each patient will have treatments specific to his or her genetic variation, and even the rarest diseases will finally have targeted therapies. We envision how genetic insights will translate into prevention. However, while there are still challenges, the progress witnessed in the field of genomics-driven drug discovery is a reason to be optimistic about how medical care might look in the future.

As was finally shown, this overview of genomics-driven drug discovery through pharmacogenomics underlined that a new era had just begun in medicine. Indeed, all the knowledge and tools developed so far by genomic studies allow better ways of treatment: more precise, effective, and personalized. While furthering the research on the human genome's role in health and disease, we come closer to making reality the promise of precision medicine. Genomic, personalized, and bright describes a bright future for drug discovery. The scientific and technological convergence on delivering therapies unique to those individuals treated is in that bright future. Let's push ahead to meet those challenges and opportunities with eyes ever-focused on improving health, minimizing suffering, and enhancing well-being through the power of genomics-driven drug discovery.

Arli Aditya Parikesit

Department of Bioinformatics, School of Life Sciences, Indonesia International Institute for Life Sciences, Jakarta, Indonesia

REFERENCES

Abul-Husn, N. S., & Kenny, E. E. (2019). Personalized medicine and the power of electronic health records. *Cell*, 177(1), 58–69. DOI: 10.1016/j.cell.2019.02.039 PMID: 30901549

Agustriawan, D., Parikesit, A. A., & Nurdiansyah, R. (2019). Big Data Computation of Drug Design: From the Natural Products to the Transcriptomic-Based Molecular Development. In *Application of Omics, AI and Blockchain in Bioinformatics Research* (pp. 59–86). World Scientific. DOI: 10.1142/9789811203589_0005

Ahmed, A., Shuaib, M., Banga, A., & Ahmad, R. (2024). Innovation in Bioinformatics: Recent tools, Database and Involvement of Artificial Intelligence. *Indonesian Journal of Medical Chemistry and Bioinformatics*, 2(2), 4. DOI: 10.7454/ijmcb.v2i2.1026

Alberto, I. R. I., Alberto, N. R. I., Altinel, Y., Blacker, S., Binotti, W. W., Celi, L. A., Chua, T., Fiske, A., Griffin, M., Karaca, G., Mokolo, N., Naawu, D. K. N., Patscheider, J., Petushkov, A., Quion, J., Senteio, C., Taisbak, S., Tırnova, İ., Tokashiki, H., & Yap, K. (2023). Who does the fairness in health AI community represent? *MedRxiv*, 2023.03.20.23287471. DOI: 10.1101/2023.03.20.23287471

Aljabali, A. A. A., El-Tanani, M., & Tambuwala, M. M. (2024). Principles of CRISPR-Cas9 technology: Advancements in genome editing and emerging trends in drug delivery. *Journal of Drug Delivery Science and Technology*, 92, 105338. DOI: 10.1016/j.jddst.2024.105338

Ansori, A. N. M., Antonius, Y., Susilo, R. J. K., Hayaza, S., Kharisma, V. D., Parikesit, A. A., Zainul, R., Jakhmola, V., Saklani, T., Rebezov, M., & Ullah, Md. E., Maksimiuk, N., Derkho, M., & Burkov, P. (2023). Application of CRISPR-Cas9 genome editing technology in various fields: A review. In *Narra J* (Vol. 3, Issue 2, pp. e184–e184). DOI: 10.52225/narra.v3i2.184

Anurogo, D., Liu, C. L., Chang, Y. C., Chang, Y. H., & Qiu, J. T. (2024). Discovery of differentially expressed proteins for CAR-T therapy of ovarian cancers with a bioinformatics analysis. *Aging (Albany NY)*, 16(14), 11409–11433. DOI: 10.18632/aging.206024 PMID: 39033780

Barik, S. (2010). SiRNA for influenza therapy. In *Viruses* (Vol. 2, Issue 7, pp. 1448–1457). Multidisciplinary Digital Publishing Institute (MDPI). DOI: 10.3390/v2071448

Blanco-Gonzalez, A., Cabezon, A., Seco-Gonzalez, A., Conde-Torres, D., Antelo-Riveiro, P., Pineiro, A., & Garcia-Fandino, R. (2022). *The Role of AI in Drug Discovery: Challenges, Opportunities, and Strategies*. https://doi.org//arxiv.2212.08104DOI: 10.48550

Blaschke, T., Arús-Pous, J., Chen, H., Margreitter, C., Tyrchan, C., Engkvist, O., Papadopoulos, K., & Patronov, A. (2020). REINVENT 2.0: An AI Tool for de Novo Drug Design. *Journal of Chemical Information and Modeling*, 60(12), 5918–5922. DOI: 10.1021/acs.jcim.0c00915 PMID: 33118816

Braverman, G., Shapiro, Z. E., & Bernstein, J. A. (2018). Ethical Issues in Contemporary Clinical Genetics. *Mayo Clinic Proceedings. Innovations, Quality & Outcomes*, 2(2), 81–90. DOI: 10.1016/j.mayocpiqo.2018.03.005 PMID: 30225437

Carthew, R. W., & Sontheimer, E. J. (2009). Origins and Mechanisms of miRNAs and siRNAs. In *Cell* (Vol. 136, Issue 4, pp. 642–655). NIH Public Access. DOI: 10.1016/j.cell.2009.01.035

Chang, Z., & Hu, J. (2006). RNAi therapeutics: Can siRNAs conquer SARS? In *Gene Therapy* (Vol. 13, Issue 11, pp. 871–872). Nature Publishing Group. DOI: 10.1038/sj.gt.3302682

Grimm, D. (2009). Asymmetry in siRNA design. *Gene Therapy*, 16(7), 827–829. DOI: 10.1038/gt.2009.45 PMID: 19404320

Heerlie, D. M., Margaretha, F., Fugaha, D. R., & Parikesit, A. A. (2024). The Effects and Treatments for Usher Syndrome: A Review. *Indonesian Journal of Life Sciences*, 6(1), 68–82. DOI: 10.54250/IJLS.V6I01.184

Hongladarom, S. (2006). Ethics of bioinformatics: A convergence between bioethics and computer ethics. *Asian Biotechnology and Development Review*, 9(1), 37–44.

Ikrar, T., & Anurogo, D. (2021). The Art of Oncoimmunovaccinomics. *World Journal of Vaccines*, 11(04), 50–66. DOI: 10.4236/wjv.2021.114007

Jordan, I. K., Sharma, S., & Mariño-Ramírez, L. (2023). Population Pharmacogenomics for Health Equity. *Genes*, 14(10), 1840. DOI: 10.3390/genes14101840 PMID: 37895188

Khaitov, M., Nikonova, A., Shilovskiy, I., Kozhikhova, K., Kofiadi, I., Vishnyakova, L., Nikolskii, A., Gattinger, P., Kovchina, V., Barvinskaia, E., Yumashev, K., Smirnov, V., Maerle, A., Kozlov, I., Shatilov, A., Timofeeva, A., Andreev, S., Koloskova, O., Kuznetsova, N., & Skvortsova, V. (2021). Silencing of SARS-CoV-2 with modified siRNA-peptide dendrimer formulation. *Allergy*, 00(9), 1–15. DOI: 10.1111/all.14850 PMID: 33837568

Koli, U., Krishnan, R. A., Pofali, P., Jain, R., & Dandekar, P. (2014). SiRNA-based therapies for pulmonary diseases. *Journal of Biomedical Nanotechnology*, 10(9), 1953–1997. https://www.ncbi.nlm.nih.gov/pubmed/25992446. DOI: 10.1166/jbn.2014.1928 PMID: 25992446

Konda, P., Garinet, S., Van Allen, E. M., & Viswanathan, S. R. (2023). Genome-guided discovery of cancer therapeutic targets. *Cell Reports*, 42(8), 112978. Advance online publication. DOI: 10.1016/j.celrep.2023.112978 PMID: 37572322

Margaretha, F., Subagja, M., Putri, A. D., Salam, C. Q., Valenthenardo, L., Adhiwijaya, P. K., Kaitlyn, G. Z., & Parikesit, A. (2023). Comprehensive analysis of protein-protein interactions (PPIs) with structure prediction program for breast cancer determination. *Berkala Penelitian Hayati*, 29(3), 87–94. DOI: 10.23869/bphjbr.29.3.20232

NIH. (2024). *Genetic Information Nondiscrimination Act (GINA)*. https://www.genome.gov/genetics-glossary/Genetic-Information-Nondiscrimination-Act-GINA

Oates, J., & Lopez, D. (2018). Pharmacogenetics: An Important Part of Drug Development with A Focus on Its Application. *International Journal of Biomedical Investigation*, 1(2), 111. DOI: 10.31531/2581-4745.1000111 PMID: 32467882

Parikesit, A. A. (2018). Introductory Chapter: The Contribution of Bioinformatics as Blueprint Lead for Drug Design. In Glavic, I. (Ed.), *Molecular Insight of Drug Design* (p. 7). InTech., DOI: 10.5772/intechopen.79664

Parikesit, A. A. (2021). Introductory Chapter: The Emerging Corner of the Omics Studies for Rational Drug Design. In Parikesit, A. A. (Ed.), *Drug Design: Novel Advances in the Omics Field and Applications* (p. 4). IntechOpen., DOI: 10.5772/intechopen.95544

Parikesit, A. A., Agustriawan, D., Valeska, M. D., Whisnu, A., Firmansyah, M., Veneqe, I., & Crystalia, A. A. (2022). Workshop And Training Of The Drug-Drug Interaction Database Using Indonesian Drug Brand Names. *Jurdimas (Jurnal Pengabdian Kepada Masyarakat). Royal*, 5(3), 333–340. DOI: 10.33330/JURDIMAS.V5I3.1574

Parikesit, A. A., Ansori, A. N. M., & Kharisma, V. D. (2022). A Computational Design of siRNA in SARS-CoV-2 Spike Glycoprotein Gene and Its Binding Capability toward mRNA. *Indonesian Journal of Chemistry*, 22(5), 1163. DOI: 10.22146/ijc.68415

Parikesit, A. A., & Nurdiansyah, R. (2018). Generating Two-Dimensional Repertoire of siRNA Linc-ROR and siRNA mRNA ARF6 from the lincRNA-RoR/miR-145/ARF6 expression Pathway that involved in the progression of Triple Negative Breast Cancer. *IOP Conference Series. Materials Science and Engineering*, 299(1), 012059. DOI: 10.1088/1757-899X/299/1/012059

Parikesit, A. A., & Nurdiansyah, R. (2020). The Predicted Structure for the Anti-Sense siRNA of the RNA Polymerase Enzyme (RdRp) gene of the SARS-CoV-2. *BERITA BIOLOGI*, 19(1), 97–108. DOI: 10.14203/beritabiologi.v19i1.3849

Parikesit, A. A., Ratnasari, N. R. P., & Anurogo, D. (2020). Application of Artificial Intelligence-Based Computation in the Health Sciences to Ward off the COVID-19 Pandemic. [IJHHS]. *International Journal of Human and Health Sciences*, 5(2), 177. DOI: 10.31344/ijhhs.v5i2.256

Parikesit, A. A., Utomo, D. H., & Karimah, N. (2018). Determination of secondary and tertiary structures of cervical cancer lncRNA diagnostic and siRNA therapeutic biomarkers. *Indonesian Journal of Biotechnology*, 23(1), 1. DOI: 10.22146/ijbiotech.28508

Pearce, A., Terrill, B., Alffenaar, J. W., Patanwala, A. E., Kummerfeld, S., Day, R., Young, M. A., & Stocker, S. L. (2022). Pharmacogenomic testing: Perception of clinical utility, enablers and barriers to adoption in Australian hospitals. *Internal Medicine Journal*, 52(7), 1135–1143. DOI: 10.1111/imj.15719 PMID: 35191159

Piccart-Gebhart, M. J., Procter, M., Leyland-Jones, B., Goldhirsch, A., Untch, M., Smith, I., Gianni, L., Baselga, J., Bell, R., Jackisch, C., Cameron, D., Dowsett, M., Barrios, C. H., Steger, G., Huang, C.-S., Andersson, M., Inbar, M., Lichinitser, M., Láng, I., & Gelber, R. D. (2005). Trastuzumab after Adjuvant Chemotherapy in HER2-Positive Breast Cancer. *The New England Journal of Medicine*, 353(16), 1659–1672. DOI: 10.1056/NEJMoa052306 PMID: 16236737

Ramanto, K. N., Agustriawan, D., Parikesit, A. A., Nurdiansyah, R., & Nasution, M. Z. A. (2019). Correlation and Meta-Analysis of HER2 in Each Stage of Breast Cancer. *IOP Conference Series. Materials Science and Engineering*, 546(6), 062024. DOI: 10.1088/1757-899X/546/6/062024

Resurreccion, E. P., & Fong, K. W. (2022). The Integration of Metabolomics with Other Omics: Insights into Understanding Prostate Cancer. *Metabolites*, 12(6), 488. DOI: 10.3390/metabo12060488 PMID: 35736421

Senft, D., Leiserson, M. D. M., Ruppin, E., & Ronai, Z. A. (2017). Precision Oncology: The Road Ahead. *Trends in Molecular Medicine*, 23(10), 874–898. DOI: 10.1016/j.molmed.2017.08.003 PMID: 28887051

Sparmann, A., & Vogel, J. (2023). RNA-based medicine: From molecular mechanisms to therapy. *The EMBO Journal*, 42(21), e114760. DOI: 10.15252/embj.2023114760 PMID: 37728251

Surendiran, A., Pradhan, S., & Adithan, C. (2008). Role of pharmacogenomics in drug discovery and development. *Indian Journal of Pharmacology*, 40(4), 137. DOI: 10.4103/0253-7613.43158 PMID: 20040945

Taherdoost, H., & Ghofrani, A. (2024). AI's role in revolutionizing personalized medicine by reshaping pharmacogenomics and drug therapy. *Intelligent Pharmacy*, 2(5), 643–650. DOI: 10.1016/j.ipha.2024.08.005

Ventola, C. L. (2013). The Role of Pharmacogenomic Biomarkers in Predicting and Improving Drug Response: Part 2: Challenges Impeding Clinical Implementation. *P&T*, 38(10), 624. https://pmc.ncbi.nlm.nih.gov/articles/PMC3875251/ PMID: 24391381

Compilation of References

Abad-Santos, F., Aliño, S. F., Borobia, A. M., García-Martín, E., Gassó, P., Maroñas, O., & Agúndez, J. A. (2024). Developments in pharmacogenetics, pharmacogenomics, and personalized medicine. *Pharmacological Research*, 200, 107061. DOI: 10.1016/j.phrs.2024.107061 PMID: 38199278

Abraham, G., Rutten-Jacobs, L., & Inouye, M. (2021). Risk prediction using polygenic risk scores for prevention of stroke and other cardiovascular diseases. *Stroke*, 52(9), 2983–2991. DOI: 10.1161/STROKEAHA.120.032619 PMID: 34399584

Abushanab, R. (2024). Role of Artificial Intelligence and Big Data in Sustainable Entrepreneurship. *Journal of Artificial Intelligence General science (JAIGS) ISSN: 3006-4023, 5*(1), 275-294.

Adesola, R. O., Opuni, E., Idris, I., Okesanya, O. J., Igwe, O., Abdulazeez, M. D., & Lucero-Prisno, D. E.III. (2024). Navigating Nigeria's Health Landscape: Population Growth and Its Health Implications. *Environmental Health Insights*, 18, 11786302241250211. DOI: 10.1177/11786302241250211 PMID: 38698838

Aditya Parikesit, A. (2023). Introductory Chapter: Current Landscape of Animal and Human Virus Treatments. In Parikesit, A. A. (Ed.), *Antiviral Strategies in the Treatment of Human and Animal Viral Infections as*. IntechOpen., DOI: 10.5772/intechopen.111733

Aebersold, R., & Mann, M. (2016). Mass spectrometry-based proteomics. *Nature*, 537(7620), 347–355. DOI: 10.1038/nature19949 PMID: 27629641

Afdhal, D., Ananta, K. W., & Hartono, W. S. (2020). Adverse Drug Reactions Prediction Using Multi-label Linear Discriminant Analysis and Multi-label Learning. *2020 International Conference on Advanced Computer Science and Information Systems (ICACSIS)*, 69–76. DOI: 10.1109/ICACSIS51025.2020.9263166

Afgan, E., Baker, D., Batut, B., Van Den Beek, M., Bouvier, D., Čech, M., Chilton, J., Clements, D., Coraor, N., Grüning, B. A., Guerler, A., Hillman-Jackson, J., Hiltemann, S., Jalili, V., Rasche, H., Soranzo, N., Goecks, J., Taylor, J., Nekrutenko, A., & Blankenberg, D. (2018). The Galaxy platform for accessible, reproducible and collaborative biomedical analyses: 2018 update. *Nucleic Acids Research*, 46(W1), W537–W544. DOI: 10.1093/nar/gky379 PMID: 29790989

Afolabi, B. L., Mazhindu, T., Zedias, C., Borok, M., Ndlovu, N., & Masimirembwa, C. (2023). Pharmacogenetics and Adverse Events in the Use of Fluoropyrimidine in a Cohort of Cancer Patients on Standard of Care Treatment in Zimbabwe. *Journal of Personalized Medicine*, 13(4), 588. DOI: 10.3390/jpm13040588 PMID: 37108974

Ahlberg, C., & Shneiderman, B. (1994, April). Visual information seeking: Tight coupling of dynamic query filters with starfield displays. In *Proceedings of the SIGCHI conference on Human factors in computing systems* (pp. 313-317). DOI: 10.1145/191666.191775

Ahluwalia, J. S., Okuyemi, K., Nollen, N., Choi, W. S., Kaur, H., Pulvers, K., & Mayo, M. S. (2006). The effects of nicotine gum and counseling among African American light smokers: A 2× 2 factorial design. *Addiction (Abingdon, England)*, 101(6), 883–891. DOI: 10.1111/j.1360-0443.2006.01461.x PMID: 16696632

Ahmed, S., Zhou, Z., Zhou, J., & Chen, S. Q. (2016). Pharmacogenomics of Drug Metabolizing Enzymes and Transporters: Relevance to Precision Medicine. *Genomics, Proteomics & Bioinformatics*, 14(5), 298–313. DOI: 10.1016/j.gpb.2016.03.008 PMID: 27729266

Akhtari, J., Tafazoli, A., Mehrad-Majd, H., & Mahrooz, A. (2018). Nanovehicle-based Small Interfering RNA (siRNA) Delivery for Therapeutic Purposes: A New Molecular Approach in Pharmacogenomics. *Current Clinical Pharmacology*, 13(3), 173–182. DOI: 10.2174/1574884713666180709152610 PMID: 29992895

Akram, F., Haq, I., Sahreen, S., Nasir, N., Naseem, W., Imitaz, M., & Aqeel, A. (2022). CRISPR/Cas9: A revolutionary genome editing tool for human cancers treatment. *Technology in Cancer Research & Treatment*, 21, 15330338221132078. Advance online publication. DOI: 10.1177/15330338221132078 PMID: 36254536

Alanazi, R. J. (2024). Role of Artificial Intelligence in Pharmacy Practice: A Systematic Review. *Archives of Pharmacy Practice (Mumbai)*, 15(2), 34–42. DOI: 10.51847/FMHsMFwgTP

Aldewachi, H., Al-Zidan, R. N., Conner, M. T., & Salman, M. M. (2021). High-throughput screening platforms in the discovery of novel drugs for neurodegenerative diseases. *Bioengineering (Basel, Switzerland)*, 8(2), 30. DOI: 10.3390/bioengineering8020030 PMID: 33672148

Aldoseri, A., Al-Khalifa, K. N., & Hamouda, A. M. (2023). Re-thinking data strategy and integration for artificial intelligence: Concepts, opportunities, and challenges. *Applied Sciences (Basel, Switzerland)*, 13(12), 7082. DOI: 10.3390/app13127082

AL-Eitan, L. N., Almasri, A. Y., & Khasawneh, R. H.AL-Eitan. (2018). Impact of CYP2C9 and VKORC1 Polymorphisms on Warfarin Sensitivity and Responsiveness in Jordanian Cardiovascular Patients during the Initiation Therapy. *Genes*, 9(12), 578. Advance online publication. DOI: 10.3390/genes9120578 PMID: 30486437

Alhajj, M., Zubair, M., & Farhana, A. (2023). *Enzyme linked immunosorbent assay*. StatPearls.

Ali, A. M., & Mohammed, M. A. (2024). A comprehensive review of artificial intelligence approaches in omics data processing: Evaluating progress and challenges. *International Journal of Mathematics, Statistics, and Computer Science*, 2, 114–167. DOI: 10.59543/ijmscs.v2i.8703

Alqahtani, A. A., Ahmed, M. M., Mohammed, A. A., & Ahmad, J. (2023). 3D Printed Pharmaceutical Systems for Personalized Treatment in Metabolic Syndrome. *Pharmaceutics*, 15(4), 1152. DOI: 10.3390/pharmaceutics15041152 PMID: 37111638

Althati, C., Tomar, M., & Shanmugam, L. (2024). Enhancing Data Integration and Management: The Role of AI and Machine Learning in Modern Data Platforms. *Journal of Artificial Intelligence General science (JAIGS) ISSN: 3006-4023,* 2(1), 220-232.

Alvarellos, M., Sheppard, H. E., Knarston, I., Davison, C., Raine, N., Seeger, T., Prieto Barja, P., & Chatzou Dunford, M. (2023). Democratizing clinical-genomic data: How federated platforms can promote benefits sharing in genomics. *Frontiers in Genetics*, 13, 1045450. Advance online publication. DOI: 10.3389/fgene.2022.1045450 PMID: 36704354

Alvarez, M. J. R., Griessler, E., & Starkbaum, J. (2022). Ethical, Legal and Social Aspects of Precision Medicine. In *Precision Medicine in Clinical Practice* (pp. 179–196). Springer Nature Singapore. DOI: 10.1007/978-981-19-5082-7_11

Amiri-Dashatan, N., Koushki, M., Abbaszadeh, H. A., Rostami-Nejad, M., & Rezaei-Tavirani, M. (2018, Fall). Proteomics Applications in Health: Biomarker and Drug Discovery and Food Industry. *Iranian Journal of Pharmaceutical Research*, 17(4), 1523–1536. PMID: 30568709

Anighoro, A., Bajorath, J., & Rastelli, G. (2014). Polypharmacology: Challenges and opportunities in drug discovery. In *Journal of Medicinal Chemistry* (Vol. 57, Issue 19, pp. 7874–7887). DOI: 10.1021/jm5006463

Ansori, A. N. M., Antonius, Y., Susilo, R. J. K., Hayaza, S., Kharisma, V. D., Parikesit, A. A., Zainul, R., Jakhmola, V., Saklani, T., Rebezov, M., & Ullah, Md. E., Maksimiuk, N., Derkho, M., & Burkov, P. (2023). Application of CRISPR-Cas9 genome editing technology in various fields: A review. In *Narra J* (Vol. 3, Issue 2, pp. e184–e184). DOI: 10.52225/narra.v3i2.184

Ansori, A. N. M., Chandara, N., Sucipto, T. H., Kharisma, V. D., Murtadlo, A. A. A., Kolesnik, E., Rebezov, M., Parikesit, A. A., Wiradana, P. A., Widhiantara, I. G., Sahadewa, S., Durry, F. D., Jakhmola, V., & Zainul, R. (2024). Computational analysis of Indonesian rabies vaccine enhanced by myeloid differentiation factor 88 (Myd88) genetic adjuvant. *Journal of Medicinal and Pharmaceutical Chemistry Research*, 6(9), 1327–1339. DOI: 10.48309/JMPCR.2024.448250.1145

Antol, D. D., Schwartz, R., Caplan, A., Casebeer, A., Erwin, C. J., Shrank, W. H., & Powers, B. W. (2022, October). Comparison of health care utilization by Medicare Advantage and traditional Medicare beneficiaries with complex care needs. []. American Medical Association.]. *JAMA Health Forum*, 3(10), e223451–e223451. DOI: 10.1001/jamahealthforum.2022.3451 PMID: 36206006

Arabnia, H. R., & Tran, Q. N. (2015). *Emerging trends in computational biology, bioinformatics, and system biology: Algorithms and software tools*. Elsevier. https://books.google.co.id/books?id=VuGoBAAAQBAJ

Arafah, A., Ali, S., Majid, S., Rashid, S., Rasool, S., Wani, H. A., Rasool, I., & Rehman, M. U. (2021). Single Nucleotide Polymorphisms and Pharmacogenomics. *Genetic Polymorphism and cancer susceptibility*, 23-52.

Arbitrio, M., Scionti, F., Di Martino, M. T., Caracciolo, D., Pensabene, L., Tassone, P., & Tagliaferri, P. (2021). Pharmacogenomics Biomarker Discovery and Validation for Translation in Clinical Practice. *Clinical and Translational Science*, 14(1), 113–119. DOI: 10.1111/cts.12869 PMID: 33089968

Argueta, N., Notari, E., & Szigeti, K. (2022). Role of pharmacogenomics in individualizing treatment for Alzheimer's disease. *CNS Drugs*, 36(4), 365–376. DOI: 10.1007/s40263-022-00915-3 PMID: 35352296

Arnold, G. L. (2018). Inborn errors of metabolism in the 21st century: Past to present. *Annals of Translational Medicine*, 6(24), 467. Advance online publication. DOI: 10.21037/atm.2018.11.36 PMID: 30740398

Aronson, J. K., & Ferner, R. E. (2017). Biomarkers—A general review. *Current Protocols in Pharmacology*, 76(1), 9–23. DOI: 10.1002/cpph.19 PMID: 28306150

Ashley, E. A. (2016). Towards precision medicine. *Nature Reviews. Genetics*, 17(9), 507–522. DOI: 10.1038/nrg.2016.86 PMID: 27528417

Asia, S., Immunoinf, A., Amin, A. M., Nur, A., Ansori, M., & Kharisma, V. D. (2024). T-Cell Epitope Vaccine Prediction Analysis Targeting Phosphoprotein (P) Rabies Virus Based on the Presence of HLA-I Alleles A, B, and C Loci Throughout T-Cell Epitope Vaccine Prediction Analysis Targeting Phosphoprotein (P) Rabies Virus Based on the. *Research Journal of Pharmacy and Technology*, 17(May), 2001–2008. DOI: 10.52711/0974-360X.2024.00317

Asiimwe, I. G., & Pirmohamed, M. (2022). Ethnic Diversity and Warfarin Pharmacogenomics. *Frontiers in Pharmacology*, 13, 866058. DOI: 10.3389/fphar.2022.866058 PMID: 35444556

Astras, G., Papagiannopoulos, C. I., Kyritsis, K. A., Markitani, C., & Vizirianakis, I. S. (2020). Pharmacogenomic testing to guide personalized cancer medicine decisions in private oncology practice: A case study. *Frontiers in Oncology*, 10, 521. DOI: 10.3389/fonc.2020.00521 PMID: 32411592

Atçeken, N., Yigci, D., Ozdalgic, B., & Tasoglu, S. (2022). CRISPR-Cas-Integrated LAMP. *Biosensors (Basel)*, 12(11), 1035. Advance online publication. DOI: 10.3390/bios12111035 PMID: 36421156

Atkins, D., & McCann, J. P. (2021). Validation of genomic biomarkers in clinical medicine. *The Lancet. Oncology*, 22(6), 847–857.

Attapong, J., Kaset, C., Nakkam, N., Tassaneeyakul, W., Wichukchinda, N., & Chomean, S. (2024). Dual Approaches in Pharmacogenetics: Developing PCR-SSP and RT-PCR Methods for HLA-B* 13: 01 Screening to Prevent Dapsone and Co-Trimoxazole SCARs. *Heliyon*, 10(15), e34977. DOI: 10.1016/j.heliyon.2024.e34977 PMID: 39144953

Attene-Ramos, M. S., Miller, N., Huang, R., Michael, S., Itkin, M., Kavlock, R. J., Austin, C. P., Shinn, P., Simeonov, A., Tice, R. R., & Xia, M. (2013). The Tox21 robotic platform for the assessment of environmental chemicals–from vision to reality. *Drug Discovery Today*, 18(15-16), 716–723. DOI: 10.1016/j.drudis.2013.05.015 PMID: 23732176

Auwerx, C., Sadler, M. C., Reymond, A., & Kutalik, Z. (2022). From pharmacogenetics to pharmaco-omics: Milestones and future directions. *HGG Advances*, 3(2), 100100. DOI: 10.1016/j.xhgg.2022.100100 PMID: 35373152

Aworunse, O. S., Adeniji, O., Oyesola, O. L., Isewon, I., Oyelade, J., & Obembe, O. O. (2018). Genomic Interventions in Medicine. *Bioinformatics and Biology Insights*, 12, 1177932218816100. DOI: 10.1177/1177932218816100 PMID: 30546257

Awwad, O., Ahram, M., Coperchini, F., & Jalil, M. A. (2024). Precision medicine: Recent advances, current challenges and future perspectives. *Frontiers in Pharmacology*, 15, 1439276. DOI: 10.3389/fphar.2024.1439276 PMID: 38910892

Ayon, N. J. (2023). High-throughput screening of natural product and synthetic molecule libraries for antibacterial drug discovery. *Metabolites*, 13(5), 625. DOI: 10.3390/metabo13050625 PMID: 37233666

Azvolinsky, A. (2013). Supreme Court Ruling Broadens BRCA Testing Options. *Journal of the National Cancer Institute*, 105(22), 1671–1672. DOI: 10.1093/jnci/djt342 PMID: 24198329

Bai1, , FLi, , SLi, , H. (2024). AI enhances drug discovery and development. *National Science Review*, 11(303), 1–3.

Balogun, O. D., Ayo-Farai, O., Ogundairo, O., Maduka, C. P., Okongwu, C. C., Babarinde, A. O., & Sodamade, O. T. (2024). The role of pharmacists in personalised medicine: A review of integrating pharmacogenomics into clinical practice. *International Medical Science Research Journal*, 4(1), 19–36. DOI: 10.51594/imsrj.v4i1.697

Baltekin, Ö., Boucharin, A., Tano, E., Andersson, D. I., & Elf, J. (2017). Antibiotic susceptibility testing in less than 30 min using direct single-cell imaging. *Proceedings of the National Academy of Sciences of the United States of America*, 114(34), 9170–9175. DOI: 10.1073/pnas.1708558114 PMID: 28790187

Barabási, A. L., & Oltvai, Z. N. (2011). Network biology: Understanding the cell's functional organization. *Nature Reviews. Genetics*, 5(2), 101–113. DOI: 10.1038/nrg1272 PMID: 14735121

Bastikar, V., Bastikar, A., & Gupta, P. (2022). Quantitative structure–activity relationship-based computational approaches. In *Computational Approaches for Novel Therapeutic and Diagnostic Designing to Mitigate SARS-CoV-2 Infection* (pp. 191–205). Academic Press.

Behl, T., Kaur, D., Sehgal, A., Singh, S., Sharma, N., Zengin, G., Andronie-Cioara, F. L., Toma, M. M., Bungau, S., & Bumbu, A. G. (2021). Role of monoamine oxidase activity in Alzheimer's disease: An insight into the therapeutic potential of inhibitors. *Molecules (Basel, Switzerland)*, 26(12), 3724. DOI: 10.3390/molecules26123724 PMID: 34207264

Behl, T., Kaur, I., Bungau, S., Kumar, A., Uddin, M. S., Kumar, C., & Arora, S. (2020). The dual impact of ACE2 in COVID-19 and ironical actions in geriatrics and pediatrics with possible therapeutic solutions. *Life Sciences*, 257, 118075. DOI: 10.1016/j.lfs.2020.118075 PMID: 32653522

Benowitz, N. L., Hukkanen, J., & Jacob, P.III. (2009). Nicotine chemistry, metabolism, kinetics and biomarkers. *Handbook of Experimental Pharmacology*, 192, 29–60. DOI: 10.1007/978-3-540-69248-5_2 PMID: 19184645

Bentley, A. R. (2020). The genetic architecture of human disease: A review of the history and challenges. *Trends in Genetics*, 36(1), 21–31. PMID: 32993968

Berthold, M. R., Cebron, N., Dill, F., Gabriel, T. R., Kötter, T., Meinl, T., Ohl, P., Thiel, K., & Wiswedel, B. (2009). KNIME-the Konstanz information miner: Version 2.0 and beyond. *SIGKDD Explorations*, 11(1), 26–31. DOI: 10.1145/1656274.1656280

Beskow, L. M., & Wolf, L. E. (2022). "Choice of law" in precision medicine research. *American Journal of Human Genetics*, 109(8), 1347–1352. DOI: 10.1016/j.ajhg.2022.06.009 PMID: 35931047

Bhagat, R. T., Butle, S. R., Khobragade, D. S., Wankhede, S. B., Prasad, C. C., Mahure, D. S., & Armarkar, A. V. (2021). Molecular docking in drug discovery. *Journal of Pharmaceutical Research International*, 33(30B), 46–58. DOI: 10.9734/jpri/2021/v33i30B31639

Bhatia, S. N., & Ingber, D. E. (2014). Microfluidic organs-on-chips. *Nature Biotechnology*, 32(8), 760–772. DOI: 10.1038/nbt.2989 PMID: 25093883

Bhattacharya, S., Heidler, P., & Varshney, S. (2022). Incorporating neglected non-communicable diseases into the national health program—A review. *Frontiers in Public Health*, 10, 1093170. Advance online publication. DOI: 10.3389/fpubh.2022.1093170 PMID: 36703821

Bienfait, K., Chhibber, A., Marshall, J. C., Armstrong, M., Cox, C., Shaw, P. M., & Paulding, C. (2022). Current challenges and opportunities for pharmacogenomics: Perspective of the Industry Pharmacogenomics Working Group (I-PWG). *Human Genetics*, 141(6), 1165–1173. DOI: 10.1007/s00439-021-02282-3 PMID: 34081195

Bini, S., Tramontano, D., Minicocci, I., Di Costanzo, A., Tambaro, F., D'Erasmo, L., & Arca, M. (2023). How ANGPTL3 Inhibition Will Help Our Clinical Practice? In *Current Atherosclerosis Reports* (Vol. 25, Issue 1). DOI: 10.1007/s11883-022-01076-w

Bjornsson, H. T. (2015). The Mendelian disorders of the epigenetic machinery. *Genome Research*, 25(10), 1473–1481. DOI: 10.1101/gr.190629.115 PMID: 26430157

Blanco-Gonzalez, A., Cabezon, A., Seco-Gonzalez, A., Conde-Torres, D., Antelo-Riveiro, P., Pineiro, A., & Garcia-Fandino, R. (2022). *The Role of AI in Drug Discovery: Challenges, Opportunities, and Strategies*. https://doi.org//arxiv.2212.08104DOI: 10.48550

Blanco-González, A., Cabezón, A., Seco-González, A., Daniel, C.-T. P. A.-R., Piñeiro, Á., & Garcia-Fandino, R. (2023). The Role of AI in Drug Discovery: Challenges, Opportunities, and Strategies. *Pharmaceuticals (Basel, Switzerland)*, 16(6), 891. DOI: 10.3390/ph16060891 PMID: 37375838

Blaschke, T., Arús-Pous, J., Chen, H., Margreitter, C., Tyrchan, C., Engkvist, O., Papadopoulos, K., & Patronov, A. (2020). REINVENT 2.0: An AI Tool for de Novo Drug Design. *Journal of Chemical Information and Modeling*, 60(12), 5918–5922. DOI: 10.1021/acs.jcim.0c00915 PMID: 33118816

Bock, C., Datlinger, P., Chardon, F., Coelho, M. A., Dong, M. B., Lawson, K. A., Lu, T., Maroc, L., Norman, T. M., Song, B., Stanley, G., Chen, S., Garnett, M., Li, W., Moffat, J., Qi, L. S., Shapiro, R. S., Shendure, J., Weissman, J. S., & Zhuang, X. (2022). High-content CRISPR screening. *Nature Reviews. Methods Primers*, 2(1), 8. DOI: 10.1038/s43586-021-00093-4 PMID: 37214176

Bodaghi, A., Fattahi, N., & Ramazani, A. (2023). Biomarkers: Promising and valuable tools towards diagnosis, prognosis and treatment of Covid-19 and other diseases. *Heliyon*, 9(2), e13323. DOI: 10.1016/j.heliyon.2023.e13323 PMID: 36744065

Bolognesi, M. L. (2019). Harnessing Polypharmacology with Medicinal Chemistry. *ACS Medicinal Chemistry Letters*, 10(3), 273–275. Advance online publication. DOI: 10.1021/acsmedchemlett.9b00039 PMID: 30891125

Boyle, E. A., Li, Y., & Pritchard, J. K. (2017). An expanded view of complex traits: From polygenic to omnigenic. *Cell*, 169(7), 1177–1186. DOI: 10.1016/j.cell.2017.05.038 PMID: 28622505

Božina, N., Vrkić Kirhmajer, M., Šimičević, L., Ganoci, L., Mirošević Skvrce, N., Klarica Domjanović, I., & Merćep, I. (2020). Use of pharmacogenomics in elderly patients treated for cardiovascular diseases. *Croatian Medical Journal*, 61(3), 147–158. DOI: 10.3325/cmj.2020.61.147 PMID: 32378381

Branch, L. E., Eller, W. S., Bias, T. K., McCawley, M. A., Myers, D. J., Gerber, B. J., & Bassler, J. R. (2019). Trends in Malware Attacks against United States Healthcare Organizations, 2016-2017. *Global Biosecurity*, 1(1), 15. Advance online publication. DOI: 10.31646/gbio.7

Brandão, M., Pondé, N. F., Poggio, F., Kotecki, N., Salis, M., Lambertini, M., & de Azambuja, E. (2018). Combination therapies for the treatment of HER2-positive breast cancer: Current and future prospects. *Expert Review of Anticancer Therapy*, 18(7), 629–649. DOI: 10.1080/14737140.2018.1477596 PMID: 29781317

Breslauer, D. N., Lee, P. J., & Lee, L. P. (2006). Microfluidics-based systems biology. *Molecular BioSystems*, 2(2), 97–112. DOI: 10.1039/b515632g PMID: 16880927

Brookes, A. J. (2019). The essence of SNPs. *Nature*, 414(6863), 10–11.

Brothers, K. B., & Rothstein, M. A. (2015). Ethical, legal and social implications of incorporating personalized medicine into healthcare. *Personalized Medicine*, 12(1), 43–51. DOI: 10.2217/pme.14.65 PMID: 25601880

Buchanan, A., Califano, A., Kahn, J., McPherson, E., Robertson, J., & Brody, B. (2017). Pharmacogenetics: ethical issues and policy options. In *Genetics and Gene Therapy* (pp. 327-341). Routledge. DOI: 10.4324/9781315254517-12

Budiani, A., Nugroho, I. B., Sari, D. A., Palupi, I., & Putranto, R. A. (2015). CRISPR / Cas9-mediated Knockout of An Oil Palm Defense-related Gene to The Pathogenic Fungi Ganoderma boninense. *Indonesian Journal of Biotechnology*, 24(1), 1–10. DOI: 10.22146/ijbiotech.52170

Bull, L. (2013). Genetics, mutations, and polymorphisms. In *Madame Curie Bioscience Database* [Internet]. Landes Bioscience.

Butler, D. (2001). Genomics. Are you ready for the revolution? *Nature*, 409(6822), 758–760. DOI: 10.1038/35057400 PMID: 11236971

Butnariu, A., Samasca, G., & Lupan, I. (2015). Ethical implications in pharmacogenetics and pharmacogenomics. *Medical Connections.*, 10, 41–44.

Bycroft, C., Freeman, C., Petkova, D., Band, G., Elliott, L. T., Sharp, K., & Marchini, J. (2017). Genome-wide genetic data on~ 500,000 UK Biobank participants. *BioRxiv,* 166298. DOI: 10.1101/166298

Bycroft, C., Freeman, C., Petkova, D., Band, G., Elliott, L. T., Sharp, K., Motyer, A., Vukcevic, D., Delaneau, O., O'Connell, J., Cortes, A., Welsh, S., Young, A., Effingham, M., McVean, G., Leslie, S., Allen, N., Donnelly, P., & Marchini, J. (2018). The UK Biobank resource with deep phenotyping and genomic data. *Nature*, 562(7726), 203–209. DOI: 10.1038/s41586-018-0579-z PMID: 30305743

Cacabelos, R., Naidoo, V., Corzo, L., Cacabelos, N., & Carril, J. C. (2021). Genophenotypic Factors and Pharmacogenomics in Adverse Drug Reactions. *International Journal of Molecular Sciences*, 22(24), 13302. Advance online publication. DOI: 10.3390/ijms222413302 PMID: 34948113

Cai, W., & Wang, Z. (2020). Introduction and Principles of Pharmacogenomics in Precision Medicine. *Pharmacogenomics in Precision Medicine: From a Perspective of Ethnic Differences*, 1-19.

Cai, W., Liu, Z., Miao, L., & Xiang, X. (Eds.). (2020). *Pharmacogenomics in precision medicine: from a perspective of ethnic differences*. Springer. DOI: 10.1007/978-981-15-3895-7

Califf, R. M. (2018). Biomarker definitions and their applications. *Experimental Biology and Medicine*, 243(3), 213–221. DOI: 10.1177/1535370217750088 PMID: 29405771

Callier, S. L., Abudu, R., Mehlman, M. J., Singer, M. E., Neuhauser, D., Caga-Anan, C., & Wiesner, G. L. (2016). Ethical, legal, and social implications of personalized genomic medicine research: Current literature and suggestions for the future. *Bioethics*, 30(9), 698–705. DOI: 10.1111/bioe.12285 PMID: 27767224

Carlsten, C., Brauer, M., Brinkman, F., Brook, J., Daley, D., McNagny, K., Pui, M., Royce, D., Takaro, T., & Denburg, J. (2014). Genes, the environment and personalized medicine. *EMBO Reports*, 15(7), 736–739. DOI: 10.15252/embr.201438480 PMID: 24907240

Carpenter, A. E., Jones, T. R., Lamprecht, M. R., Clarke, C., Kang, I. H., Friman, O., Guertin, D. A., Chang, J. H., Lindquist, R. A., Moffat, J., Golland, P., & Sabatini, D. M. (2006). CellProfiler: Image analysis software for identifying and quantifying cell phenotypes. *Genome Biology*, 7(10), 1–11. DOI: 10.1186/gb-2006-7-10-r100 PMID: 17076895

Carpenter, A. E., & Sabatini, D. M. (2004). Systematic genome-wide screens of gene function. *Nature Reviews. Genetics*, 5(1), 11–22. DOI: 10.1038/nrg1248 PMID: 14708012

Carranza-Leon, D., Dickson, A. L., Gaedigk, A., Stein, C. M., & Chung, C. P. (2021). CYP2D6 genotype and reduced codeine analgesic effect in real-world clinical practice. *The Pharmacogenomics Journal*, 21(4), 484–490. DOI: 10.1038/s41397-021-00226-8 PMID: 33750887

Carr, D. F., Turner, R. M., & Pirmohamed, M. (2021). Pharmacogenomics of anticancer drugs: Personalising the choice and dose to manage drug response. *British Journal of Clinical Pharmacology*, 87(2), 237–255. DOI: 10.1111/bcp.14407 PMID: 32501544

Carrillo-Vázquez, J. P., Chimal-Vega, B., Zamora-López, B., Marchat, L. A., Benítez-Cardoza, C. G., Reyes-López, C. A. S., & Zamorano-Carrillo, A. (2013). Structural Consequences Of The Polymorphism Q223r In The Human Leptin Receptor: A Molecular Dynamics Study. *American Journal of Agricultural and Biological Sciences*, 8(3), 239–248. DOI: 10.3844/ajabssp.2013.239.248

Carter, M., & Shieh, J. (2015). Biochemical assays and intracellular signaling. *Guide to research techniques in neuroscience*, pp.311-343.

Caspar, S. M., Schneider, T., Meienberg, J., & Matyas, G. (2020). Added value of clinical sequencing: WGS-based profiling of pharmacogenes. *International Journal of Molecular Sciences*, 21(7), 2308. DOI: 10.3390/ijms21072308 PMID: 32225115

Caspar, S. M., Schneider, T., Stoll, P., Meienberg, J., & Matyas, G. (2021). Potential of whole-genome sequencing-based pharmacogenetic profiling. *Pharmacogenomics*, 22(3), 177–190. DOI: 10.2217/pgs-2020-0155 PMID: 33517770

Caudle, K. E. (2020). Clinical pharmacogenetics implementation consortium: Updated clinical practice guidelines for pharmacogenetic testing in the context of cancer treatment. *Clinical Pharmacology and Therapeutics*, 107(3), 548–561.

Center for Drug Evaluation and Research. (2019). *Table of Pharmacogenomic Biomarkers*. U.S. Food and Drug Administration. https://www.fda.gov/drugs/science-and-research-drugs/table-pharmacogenomic-biomarkers-drug-labeling

Chai, S. C., Goktug, A. N., & Chen, T. (2015). Assay validation in high throughput screening–from concept to application. *Drug Discov. Dev. Mol. Med*, 10, 59765. DOI: 10.5772/59765

Chandrasekaran, S. N., Ceulemans, H., Boyd, J. D., & Carpenter, A. E. (2021). Image-based profiling for drug discovery: Due for a machine-learning upgrade? *Nature Reviews. Drug Discovery*, 20(2), 145–159. DOI: 10.1038/s41573-020-00117-w PMID: 33353986

Chan, Y. T., Lu, Y., Wu, J., Zhang, C., Tan, H. Y., Bian, Z. X., Wang, N., & Feng, Y. (2022, April 11). CRISPR-Cas9 library screening approach for anti-cancer drug discovery: Overview and perspectives. *Theranostics*, 12(7), 3329–3344. DOI: 10.7150/thno.71144 PMID: 35547744

Chapman, P. B., Hauschild, A., Robert, C., Haanen, J. B., Ascierto, P., Larkin, J., Dummer, R., Garbe, C., Testori, A., Maio, M., Hogg, D., Lorigan, P., Lebbe, C., Jouary, T., Schadendorf, D., Ribas, A., O'Day, S. J., Sosman, J. A., Kirkwood, J. M., & McArthur, G. A. (2011). Improved survival with BRAF inhibition in melanoma with BRAF V600E mutation. *The New England Journal of Medicine*, 364(26), 2507–2516. DOI: 10.1056/NEJMoa1103782 PMID: 21639808

Charnysh, E., Pal, S., Reader, J. M., Uhlmann, W. R., McCain, S., Sanghavi, K., Blasco, D., Brandt, R., Feero, W. G., Ferber, R., Giri, V. N., Hendy, K., Prince, A. E. R., Lee, C., Roberts, J. S., Crumpler, N., Leader, A., Mathews, D., Ryan, K., & Saidi, F. (2024). Health care utilization and behavior changes after workplace genetic testing at a large US health care system. *Genetics in Medicine*, 26(8), 101160. DOI: 10.1016/j.gim.2024.101160 PMID: 38733246

Cheng, F., Li, W., Zhou, Y., Shen, J., Wu, Z., Liu, G., Lee, P. W., & Tang, Y. (2012). AdmetSAR: A comprehensive source and free tool for assessment of chemical ADMET properties. *Journal of Chemical Information and Modeling*, 52(11), 3099–3105. DOI: 10.1021/ci300367a PMID: 23092397

Cheng, F., Ma, Y., Uzzi, B., & Loscalzo, J. (2020). Importance of scientific collaboration in contemporary drug discovery and development: A detailed network analysis. *BMC Biology*, 18(1), 1–9. DOI: 10.1186/s12915-020-00868-3 PMID: 33050894

Chen, L. M., & Lee, J. M. (2020). BRCA1/2 mutations and the impact on personalized medicine for breast and ovarian cancer. Cancer Treatment Reviews, 88, 102040. Choi, H., & Lee, K. (2018). Data privacy and security in genomics: Challenges and solutions. *Nature Biotechnology*, 36(7), 629–636.

Chenoweth, M. J., Giacomini, K. M., Pirmohamed, M., Hill, S. L., van Schaik, R. H., Schwab, M., Shuldiner, A. R., Relling, M. V., & Tyndale, R. F. (2020). Global pharmacogenomics within precision medicine: Challenges and opportunities. *Clinical Pharmacology and Therapeutics*, 107(1), 57–61. DOI: 10.1002/cpt.1664 PMID: 31696505

Chen, R., & Snyder, M. (2013). Promise of personalized omics to precision medicine. *Wiley Interdisciplinary Reviews. Systems Biology and Medicine*, 5(1), 73–82. DOI: 10.1002/wsbm.1198 PMID: 23184638

Chen, Y., Zhang, X., Ye, Q., Zhang, X., Cao, N., Li, S. Y., Yu, J., Zhao, S. T., Zhang, J., Xu, X. M., Shi, Y. K., & Yang, L. X. (2024). Machine learning-based prediction model for myocardial ischemia under high altitude exposure: A cohort study. *Scientific Reports*, 14(1), 686. DOI: 10.1038/s41598-024-51202-8 PMID: 38182722

Chu, X., Prasad, B., Neuhoff, S., Yoshida, K., Leeder, J. S., Mukherjee, D., Taskar, K., Varma, M. V. S., Zhang, X., Yang, X., & Galetin, A. (2022). Clinical implications of altered drug transporter abundance/function and PBPK modeling in specific populations: An ITC perspective. *Clinical Pharmacology and Therapeutics*, 112(3), 501–526. DOI: 10.1002/cpt.2643 PMID: 35561140

Cilluffo, G., Fasola, S., Ferrante, G., Malizia, V., Montalbano, L., & La Grutta, S. (2021). Machine learning: An overview and applications in pharmacogenetics. *Genes*, 12(10), 1511. DOI: 10.3390/genes12101511 PMID: 34680905

Clayton, E. W., Evans, B. J., Hazel, J. W., & Rothstein, M. A. (2019). The law of genetic privacy: Applications, implications, and limitations. *Journal of Law and the Biosciences*, 6(1), 1–36. DOI: 10.1093/jlb/lsz007 PMID: 31666963

Collins, F. S. (2017). The molecular and genetic basis of human health and disease. *The New England Journal of Medicine*, 377(6), 512–519.

Collins, F. S., & Fink, L. (1995). The Human Genome Project. *Alcohol Health and Research World*, 19(3), 190–195. PMID: 31798046

Collins, F. S., Morgan, M., & Patrinos, A. (2003). The Human Genome Project: Lessons from large-scale biology. *Science*, 300(5617), 286–290. DOI: 10.1126/science.1084564 PMID: 12690187

Collins, F. S., & Varmus, H. (2015). A new initiative on precision medicine. *The New England Journal of Medicine*, 372(9), 793–795. DOI: 10.1056/NEJMp1500523 PMID: 25635347

Collins, F. S., Varmus, H., & Cook-Deegan, R. (2017). Genomic medicine - A decade of progress. *The New England Journal of Medicine*, 377(6), 512–519.

Collisson, E. A. C. J., Campbell, J., Brooks, A., Berger, A., Lee, W., Chmielecki, J., & Helman, E. (2014). Comprehensive molecular profiling of lung adenocarcinoma: The cancer genome atlas research network. *Nature*, 511(7511), 543–550. DOI: 10.1038/nature13385 PMID: 25079552

Conesa, A., & Götz, S. (2008). Blast2GO: A comprehensive suite for functional analysis in plant genomics. *International Journal of Plant Genomics*, 619832, 1–12. Advance online publication. DOI: 10.1155/2008/619832 PMID: 18483572

Cook, C. E., Bergman, M. T., Finn, R. D., Cochrane, G., Birney, E., & Apweiler, R. (2016). The European Bioinformatics Institute in 2016: Data growth and integration. *Nucleic Acids Research*, 44(D1), D20–D26. DOI: 10.1093/nar/gkv1352 PMID: 26673705

Cook, J. C., Wu, H., Aleo, M. D., & Adkins, K. (2018). Principles of precision medicine and its application in toxicology. *The Journal of Toxicological Sciences*, 43(10), 565–577. DOI: 10.2131/jts.43.565 PMID: 30298845

Cooper-DeHoff, R. M., Niemi, M., Ramsey, L. B., Luzum, J. A., Tarkiainen, E. K., Straka, R. J., Gong, L., Tuteja, S., Wilke, R. A., Wadelius, M., Larson, E. A., Roden, D. M., Klein, T. E., Yee, S. W., Krauss, R. M., Turner, R. M., Palaniappan, L., Gaedigk, A., Giacomini, K. M., & Voora, D. (2022). The Clinical Pharmacogenetics Implementation Consortium Guideline for SLCO1B1, ABCG2, and CYP2C9 genotypes and Statin-Associated Musculoskeletal Symptoms. *Clinical Pharmacology and Therapeutics*, 111(5), 1007–1021. DOI: 10.1002/cpt.2557 PMID: 35152405

Crews, K. R., Hicks, J. K., Pui, C. H., Relling, M. V., & Evans, W. E. (2012). Pharmacogenomics and individualized medicine: Translating science into practice. *Clinical Pharmacology and Therapeutics*, 92(4), 467–475. DOI: 10.1038/clpt.2012.120 PMID: 22948889

Crisafulli, C., Romeo, P. D., Calabrò, M., Epasto, L. M., & Alberti, S. (2019). Pharmacogenetic and pharmacogenomic discovery strategies. *Cancer Drug Resistance (Alhambra, Calif.)*, 2(2), 225. PMID: 35582724

Crowley, E., Warner, N., Pan, J., Khalouei, S., Elkadri, A., Fiedler, K., Foong, J., Turinsky, A. L., Bronte-Tinkew, D., Zhang, S., Hu, J., Tian, D., Li, D., Center, R. G., Horowitz, J., Siddiqui, I., Upton, J., Roifman, C. M., Church, P. C., & Muise, A. M. (2020). Prevalence and Clinical Features of Inflammatory Bowel Diseases Associated with Monogenic Variants, Identified by Whole-exome sequencing in 1000 Children at a Single Center. *Gastroenterology*, 158(8), 2208–2220. DOI: 10.1053/j.gastro.2020.02.023 PMID: 32084423

Crowther, J. R. (2000). The ELISA guidebook. *Methods in Molecular Biology (Clifton, N.J.)*, 149, III–IV, 1–413. DOI: 10.1385/1592590497 PMID: 11028258

Cunningham, B. T. (2009). Label-free optical biosensors: An introduction. Label-free biosensors: techniques and applications, 1.

D'ARVILLE, C.A.R.O.L.Y.N. (2004). Partnering: Key to early-stage biotech survival, but at what cost? *Biotechnology Healthcare*, 1(6), 26. PMID: 23424295

Dahlin, J. L., & Walters, M. A. (2014). The essential roles of chemistry in high-throughput screening triage. *Future Medicinal Chemistry*, 6(11), 1265–1290. DOI: 10.4155/fmc.14.60 PMID: 25163000

Dancey, J. E. (2012). The role of companion diagnostics in personalized medicine. *Nature Reviews. Drug Discovery*, 11(8), 565–573.

David, S. P., Devine, B., Palaniappan, L., & Houwink, E. J. (2024). State of the science of pharmacogenomics implementation in healthcare systems and communities. *Frontiers in Pharmacology*, 15, 1463384. DOI: 10.3389/fphar.2024.1463384 PMID: 39114360

Davis, A. A., McKee, A. E., Kibbe, W. A., & Villaflor, V. M. (2018). Complexity of delivering precision medicine: Opportunities and challenges. *American Society of Clinical Oncology Educational Book*, 38(38), 998–1007. DOI: 10.1200/EDBK_200279 PMID: 30231318

Davis, A. M., Engkvist, O., Fairclough, R. J., Feierberg, I., Freeman, A., & Iyer, P. (2021). Public-private partnerships: Compound and data sharing in drug discovery and development. *SLAS Discovery*, 26(5), 604–619. DOI: 10.1177/2472555220982268 PMID: 33586501

Davis, B. H., & Limdi, N. A. (2021). Translational pharmacogenomics: Discovery, evidence synthesis and delivery of race-conscious medicine. *Clinical Pharmacology and Therapeutics*, 110(4), 909–925. DOI: 10.1002/cpt.2357 PMID: 34233023

de Andrés, F., Altamirano-Tinoco, C., Ramírez-Roa, R., Montes-Mondragón, C. F., Dorado, P., Peñas-Lledó, E. M., & LLerena, A. (2021). Relationships between CYP1A2, CYP2C9, CYP2C19, CYP2D6 and CYP3A4 metabolic phenotypes and genotypes in a Nicaraguan Mestizo population. *The Pharmacogenomics Journal*, 21(2), 140–151. DOI: 10.1038/s41397-020-00190-9 PMID: 33024249

De Maria Marchiano, R., Di Sante, G., Piro, G., Carbone, C., Tortora, G., Boldrini, L., Pietragalla, A., Daniele, G., Tredicine, M., Cesario, A., Valentini, V., Gallo, D., Babini, G., D'Oria, M., & Scambia, G. (2021). Translational research in the era of precision medicine: Where we are and where we will go. *Journal of Personalized Medicine*, 11(3), 216. DOI: 10.3390/jpm11030216 PMID: 33803592

De Stefano, P., Bianchi, E., & Dubini, G. (2022). The impact of microfluidics in high-throughput drug-screening applications. *Biomicrofluidics*, 16(3), 031501. DOI: 10.1063/5.0087294 PMID: 35646223

Deac, A. L., Pop, R. M., Burz, C. C., Bocşan, I. C., Militaru, C., Suciu, Ş., & Buzoianu, A. D. (2023). 5-fluorouracil therapeutic drug monitoring and adverse events in a Romanian population. *Medicine and Pharmacy Reports*, 96(4), 413–419. DOI: 10.15386/mpr-2643 PMID: 37970190

Delpierre, C., & Lefèvre, T. (2023). Precision and personalized medicine: What their current definition says and silences about the model of health they promote. Implication for the development of personalized health. *Frontiers in Sociology*, 8, 1112159. DOI: 10.3389/fsoc.2023.1112159 PMID: 36895332

Demkow, U. (2016). Next generation sequencing in pharmacogenomics. In *Clinical Applications for Next-Generation Sequencing* (pp. 217–240). Elsevier. DOI: 10.1016/B978-0-12-801739-5.00011-8

Dhudum, R., Ganeshpurkar, A., & Pawar, A. (2024). Revolutionizing Drug Discovery: A Comprehensive Review of AI Applications. *Drugs and Drug Candidates*, 3(1), 148–171. DOI: 10.3390/ddc3010009

Diansyah, M. R., Annisa, A., & Kusuma, W. A. (2021). Analysis using top-k skyline query of protein-protein interaction reveals alpha-synuclein as the most important protein in Parkinson's disease. *Indonesian Journal of Biotechnology*, 26(4), 197–205. DOI: 10.22146/ijbiotech.63023

Diaz, L. A. (2012). The molecular evolution of the cancer genome. *Journal of Clinical Oncology*, 30(16), 1887–1891.

Dinse, G. E. (2019). Regulatory oversight of genomic diagnostics: Frameworks and challenges. *Genomics*, 112(5), 347–352.

Dirks, R. A., Stunnenberg, H. G., & Marks, H. (2016, December). Genome-wide epigenomic profiling for biomarker discovery. *Clinical Epigenetics*, 8(1), 1–7. DOI: 10.1186/s13148-016-0284-4 PMID: 27895806

Dittrich, P. S., & Manz, A. (2006). Lab-on-a-chip: Microfluidics in drug discovery. *Nature Reviews. Drug Discovery*, 5(3), 210–218. DOI: 10.1038/nrd1985 PMID: 16518374

Dixon Omere, Z. O., & King, J. (2022). America's History of Health Disparities: How Does this Affect the Future of Healthcare In Mississippi? Retrieved from https://egrove.olemiss.edu/hon_thesis/2708

Doench, J. G. (2018). The CRISPR–Cas9 genome editing system: A new era in functional genomics. *Nature Reviews. Genetics*, 19(2), 97–111. PMID: 29199283

Dolz, M., Monterrey, D. T., Beltrán-Nogal, A., Menés-Rubio, A., Keser, M., González-Pérez, D., de Santos, P. G., Viña-González, J., & Alcalde, M. (2023). The colors of peroxygenase activity: Colorimetric high-throughput screening assays for directed evolution. *Methods in Enzymology*, 693, 73–109. DOI: 10.1016/bs.mie.2023.09.006 PMID: 37977739

Dominic, F., Ekpan, M., Ori, M. O., Samuel, H. S., & Egwuatu, O. P. (2024). The Synergy of AI and Drug Delivery: A Revolution in Healthcare. *International Journal of Advanced Biological and Biomedical Research*, 12(1), 44–66.

Dong, C., Ly, C., Dunlap, L. E., Vargas, M. V., Sun, J., Hwang, I. W., Azinfar, A., Oh, W. C., Wetsel, W. C., Olson, D. E., & Tian, L. (2021). Psychedelic-inspired drug discovery using an engineered biosensor. *Cell*, 184(10), 2779–2792. DOI: 10.1016/j.cell.2021.03.043 PMID: 33915107

Doudna, J. A., & Charpentier, E. (2014). The new frontier of genome engineering with CRISPR-Cas9. *Science*, 346(6213), 1258096. DOI: 10.1126/science.1258096 PMID: 25430774

Dove, E. S. (2018). The EU General Data Protection Regulation: Implications for International Scientific Research in the Digital Era. *The Journal of Law, Medicine & Ethics*, 46(4), 1013–1030. DOI: 10.1177/1073110518822003

Drews, J. (2000, March 17). Drug discovery: A historical perspective. *Science*, 287(5460), 1960–1964. DOI: 10.1126/science.287.5460.1960 PMID: 10720314

Drucker, E., & Krapfenbauer, K. (2013). Pitfalls and limitations in translation from biomarker discovery to clinical utility in predictive and personalised medicine. *The EPMA Journal*, 4(1), 1–10. DOI: 10.1186/1878-5085-4-7 PMID: 23442211

Duan, X. P., Qin, B. D., Jiao, X. D., Liu, K., Wang, Z., & Zang, Y. S. (2024). New clinical trial design in precision medicine: Discovery, development and direction. *Signal Transduction and Targeted Therapy*, 9(1), 57. DOI: 10.1038/s41392-024-01760-0 PMID: 38438349

Duarte, J. D., & Cavallari, L. H. (2021). Pharmacogenetics to guide cardiovascular drug therapy. *Nature Reviews. Cardiology*, 18(9), 649–665. Advance online publication. DOI: 10.1038/s41569-021-00549-w PMID: 33953382

Dueñas, M. E., Peltier-Heap, R. E., Leveridge, M., Annan, R. S., Büttner, F. H., & Trost, M. (2023, January 11). Advances in high-throughput mass spectrometry in drug discovery. *EMBO Molecular Medicine*, 15(1), e14850. DOI: 10.15252/emmm.202114850 PMID: 36515561

Dutescu, I. A., & Hillier, S. A. (2021). Encouraging the development of new antibiotics: Are financial incentives the right way forward? A systematic review and case study. *Infection and Drug Resistance*, 14, 415–434. DOI: 10.2147/IDR.S287792 PMID: 33574682

Dzau, V. J., & Ginsburg, G. S. (2016). Realizing the Full Potential of Precision Medicine in Health and Health Care. *Journal of the American Medical Association*, 316(16), 1659–1660. DOI: 10.1001/jama.2016.14117 PMID: 27669484

Edmondson, R., Broglie, J. J., Adcock, A. F., & Yang, L. (2014, May). Three-dimensional cell culture systems and their applications in drug discovery and cell-based biosensors. *Assay and Drug Development Technologies*, 12(4), 207–218. DOI: 10.1089/adt.2014.573 PMID: 24831787

Ege, N., Bouguenina, H., Tatari, M., & Chopra, R. (2021). Phenotypic screening with target identification and validation in the discovery and development of E3 ligase modulators. *Cell Chemical Biology*, 28(3), 283–299. DOI: 10.1016/j.chembiol.2021.02.011 PMID: 33740433

Eichler, H. G., Pignatti, F., Schwarzer-Daum, B., Hidalgo-Simon, A., Eichler, I., Arlett, P., Humphreys, A., Vamvakas, S., Brun, N., & Rasi, G. (2021). Randomized controlled trials versus real world evidence: Neither magic nor myth. *Clinical Pharmacology and Therapeutics*, 109(5), 1212–1218. DOI: 10.1002/cpt.2083 PMID: 33063841

Eke, A. C. (2022). An update on the physiologic changes during pregnancy and their impact on drug pharmacokinetics and pharmacogenomics. *Journal of Basic and Clinical Physiology and Pharmacology*, 33(5), 581–598. DOI: 10.1515/jbcpp-2021-0312 PMID: 34881531

Elgarhy, F. M., Borham, A., Alziny, N., AbdElaal, K. R., Shuaib, M., Musaibah, A. S., Hussein, M. A., & Abdelnaser, A. (2024). From Drug Discovery to Drug Approval: A Comprehensive Review of the Pharmacogenomics Status Quo with a Special Focus on Egypt. *Pharmaceuticals (Basel, Switzerland)*, 17(7), 881. DOI: 10.3390/ph17070881 PMID: 39065732

Ellegren, H. (2019). The study of STRs in population genetics. *Nature Reviews. Genetics*, 20(2), 129–138.

Elvas, L. B., Nunes, M., Ferreira, J. C., Dias, M. S., & Rosário, L. B. (2023). AI-Driven Decision Support for Early Detection of Cardiac Events: Unveiling Patterns and Predicting Myocardial Ischemia. *Journal of Personalized Medicine*, 13(9), 1421. DOI: 10.3390/jpm13091421 PMID: 37763188

Enko, D., Michaelis, S., Schneider, C., Schaflinger, E., Baranyi, A., Schnedl, W. J., & Muller, D. J. (2023). The Use of Next-Generation Sequencing in Pharmacogenomics. *Clinical Laboratory*, 69(8). Advance online publication. DOI: 10.7754/Clin.Lab.2023.230103 PMID: 37560847

Erceg, D. (2024). Pharmacogenomics in Drug Research and Development. In *Pharmacogenomics in Clinical Practice* (pp. 439–461). Springer International Publishing.

Esteva, A., Robicquet, A., Ramsundar, B., Kuleshov, V., DePristo, M., Chou, K., Cui, C., Corrado, G., Thrun, S., & Dean, J. (2019). A guide to deep learning in healthcare. *Nature Medicine*, 25(1), 24–29. DOI: 10.1038/s41591-018-0316-z PMID: 30617335

Fachrul, M., Utomo, D. H., & Parikesit, A. A. (2018). lncRNA-based study of epigenetic regulations in diabetic peripheral neuropathy. *In Silico Pharmacology*, 6(1), 7. DOI: 10.1007/s40203-018-0042-8 PMID: 30607320

Fan, F., & Wood, K. V. (2007, February). Bioluminescent assays for high-throughput screening. *Assay and Drug Development Technologies*, 5(1), 127–136. DOI: 10.1089/adt.2006.053 PMID: 17355205

Fang, X., Zheng, Y., Duan, Y., Liu, Y., & Zhong, W. (2018). Recent advances in design of fluorescence-based assays for high-throughput screening. *Analytical Chemistry*, 91(1), 482–504. DOI: 10.1021/acs.analchem.8b05303 PMID: 30481456

Fekete, F., Mangó, K., Déri, M., Incze, E., Minus, A., & Monostory, K. (2021). Impact of genetic and non-genetic factors on hepatic CYP2C9 expression and activity in Hungarian subjects. *Scientific Reports*, 11(1), 17081. DOI: 10.1038/s41598-021-96590-3 PMID: 34429480

Feldman, E. A. (2012). The Genetic Information Non-discrimination Act (GINA): Public Policy and Medical Practice in the Age of Personalized Medicine. *Journal of General Internal Medicine*, 27(6), 743–746. DOI: 10.1007/s11606-012-1988-6 PMID: 22314637

Fernald, G. H., Capriotti, E., Daneshjou, R., Karczewski, K. J., & Altman, R. B. (2011). Bioinformatics challenges for personalized medicine. *Bioinformatics (Oxford, England)*, 27(13), 1741–1748. DOI: 10.1093/bioinformatics/btr295 PMID: 21596790

Feuk, L., Carson, A. R., & Scherer, S. W. (2006). Structural variation in the human genome. *Nature Reviews. Genetics*, 7(2), 85–97. DOI: 10.1038/nrg1767 PMID: 16418744

Fidai, S. S., Sharma, A. E., Johnson, D. N., Segal, J. P., & Lastra, R. R. (2018). Dihydropyrimidine dehydrogenase deficiency as a cause of fatal 5-Fluorouracil toxicity. *Autopsy and Case Reports*, 8(4), e2018049. DOI: 10.4322/acr.2018.049 PMID: 30775324

Ford, N., Meintjes, G., Pozniak, A., Bygrave, H., Hill, A., Peter, T., Davies, M.-A., Grinsztejn, B., Calmy, A., Kumarasamy, N., Phanuphak, P., deBeaudrap, P., Vitoria, M., Doherty, M., Stevens, W., & Siberry, G. K. (2015). The future role of CD4 cell count for monitoring antiretroviral therapy. *The Lancet. Infectious Diseases*, 15(2), 241–247. DOI: 10.1016/S1473-3099(14)70896-5 PMID: 25467647

Foulkes, A. L., Soda, T., Farrell, M., Giusti-Rodríguez, P., & Lázaro-Muñoz, G. (2019). Legal and ethical implications of CRISPR applications in psychiatry. *North Carolina Law Review*, 97(5), 1359. PMID: 31871365

Founou, R. C., Founou, L. L., & Essack, S. Y. (2017). Clinical and economic impact of antibiotic resistance in developing countries: A systematic review and meta-analysis. *PLoS One*, 12(12), e0189621. DOI: 10.1371/journal.pone.0189621 PMID: 29267306

Franken, H., Mathieson, T., Childs, D., Sweetman, G. M., Werner, T., Tögel, I., Doce, C., Gade, S., Bantscheff, M., Drewes, G., Reinhard, F. B. M., Huber, W., & Savitski, M. M. (2015). Thermal proteome profiling for unbiased identification of direct and indirect drug targets using multiplexed quantitative mass spectrometry. *Nature Protocols*, 10(10), 1567–1593. DOI: 10.1038/nprot.2015.101 PMID: 26379230

French, M. T., Homer, J., Gumus, G., & Hickling, L. (2016). Key Provisions of the Patient Protection and Affordable Care Act (ACA): A Systematic Review and Presentation of Early Research Findings. *Health Services Research*, 51(5), 1735–1771. DOI: 10.1111/1475-6773.12511 PMID: 27265432

Gainor, J. F., Longo, D. L., & Chabner, B. A. (2014). Pharmacodynamic biomarkers: Falling short of the mark? *Clinical Cancer Research*, 20(10), 2587–2594. DOI: 10.1158/1078-0432.CCR-13-3132 PMID: 24831281

Galperin, M. Y., & Koonin, E. V. (2000). Who's your neighbor? New computational approaches for functional genomics. *Nature Biotechnology*, 18(6), 609–613. DOI: 10.1038/76443 PMID: 10835597

Gantz, M., Neun, S., Medcalf, E. J., van Vliet, L. D., & Hollfelder, F. (2023). Ultrahigh-throughput enzyme engineering and discovery in in vitro compartments. *Chemical Reviews*, 123(9), 5571–5611. DOI: 10.1021/acs.chemrev.2c00910 PMID: 37126602

Garcia-Agundez, A., Garcia-Martin, E., & Eickhoff, C. (2023). The Potential of Machine-learning in Pharmacogenetics, Pharmacogenomics and Pharmacoepidemiology: Volume II. *Frontiers in Pharmacology*, 14, 1277561. DOI: 10.3389/fphar.2023.1277561 PMID: 37711174

Gardy, J. L., & Loman, N. J. (2018). Towards a genomics-informed, real-time, global pathogen surveillance system. *Nature Reviews. Genetics*, 19(1), 9–20. DOI: 10.1038/nrg.2017.88 PMID: 29129921

Gargis, A. S. (2012). Assay validation for clinical application. *Clinical Laboratory*, 58(6), 249–257.

Gaulton, A., Hersey, A., Nowotka, M., Bento, A. P., Chambers, J., Mendez, D., Mutowo, P., Atkinson, F., Bellis, L. J., Cibrián-Uhalte, E., Davies, M., Dedman, N., Karlsson, A., Magariños, M. P., Overington, J. P., Papadatos, G., Smit, I., & Leach, A. R. (2017). The ChEMBL database in 2017. *Nucleic Acids Research*, 45(D1), D945–D954. DOI: 10.1093/nar/gkw1074 PMID: 27899562

Getahun, K. A., Angaw, D. A., Asres, M. S., Kahaliw, W., Petros, Z., Abay, S. M., Yimer, G., & Berhane, N. (2024). The Role of Pharmacogenomics Studies for Precision Medicine Among Ethiopian Patients and Their Clinical Implications: A Scoping Review. *Pharmacogenomics and Personalized Medicine*, 17, 347–361. DOI: 10.2147/PGPM.S454328 PMID: 38974617

Ghadessi, M., Di, J., Wang, C., Toyoizumi, K., Shao, N., Mei, C., Demanuele, C., Tang, R. S., McMillan, G., & Beckman, R. A. (2023). Decentralized clinical trials and rare diseases: A Drug Information Association Innovative Design Scientific Working Group (DIA-IDSWG) perspective. *Orphanet Journal of Rare Diseases*, 18(1), 79. DOI: 10.1186/s13023-023-02693-7 PMID: 37041605

Giannopoulou, E., Katsila, T., Mitropoulou, C., Tsermpini, E.-E., & Patrinos, G. P. (2019). Integrating next-generation sequencing in the clinical pharmacogenomics workflow. *Frontiers in Pharmacology*, 10, 384. DOI: 10.3389/fphar.2019.00384 PMID: 31024324

Giansanti, D. (2024). Joint Expedition: Exploring the Intersection of Digital Health and AI in Precision Medicine with Team Integration. *Journal of Personalized Medicine*, 14(4), 388. Advance online publication. DOI: 10.3390/jpm14040388 PMID: 38673015

Ginsburg, G. S. (2019). A global collaborative to advance genomic medicine. *American Journal of Human Genetics*, 104(3), 407–409. DOI: 10.1016/j.ajhg.2019.02.010 PMID: 30849327

Ginsburg, G. S., & Phillips, K. A. (2018). Precision medicine: From science to value. *Health Affairs*, 37(5), 746–754. DOI: 10.1377/hlthaff.2017.1624 PMID: 29733705

Giuliano, K. A., & Taylor, D. L. (1998). Fluorescent-protein biosensors: New tools for drug discovery. *Trends in Biotechnology*, 16(3), 135–140. DOI: 10.1016/S0167-7799(97)01166-9 PMID: 9523461

Gladding, P. A., Legget, M., Fatkin, D., Larsen, P., & Doughty, R. (2020). Polygenic risk scores in coronary artery disease and atrial fibrillation. *Heart Lung and Circulation*, 29(4), 634–640. DOI: 10.1016/j.hlc.2019.12.004 PMID: 31974023

Goetz, L. H., & Schork, N. J. (2018). Personalized medicine: Motivation, challenges, and progress. *Fertility and Sterility*, 109(6), 952–963. DOI: 10.1016/j.fertnstert.2018.05.006 PMID: 29935653

Goetz, M. P. (2018). Pharmacogenetic testing in breast cancer: Focus on tamoxifen therapy. *Clinical Pharmacology and Therapeutics*, 104(2), 229–238.

Gold, M. (1985). *A Conspiracy of Cells: One woman's immortal legacy-and the medical scandal it caused*. State University of New York Press.

Gong, L., Whirl-Carrillo, M., & Klein, T. E. (2021). PharmGKB, an Integrated Resource of Pharmacogenomic Knowledge. *Current Protocols*, 1(8), e226. DOI: 10.1002/cpz1.226 PMID: 34387941

Gonzalez, F. J., Skodat, R. C., Kimura, S., Umeno, M., Zanger, U. M., Nebert, D. W., Gelboin, H. V., Hardwick, J. P., & Meyer, U. A. (1988). Characterization of the common genetic defect in humans deficient in debrisoquine metabolism. *Nature*, 331(6155), 442–446. DOI: 10.1038/331442a0 PMID: 3123997

Goodwin, S., McPherson, J. D., & McCombie, W. R. (2016, May 17). Coming of age: Ten years of next-generation sequencing technologies. *Nature Reviews. Genetics*, 17(6), 333–351. DOI: 10.1038/nrg.2016.49 PMID: 27184599

Graansma, L. J., Zhai, Q., Busscher, L., Menafra, R., van den Berg, R. R., Kloet, S. L., & van der Lee, M. (2023). From gene to dose: Long-read sequencing and *-allele tools to refine phenotype predictions of CYP2C19. *Frontiers in Pharmacology*, 14, 1076574. DOI: 10.3389/fphar.2023.1076574 PMID: 36937863

Grant, D. M., & Tyndale, R. F. (2008). In memoriam: Werner Kalow, MD (1917–2008). *Pharmacogenetics and Genomics*, 18(10), 835–836. DOI: 10.1097/FPC.0b013e32830d102d

Grapov, D., Fahrmann, J., Wanichthanarak, K., & Khoomrung, S. (2018). Rise of Deep Learning for Genomic, Proteomic, and Metabolomic Data Integration in Precision Medicine. *OMICS: A Journal of Integrative Biology*, 22(10), 630–636. DOI: 10.1089/omi.2018.0097 PMID: 30124358

Greely, H. T. (2019). CRISPR'd babies: Human germline genome editing in the 'He Jiankui affair'. *Journal of Law and the Biosciences*, 6(1), 111–183. DOI: 10.1093/jlb/lsz010 PMID: 31666967

Green, R. M. (2007). *Babies by Design: The Ethics of Genetic Choice*. Yale University Press.

Grimberg, F., Asprion, P. M., Schneider, B., Miho, E., Babrak, L., & Habbabeh, A. (2021). The real-world data challenges radar: A review on the challenges and risks regarding the use of real-world data. *Digital Biomarkers*, 5(2), 148–157. DOI: 10.1159/000516178 PMID: 34414352

Grisanti, L. A. (2023). Radioligand Binding to Quantify Adrenergic Receptor Expression in the Heart. *Current Protocols*, 3(1), e649. DOI: 10.1002/cpz1.649 PMID: 36602296

Gronde, T. V. D., Uyl-de Groot, C. A., & Pieters, T. (2017). Addressing the challenge of high-priced prescription drugs in the era of precision medicine: A systematic review of drug life cycles, therapeutic drug markets and regulatory frameworks. *PLoS One*, 12(8), e0182613. DOI: 10.1371/journal.pone.0182613 PMID: 28813502

Gummadi, A. C., & Guddati, A. K. (2021). Genetic Polymorphisms in Pharmaceuticals and Chemotherapy. *World Journal of Oncology*, 12(5), 149–154. DOI: 10.14740/wjon1405 PMID: 34804277

Gunawan, A., Nadia, J., & Parikesit, A. A. (2024). The Molecular Docking Approach Toward Targeting Transcription Factors in Prostate Cancer. In *A Closer Look at Cancer Biomarkers* (p. Chapter 3). Nova Science Publishers. https://www.researchgate.net/publication/380032683_The_Molecular_Docking_Approach_Toward_Targeting_Transcription_Factors_in_Prostate_Cancer

Guo, Z., & Priefer, R. (2021). Current progress in pharmacogenomics of Type 2 diabetes: A systemic overview. *Diabetes & Metabolic Syndrome*, 15(5), 102239. DOI: 10.1016/j.dsx.2021.102239 PMID: 34371302

Gupta, G., Jha, K., & Chaudhary, S. (2023, July). Biosensor Assays Types and Their Roles Toward Ligand-Receptor Interactions in Drug Discovery. *Assay and Drug Development Technologies*, 21(5), 190–201. DOI: 10.1089/adt.2023.003 PMID: 37285190

Gupta, S., & Jhawat, V. (2017). Quality by design (QbD) approach of pharmacogenomics in drug designing and formulation development for optimization of drug delivery systems. *Journal of Controlled Release*, 245, 15–26. DOI: 10.1016/j.jconrel.2016.11.018 PMID: 27871989

Gymrek, M. (2017). The role of short tandem repeats in disease. *Nature Reviews. Genetics*, 18(3), 139–153.

Haldane, J. B. S. (1949). Disease and evolution. *La Ricerca Scientifica*, 19, 68–75.

Hamburg, M. A., & Collins, F. S. (2010). The path to personalized medicine. *The New England Journal of Medicine*, 363(4), 301–304. DOI: 10.1056/NEJMp1006304 PMID: 20551152

Hamilton, J. G., Banerjee, S. C., Carlsson, S. V., Vera, J., Lynch, K. A., Sar-Graycar, L., Martin, C. M., Parker, P. A., & Hay, J. L. (2021). Clinician perspectives on communication and implementation challenges in precision oncology. *Personalized Medicine*, 18(6), 559–572. DOI: 10.2217/pme-2021-0048 PMID: 34674550

Hariharan, S., Lorintiu, O., Lee, C. C., Duchemin-Pelletier, E., Li, X., Healy, A., . . . Christoforou, N. (2023). Development and validation of a high throughput screening platform to enable target identification in skeletal muscle cells from Duchenne Muscular Dystrophy (DMD) patients. bioRxiv, 2023-05.

Hartl, D., de Luca, V., Kostikova, A., Laramie, J., Kennedy, S., Ferrero, E., Siegel, R., Fink, M., Ahmed, S., Millholland, J., Schuhmacher, A., Hinder, M., Piali, L., & Roth, A. (2021). Translational precision medicine: An industry perspective. *Journal of Translational Medicine*, 19(1), 245. DOI: 10.1186/s12967-021-02910-6 PMID: 34090480

Hart, T., Chandrashekhar, M., Aregger, M., Steinhart, Z., Brown, K. R., MacLeod, G., Mis, M., Zimmermann, M., Fradet-Turcotte, A., Sun, S., Mero, P., Dirks, P., Sidhu, S., Roth, F. P., Rissland, O. S., Durocher, D., Angers, S., & Moffat, J. (2015). High-resolution CRISPR screens reveal fitness genes and genotype-specific cancer liabilities. *Cell*, 163(6), 1515–1526. DOI: 10.1016/j.cell.2015.11.015 PMID: 26627737

Hasanzad, M., Sarhangi, N., Ehsani Chimeh, S., Ayati, N., Afzali, M., Khatami, F., Nikfar, S., & Aghaei Meybodi, H. R. (2022). Precision medicine journey through omics approach. *Journal of Diabetes and Metabolic Disorders*, 21(1), 881–888. DOI: 10.1007/s40200-021-00913-0 PMID: 35673436

Hasin, Y., Seldin, M., & Lusis, A. (2017). Multi-omics approaches to disease. *Genome Biology*, 18(1), 1–15. DOI: 10.1186/s13059-017-1215-1 PMID: 28476144

Hassan, R., Allali, I., Agamah, F. E., Elsheikh, S. S. M., Thomford, N. E., Dandara, C., & Chimusa, E. R. (2021). Drug response in association with pharmacogenomics and pharmacomicrobiomics: Towards a better personalized medicine. *Briefings in Bioinformatics*, 22(4), bbaa292. Advance online publication. DOI: 10.1093/bib/bbaa292 PMID: 33253350

Healing, E., Charlier, C. F., Meira, L. B., & Elliott, R. M. (2019). A panel of colorimetric assays to measure enzymatic activity in the base excision DNA repair pathway. *Nucleic Acids Research*, 47(11), e61–e61. DOI: 10.1093/nar/gkz171 PMID: 30869144

Heerlie, D. M., Margaretha, F., Fugaha, D. R., & Parikesit, A. A. (2024). The Effects and Treatments for Usher Syndrome: A Review. *Indonesian Journal of Life Sciences*, 6(1), 68–82. DOI: 10.54250/IJLS.V6I01.184

Henderson, L. M., Claw, K. G., Woodahl, E. L., Robinson, R. F., Boyer, B. B., Burke, W., & Thummel, K. E. (2018). P450 pharmacogenetics in indigenous North American populations. *Journal of Personalized Medicine*, 8(1), 9. DOI: 10.3390/jpm8010009 PMID: 29389890

Hensing, W., Santa-Maria, C. A., Peterson, L. L., & Sheng, J. Y. (2020, October). Landmark trials in the medical oncology management of early stage breast cancer. [). WB Saunders.]. *Seminars in Oncology*, 47(5), 278–292. DOI: 10.1053/j.seminoncol.2020.08.001 PMID: 32933761

Hevener, K. E., Pesavento, R., Ren, J. H., Lee, H., Ratia, K., & Johnson, M. E. (2018). Hit-to-Lead: hit validation and assessment. In *Methods in Enzymology* (Vol. 610, pp. 265–309). Academic Press.

Hewitt, J. E. (2018). The ethical, legal and regulatory issues associated with pharmacogenomics: Systematically quantifying the literature. *Journal of Law and Medicine*, 25(3), 782–793. PMID: 29978667

Hippman, C., & Nislow, C. (2019). Pharmacogenomic Testing: Clinical Evidence and Implementation Challenges. *Journal of Personalized Medicine*, 9(3), 40. Advance online publication. DOI: 10.3390/jpm9030040 PMID: 31394823

Hirsch, F. R. (2017). PD-L1 testing in cancer: The challenges of biomarker testing and interpretation. *Nature Reviews. Clinical Oncology*, 14(9), 508–517.

Hirschhorn, J. N., & Daly, M. J. (2005). Genome-wide association studies for common diseases and complex traits. *Nature Reviews. Genetics*, 6(2), 95–108. DOI: 10.1038/nrg1521 PMID: 15716906

Hobbs, B. P., Pestana, R. C., Zabor, E. C., Kaizer, A. M., & Hong, D. S. (2022). Basket trials: Review of current practice and innovations for future trials. *Journal of Clinical Oncology*, 40(30), 3520–3528. DOI: 10.1200/JCO.21.02285 PMID: 35537102

Ho, D., Quake, S. R., McCabe, E. R., Chng, W. J., Chow, E. K., Ding, X., Gelb, B. D., Ginsburg, G. S., Hassenstab, J., Ho, C.-M., Mobley, W. C., Nolan, G. P., Rosen, S. T., Tan, P., Yen, Y., & Zarrinpar, A. (2020). Enabling technologies for personalized and precision medicine. *Trends in Biotechnology*, 38(5), 497–518. DOI: 10.1016/j.tibtech.2019.12.021 PMID: 31980301

Hongladarom, S. (2006). Ethics of bioinformatics: A convergence between bioethics and computer ethics. *Asian Biotechnology and Development Review*, 9(1), 37–44.

Horgan, D., Ciliberto, G., Conte, P., Curigliano, G., Seijo, L., Montuenga, L. M., Garassino, M., Penault-Llorca, F., Galli, F., Ray-Coquard, I., Querleu, D., Riegman, P., Kerr, K., Van Poppel, H., Bjartell, A., Codacci-Pisanelli, G., Koeva-Balabanova, J., Paradiso, A., Maravic, Z., & Capoluongo, E. (2021). Bringing onco-innovation to Europe's healthcare systems: The potential of biomarker testing, real world evidence, tumour agnostic therapies to empower personalised medicine. *Cancers (Basel)*, 13(3), 583. DOI: 10.3390/cancers13030583 PMID: 33540773

Horn, S. (2018). TERT promoter mutations in telomere-driven diseases. *Nature Genetics*, 50(4), 523–529.

Horwitz, R. I., Lobitz, G., Mawn, M., Conroy, A. H., Cullen, M. R., Sim, I., & Singer, B. H. (2021). Biosocial medicine: Biology, biography, and the tailored care of the patient. *SSM - Population Health*, 15, 100863. DOI: 10.1016/j.ssmph.2021.100863 PMID: 34430699

Hossain, M. A. (2023). Precision Medicine: An emerging approach for patient care. *Bangladesh Journal of Medicine*, 201-201.

Houle, D., Govindaraju, D. R., & Omholt, S. W. (2010). Phenomics: The next challenge. *Nature Reviews. Genetics*, 11(12), 855–866. DOI: 10.1038/nrg2897 PMID: 21085204

Huang, D., Yang, M., Wen, X., Xia, S., & Yuan, B. (2024). AI-DRIVEN DRUG DISCOVERY: ACCELERATING THE DEVELOPMENT OF NOVEL THERAPEUTICS IN BIOPHARMACEUTICALS. *Journal of Knowledge Learning and Science Technology ISSN: 2959-6386 (online), 3*(3), 206-224.

Huang, D. W. (2009). DAVID: A web-based tool for functional annotation of gene lists. *Bioinformatics (Oxford, England)*, 16(9), 525–527.

Huang, J. A., Wu, Y. H., Chen, P. L., Weng, Y. C., Chiang, I. C., Huang, Y. T., & Chou, W. H. (2024). MMP-9 upregulation may predict hemorrhagic transformation after endovascular thrombectomy. *Frontiers in Neurology*, 15, 1400270. DOI: 10.3389/fneur.2024.1400270 PMID: 38798706

Huber, W., Carey, V. J., Gentleman, R., Anders, S., Carlson, M., Carvalho, B. S., Bravo, H. C., Davis, S., Gatto, L., Girke, T., Gottardo, R., Hahne, F., Hansen, K. D., Irizarry, R. A., Lawrence, M., Love, M. I., MacDonald, J., Obenchain, V., Oleś, A. K., & Morgan, M. (2015). Orchestrating high-throughput genomic analysis with Bioconductor. *Nature Methods*, 12(2), 115–121. DOI: 10.1038/nmeth.3252 PMID: 25633503

Hughes, J. P., Rees, S., Kalindjian, S. B., & Philpott, K. L. (2011). Principles of early drug discovery. *British Journal of Pharmacology*, 162(6), 1239–1249. DOI: 10.1111/j.1476-5381.2010.01127.x PMID: 21091654

Hukerikar, N., Hingorani, A. D., Asselbergs, F. W., Finan, C., & Schmidt, A. F. (2024). Prioritising genetic findings for drug target identification and validation. *Atherosclerosis*, 390, 117462. Advance online publication. DOI: 10.1016/j.atherosclerosis.2024.117462 PMID: 38325120

Hulme, E. C., & Trevethick, M. A. (2010, November). Ligand binding assays at equilibrium: Validation and interpretation. *British Journal of Pharmacology*, 161(6), 1219–1237. DOI: 10.1111/j.1476-5381.2009.00604.x PMID: 20132208

Huttlin, E. L. (2021). The BioPlex network: A comprehensive map of the human interactome. *Science*, 360(6393), 181–187.

Hyman, D. M. (2017). Implementing genome-driven oncology. *American Journal of Clinical Oncology*, 40(1), 1–11. PMID: 28059930

Ingelman-Sundberg, M., Nebert, D. W., & Lauschke, V. M. (2023). Emerging trends in pharmacogenomics: From common variant associations toward comprehensive genomic profiling. *Human Genomics*, 17(1), 105. DOI: 10.1186/s40246-023-00554-9 PMID: 37996916

Issa, J., Riad, A., Olszewski, R., & Dyszkiewicz-Konwińska, M. (2023). The Influence of Slice Thickness, Sharpness, and Contrast Adjustments on Inferior Alveolar Canal Segmentation on Cone-Beam Computed Tomography Scans: A Retrospective Study. *Journal of Personalized Medicine*, 13(10), 1518. DOI: 10.3390/jpm13101518 PMID: 37888129

Jameson, J. L., & Longo, D. L. (2015). Precision medicine—Personalized, problematic, and promising. *The New England Journal of Medicine*, 372(23), 2229–2231. DOI: 10.1056/NEJMsb1503104 PMID: 26014593

Jarvis, J. P., Peter, A. P., Keogh, M., Baldasare, V., Beanland, G. M., Wilkerson, Z. T., Kradel, S., & Shaman, J. A. (2022). Real-world impact of a pharmacogenomics-enriched comprehensive medication management program. *Journal of Personalized Medicine*, 12(3), 421. DOI: 10.3390/jpm12030421 PMID: 35330421

Jarvis, J. P., Peter, A. P., & Shaman, J. A. (2019). Consequences of CYP2D6 copy-number variation for pharmacogenomics in psychiatry. *Frontiers in Psychiatry*, 10, 432. DOI: 10.3389/fpsyt.2019.00432 PMID: 31281270

Jhawat, V., Gulia, M., Gupta, S., Maddiboyina, B., & Dutt, R. (2020). Integration of pharmacogenomics and theranostics with nanotechnology as quality by design (QbD) approach for formulation development of novel dosage forms for effective drug therapy. *Journal of Controlled Release*, 327, 500–511. DOI: 10.1016/j.jconrel.2020.08.039 PMID: 32858073

Jiang, Z., & Zhou, Y. (2005). Using bioinformatics for drug target identification from the genome. *American Journal of Pharmacogenomics*, 5(6), 387–396. DOI: 10.2165/00129785-200505060-00005 PMID: 16336003

Jin, B., Tao, Q., Peng, J., Soo, H. M., Wu, W., Ying, J., Fields, C. R., Delmas, A. L., Liu, X., Qiu, J., & Robertson, K. D. (2008). DNA methyltransferase 3B (DNMT3B) mutations in ICF syndrome lead to altered epigenetic modifications and aberrant expression of genes regulating development, neurogenesis and immune function. *Human Molecular Genetics*, 17(5), 690–709. DOI: 10.1093/hmg/ddm341 PMID: 18029387

Ji, Y., & Shaaban, S. (2024). Interrogating Pharmacogenetics Using Next-Generation Sequencing. *The Journal of Applied Laboratory Medicine*, 9(1), 50–60. DOI: 10.1093/jalm/jfad097 PMID: 38167765

Johansson, E. (2019). Analytical validation of genomic tests. *European Journal of Human Genetics*, 27(4), 475–487.

John, S., Ravichandran, N., & Khan, M. F. (2018). Electronic medical record for deliverance of effective healthcare delivery: Ethical issues and challenges of digitalization in clinical information and Electronic Medical Records (EMR) management. *IOSR Journal of Business and Management (IOSR-JBM), 20*(3), 01-06. DOI: DOI: 10.9790/487X-2003020106

Johnson, J. A., & Cavallari, L. H. (2015). Warfarin Pharmacogenetics. *Trends in Cardiovascular Medicine*, 25(1), 33–41. DOI: 10.1016/j.tcm.2014.09.001 PMID: 25282448

Johnson, J. A., & Cavallari, L. H. (2019). Pharmacogenetics of warfarin: An update of the clinical implications. *Journal of Clinical Pharmacology*, 59(6), 738–746.

Johnson, K. B., Wei, Q., Weeraratne, D., Frisse, M. E., Misulis, K., Rhee, K., Zhao, J., & Snowdon, J. L. (2020). Precision Medicine, AI, and the Future of Personalized Health Care. *Clinical and Translational Science*, 14(1), 86–93. DOI: 10.1111/cts.12884 PMID: 32961010

Jolly, B., & Scaria, V. (2024). Ethnic differences in pharmacogenomic variants: A south Asian perspective. *Pharmacogenomics*, 25(4), 171–174. DOI: 10.2217/pgs-2024-0026 PMID: 38511426

Jones, E., Michael, S., & Sittampalam, G. S. (2016). *Basics of assay equipment and instrumentation for high throughput screening*. Assay Guidance Manual. [Internet]

Jones, P. A., & Baylin, S. B. (2019). The epigenomics of cancer. *Cell*, 128(4), 683–692. DOI: 10.1016/j.cell.2007.01.029 PMID: 17320506

Jordan, I. K., Sharma, S., & Mariño-Ramírez, L. (2023). Population pharmacogenomics for health equity. *Genes*, 14(10), 1840. DOI: 10.3390/genes14101840 PMID: 37895188

Kabadi, A., McDonnell, E., Frank, C. L., & Drowley, L. (2020). Applications of Functional Genomics for Drug Discovery. In *SLAS Discovery* (Vol. 25, Issue 8). DOI: 10.1177/2472555220902092

Kabbani, D., Akika, R., Wahid, A., Daly, A. K., Cascorbi, I., & Zgheib, N. K. (2023). Pharmacogenomics in practice: A review and implementation guide. *Frontiers in Pharmacology*, 14, 1189976. DOI: 10.3389/fphar.2023.1189976 PMID: 37274118

Kachhadiya, D. K., Pooja, K., Mishra, S. K., & George, J. J. (2024). In Silico Based Identification Of Novel Inhibitors For Selected MDR Protein From Shigella Species: A Validation Through Molecular Docking Analysis. *Educational Administration: Theory And Practice, 30*(6 (S)), 309-316.

Kalinin, A. A., Higgins, G. A., Reamaroon, N., Soroushmehr, S., Allyn-Feuer, A., Dinov, I. D., Najarian, K., & Athey, B. D. (2018). Deep learning in pharmacogenomics: From gene regulation to patient stratification. *Pharmacogenomics*, 19(7), 629–650. DOI: 10.2217/pgs-2018-0008 PMID: 29697304

Kantae, V., Krekels, E. H., Esdonk, M. J. V., Lindenburg, P., Harms, A. C., Knibbe, C. A., Van der Graaf, P. H., & Hankemeier, T. (2017). Integration of pharmacometabolomics with pharmacokinetics and pharmacodynamics: Towards personalized drug therapy. *Metabolomics*, 13(1), 1–11. DOI: 10.1007/s11306-016-1143-1 PMID: 28058041

Karczewski, K. J., & Snyder, M. P. (2018). Integrating clinical and genomic data for precision medicine. *Genome Research*, 28(5), 749–758.

Karouzakis, E., Gay, R. E., Michel, B. A., Gay, S., & Neidhart, M. (2009). DNA hypomethylation in rheumatoid arthritis synovial fibroblasts. *Arthritis and Rheumatism*, 60(12), 3613–3622. DOI: 10.1002/art.25018 PMID: 19950268

Katara, P. (2013). Role of bioinformatics and pharmacogenomics in drug discovery and development process. *Network Modeling and Analysis in Health Informatics and Bioinformatics*, 2(4), 225–230. DOI: 10.1007/s13721-013-0039-5

Katara, P., & Yadav, A. (2019). Pharmacogenes (PGx-genes): Current understanding and future directions. *Gene*, 718, 144050. DOI: 10.1016/j.gene.2019.144050 PMID: 31425740

Kato, S., Subbiah, V., & Kurzrock, R. (2017). Counterpoint: successes in the pursuit of precision medicine: biomarkers take credit. *Journal of the National Comprehensive Cancer Network : JNCCN*, 15(7), 863–866. DOI: 10.6004/jnccn.2017.0127 PMID: 28687573

Keefe, D., & Bateman, E. (2012). Tumor control versus adverse events with targeted anticancer therapies. *Nature Reviews. Clinical Oncology*, 9(2), 98–109. DOI: 10.1038/nrclinonc.2011.192 PMID: 22182972

Keerthana, V., Mohaideen, S. S., Vigneshwaran, L. V., & Kumar, M. S. (2022). Role of artificial intelligence in drug development. *International Journal of Research in Pharmaceutical Sciences and Technology, 3*(1), 09-14.

Keller, L. (2021). Quality assurance and accreditation in genomic testing: A review of regulations and standards. *Genetic Testing and Molecular Biomarkers*, 25(7), 489–495.

Kenner, B., Chari, S. T., Kelsen, D., Klimstra, D. S., Pandol, S. J., Rosenthal, M., Rustgi, A. K., Taylor, J. A., Yala, A., Abul-Husn, N., Andersen, D. K., Bernstein, D., Brunak, S., Canto, M. I., Eldar, Y. C., Fishman, E. K., Fleshman, J., Go, V. L. W., Holt, J. M., & Wolpin, B. (2021). Artificial intelligence and early detection of pancreatic cancer: 2020 summative review. *Pancreas*, 50(3), 251–279. DOI: 10.1097/MPA.0000000000001762 PMID: 33835956

Keskin, O., & Gursoy, A. (2018). Role of protein-protein interactions in human disease. *Nature Reviews. Molecular Cell Biology*, 19(4), 220–232.

Khedr, S., Donia, T., Salim, E., & Hessien, M. (2020). CYP2D6 expression is cumulatively up regulated in multidrug treated hepatoma cells: A predective pharmacogenetics in vitro model. *The Egyptian Journal of Biochemistry and Molecular Biology*, 38(1), 39–55. DOI: 10.21608/ejb.2020.136526

Kiani, A. K., Pheby, D., Henehan, G., Brown, R., Sieving, P., Sykora, P., & Marks, R.. (2022). Ethical considerations regarding animal experimentation. *Journal of Preventive Medicine and Hygiene*, 63(2, Suppl 3), E255. PMID: 36479489

Kim, I. W., Han, N., Kim, M. G., Kim, T., & Oh, J. M. (2015). Copy number variability analysis of pharmacogenes in patients with lymphoma, leukemia, hepatocellular, and lung carcinoma using The Cancer Genome Atlas data. *Pharmacogenetics and Genomics*, 25(1), 1–7. DOI: 10.1097/FPC.0000000000000097 PMID: 25379720

Kim, J. A., Ceccarelli, R., & Lu, C. Y. (2021). Pharmacogenomic Biomarkers in US FDA-Approved Drug Labels (2000-2020). *Journal of Personalized Medicine*, 11(3), 179. Advance online publication. DOI: 10.3390/jpm11030179 PMID: 33806453

Kim, J., Basak, J. M., & Holtzman, D. M. (2009). The Role of Apolipoprotein E in Alzheimer's Disease. *Neuron*, 63(3), 287–303. DOI: 10.1016/j.neuron.2009.06.026 PMID: 19679070

Kim, S., Chen, J., Cheng, T., Gindulyte, A., He, J., He, S., Li, Q., Shoemaker, B. A., Thiessen, P. A., Yu, B., Zaslavsky, L., Zhang, J., & Bolton, E. E. (2019). PubChem 2019 update: Improved access to chemical data. *Nucleic Acids Research*, 47(D1), D1102–D1109. DOI: 10.1093/nar/gky1033 PMID: 30371825

Kim, W. Y., Kim, H. S., Oh, M., & Shin, J. G. (2020). Survey of physicians' views on the clinical implementation of pharmacogenomics-based personalized therapy. *Translational and Clinical Pharmacology*, 28(1), 34–42. DOI: 10.12793/tcp.2020.28.e6 PMID: 32274379

Kim, Y., & Lee, H.-M. (2023). CRISPR-Cas System Is an Effective Tool for Identifying Drug Combinations That Provide Synergistic Therapeutic Potential in Cancers. *Cells*, 12(22), 2593. DOI: 10.3390/cells12222593 PMID: 37998328

Kircher, M., Witten, D. M., Jain, P., O'Roak, B. J., Cooper, G. M., & Shendure, J. (2014). A general framework for estimating the relative pathogenicity of human genetic variants. *Nature Genetics*, 46(3), 310–315. DOI: 10.1038/ng.2892 PMID: 24487276

Kivipelto, M., Mangialasche, F., Snyder, H. M., Allegri, R., Andrieu, S., Arai, H., Baker, L., Belleville, S., Brodaty, H., Brucki, S. M., Calandri, I., Caramelli, P., Chen, C., Chertkow, H., Chew, E., Choi, S. H., Chowdhary, N., Crivelli, L., La Torre, R. D., & Carrillo, M. C. (2020). World-Wide FINGERS Network: A global approach to risk reduction and prevention of dementia. *Alzheimer's & Dementia*, 16(7), 1078–1094. DOI: 10.1002/alz.12123 PMID: 32627328

Klann, T. (2020). CRISPR-Cas9 genome editing: Current challenges and future prospects. *Molecular Therapy*, 28(5), 1120–1131.

Klein, M. E., Parvez, M. M., & Shin, J. G. (2017). Clinical implementation of pharmacogenomics for personalized precision medicine: Barriers and solutions. *Journal of Pharmaceutical Sciences*, 106(9), 2368–2379. DOI: 10.1016/j.xphs.2017.04.051 PMID: 28619604

Knez, D., Gobec, S., & Hrast, M. (2022). Screening of Big Pharma's Library against Various *in-house* Biological Targets. *Molecules (Basel, Switzerland)*, 27(14), 4484. DOI: 10.3390/molecules27144484 PMID: 35889355

Knoppers, B. M. (2014). International ethics harmonization and the global alliance for genomics and health. *Genome Medicine*, 6(2), 13. DOI: 10.1186/gm530 PMID: 25031613

Knoppers, B. M. (2018). Genomic privacy and ethical governance. *Human Genetics*, 137(6), 799–811.

Knowles, L., Luth, W., & Bubela, T. (2017). Paving the road to personalized medicine: Recommendations on regulatory, intellectual property and reimbursement challenges. *Journal of Law and the Biosciences*, 4(3), 453–506. DOI: 10.1093/jlb/lsx030 PMID: 29868182

Koboldt, D. C., Steinberg, K. M., Larson, D. E., Wilson, R. K., & Mardis, E. R. (2013). The next-generation sequencing revolution and its impact on genomics. *Cell*, 155(1), 27–39. DOI: 10.1016/j.cell.2013.09.006 PMID: 24074859

Kojima, T., Kawai, K., Miyazaki, J., & Nishiyama, H. (2017). Biomarkers for precision medicine in bladder cancer. *International Journal of Clinical Oncology*, 22(2), 207–213. DOI: 10.1007/s10147-016-1068-8 PMID: 27896485

Kolluri, S., Lin, J., Liu, R., Zhang, Y., & Zhang, W. (2022). Machine learning and artificial intelligence in pharmaceutical research and development: A review. *The AAPS Journal*, 24, 1–10.

König, I. R., Fuchs, O., Hansen, G., von Mutius, E., & Kopp, M. V. (2017). What is precision medicine? *The European Respiratory Journal*, 50(4), 1700391. DOI: 10.1183/13993003.00391-2017 PMID: 29051268

Krebs, K., & Milani, L. (2019). Translating pharmacogenomics into clinical decisions: Do not let the perfect be the enemy of the good. *Human Genomics*, 13(1), 39. Advance online publication. DOI: 10.1186/s40246-019-0229-z PMID: 31455423

Krzyszczyk, P., Acevedo, A., Davidoff, E. J., Timmins, L. M., Marrero-Berrios, I., Patel, M., White, C., Lowe, C., Sherba, J. J., Hartmanshenn, C., O'Neill, K. M., Balter, M. L., Fritz, Z. R., Androulakis, I. P., Schloss, R. S., & Yarmush, M. L. (2018). The growing role of precision and personalized medicine for cancer treatment. *Technology*, 6(3-4), 79–100. DOI: 10.1142/S2339547818300020 PMID: 30713991

Kulkarni, L., Nirmala, M. S., Najma, U., & Jyothi, N. M. (2023). AI-Enabled Genomic Biomarkers: The Future of Pharmaceutical Industry and Personalized Medicine. *Seybold Report Journal*, 18(08), 54–72. DOI: 10.5110/77-1024

Kumari, D., Swaroop, M., Southall, N., Huang, W., Zheng, W., & Usdin, K. (2015). High-throughput screening to identify compounds that increase fragile X mental retardation protein expression in neural stem cells differentiated from fragile X syndrome patient-derived induced pluripotent stem cells. *Stem Cells Translational Medicine*, 4(7), 800–808. DOI: 10.5966/sctm.2014-0278 PMID: 25999519

Kunkle, B. W., Grenier-Boley, B., Sims, R., Bis, J. C., Damotte, V., Naj, A. C., & Rotter, J. I. (2019). Genetic meta-analysis of diagnosed Alzheimer's disease identifies new risk loci and implicates A beta, tau, immunity and lipid processing (vol 51, pg 414, 2019). *Nature Genetics*, 51(9), 1423–1424. DOI: 10.1038/s41588-019-0495-7 PMID: 31417202

Ku, S. K., Yi, S. H., & Nam, D.-H. (2014). High-throughput screening (HTS) of anticancer drug efficacy on a micropillar/microwell chip platform. *Analytical Chemistry*, 86(1), 535–542. DOI: 10.1021/ac402546b PMID: 24199994

Lalitkumar, K. (2023). Artificial Intelligence in Pharmaceutical Technology and Drug Delivery Design. *Pharmaceutics*, 15(7), 1916. PMID: 37514102

Lal, S., Wong, Z. W., Jada, S. R., Xiang, X., Shu, X. C., Ang, P. C. S., Figg, W. D., Lee, E. J. D., & Chowbay, B. (2007). Novel SLC22A16 polymorphisms and influence on doxorubicin pharmacokinetics in Asian breast cancer patients. *Pharmacogenomics*, 8(6), 567–575. DOI: 10.2217/14622416.8.6.567 PMID: 17559346

Latella, M. C., Di Salvo, M. T., Cocchiarella, F., Benati, D., Grisendi, G., Comitato, A., Marigo, V., & Recchia, A. (2016). In vivo Editing of the Human Mutant Rhodopsin Gene by Electroporation of Plasmid-based CRISPR/Cas9 in the Mouse Retina. *Molecular Therapy. Nucleic Acids*, 5, e389. DOI: 10.1038/mtna.2016.92 PMID: 27874856

Lauschke, V. M., & Ingelman-Sundberg, M. (2020). Emerging strategies to bridge the gap between pharmacogenomic research and its clinical implementation. *NPJ Genomic Medicine*, 5(1), 9. DOI: 10.1038/s41525-020-0119-2 PMID: 32194983

Lauschke, V. M., Milani, L., & Ingelman-Sundberg, M. (2018). Pharmacogenomic biomarkers for improved drug therapy—Recent progress and future developments. *The AAPS Journal*, 20(1), 1–16. DOI: 10.1208/s12248-017-0161-x PMID: 29181807

Lauschke, V. M., Zhou, Y., & Ingelman-Sundberg, M. (2024). Pharmacogenomics beyond single common genetic variants: The way forward. *Annual Review of Pharmacology and Toxicology*, 64(1), 33–51. DOI: 10.1146/annurev-pharmtox-051921-091209 PMID: 37506333

Lawson, C. E., Manuel Martí, J., Radivojevic, T., Sai Vamshi, R. J., Gentz, R., Hillson, N. J., & Peisert, S.. (2021). Machine learning for metabolic engineering: A review. *Metabolic Engineering*, 63, 34–60. DOI: 10.1016/j.ymben.2020.10.005 PMID: 33221420

le Sage, C., Lawo, S., & Cross, B. C. S. (2020, March). CRISPR: A Screener's Guide. *SLAS Discovery*, 25(3), 233–240. DOI: 10.1177/2472555219883621 PMID: 31658850

Le Tourneau, C. (2015). Molecularly targeted therapy of cancer. *Nature Reviews. Clinical Oncology*, 12(3), 179–190. PMID: 26342236

LeBlanc, E. V., Shekhar-Guturja, T., Whitesell, L., & Cowen, L. E. (2021). Fluorescence Polarization-Based Measurement of Protein-Ligand Interaction in Fungal Cell Lysates. *Current Protocols*, 1(1), e17. DOI: 10.1002/cpz1.17 PMID: 33484500

Leitner, A., Faini, M., Stengel, F., & Aebersold, R. (2016). Crosslinking and mass spectrometry: An integrated technology to understand the structure and function of molecular machines. *Trends in Biochemical Sciences*, 41(1), 20–32. DOI: 10.1016/j.tibs.2015.10.008 PMID: 26654279

Lei, X., & Liyin, L. (2024, August). Qiutang, Wang., Bing, Pan., Leilei, Zheng., Zheng, Lin. (2024). Effect of pharmacogenomic testing on the clinical treatment of patients with depressive disorder: A randomized clinical trial. *Journal of Affective Disorders*, 359, 117–124. Advance online publication. DOI: 10.1016/j.jad.2024.05.063

Lenz, C., Narang, A., & Bousman, C. A. (2024). Pharmacogenomic allele coverage of genome-wide genotyping arrays: A comparative analysis. *Pharmacogenetics and Genomics*, 34(4), 130–134. DOI: 10.1097/FPC.0000000000000523 PMID: 38359167

Lever, S. Z., Fan, K. H., & Lever, J. R. (2017). Tactics for preclinical validation of receptor-binding radiotracers. *Nuclear Medicine and Biology*, 44, 4–30. DOI: 10.1016/j.nucmedbio.2016.08.015 PMID: 27755986

Levine, K. S., Leonard, H. L., Blauwendraat, C., Iwaki, H., Johnson, N., Bandres-Ciga, S., Ferrucci, L., Faghri, F., Singleton, A. B., & Nalls, M. A. (2023). Virus exposure and neurodegenerative disease risk across national biobanks. *Neuron*, 0(0), 1086–1093.e2. Advance online publication. DOI: 10.1016/j.neuron.2022.12.029 PMID: 36669485

Lewis, C. M., & Vassos, E. (2020). Polygenic risk scores: From research tools to clinical instruments. *The Lancet. Psychiatry*, 7(1), 5–7. PMID: 31860452

Libbrecht, M. W., & Noble, W. S. (2015). Machine learning applications in genetics and genomics. *Nature Reviews. Genetics*, 16(6), 321–332. DOI: 10.1038/nrg3920 PMID: 25948244

Liebman, I. (2022). The Role of Artificial Intelligence in Drug Discovery and Development. *Chemistry International*, 44(1), 16–19. DOI: 10.1515/ci-2022-0105

Li, H., & Durbin, R. (2009). Fast and accurate short read alignment with Burrows-Wheeler transform. *Bioinformatics (Oxford, England)*, 25(14), 1754–1760. DOI: 10.1093/bioinformatics/btp324 PMID: 19451168

Li, J., & Qiu, C. (2022). Recent advances in pharmacogenomics research of anti-asthmatic drugs: A narrative review. *Annals of Translational Medicine*, 10(6), 369. DOI: 10.21037/atm-22-291 PMID: 35434004

Lin, E., Lin, C. H., & Lane, H. Y. (2020). Precision psychiatry applications with pharmacogenomics: Artificial intelligence and machine learning approaches. *International Journal of Molecular Sciences*, 21(3), 969. DOI: 10.3390/ijms21030969 PMID: 32024055

Lionta, E., Spyrou, G. K., Vassilatis, D., & Cournia, Z. (2014). Structure-based virtual screening for drug discovery: Principles, applications and recent advances. *Current Topics in Medicinal Chemistry*, 14(16), 1923–1938. DOI: 10.2174/1568026614666140929124445 PMID: 25262799

Little, J. (2009). The Strengthening the Reporting of Genetic Association Studies (STREGA) Statement: Guidelines for reporting genetic association studies. *American Journal of Human Genetics*, 78(1), 29–41.

Liu, A., Chokshi, M., Nguyen, N., Powell, R. T., Stephan, C. C., & Bao, G. (2024). Cystic fibrosis cell models for high-throughput analysis and drug screening. *Journal of Cystic Fibrosis*, 23(4), 716–724. DOI: 10.1016/j.jcf.2024.07.001 PMID: 39060183

Liu, X., Luo, X., Jiang, C., & Zhao, H. (2019). Difficulties and challenges in the development of precision medicine. *Clinical Genetics*, 95(5), 569–574. DOI: 10.1111/cge.13511 PMID: 30653655

Liu, Y., Aryee, M. J., Padyukov, L., Fallin, M. D., Hesselberg, E., Runarsson, A., Reinius, L., Acevedo, N., Taub, M., Ronninger, M., Shchetynsky, K., Scheynius, A., Kere, J., Alfredsson, L., Klareskog, L., Ekström, T. J., & Feinberg, A. P. (2013). Epigenome-wide association data implicate DNA methylation as an intermediary of genetic risk in rheumatoid arthritis. *Nature Biotechnology*, 31(2), 142–147. DOI: 10.1038/nbt.2487 PMID: 23334450

Liu, Y., Bian, C., Huang, Y., You, X., & Shi, Q. (2017). Recent advances in bioinformatics for fish genomics. In *Bioinformarics* (p. 11). Avid Science., https://www.pacb.com/publications/recent-advances-in-bioinformatics-for-fish-genomics/

Liu, Y., Song, F., Li, Z., Chen, L., Xu, Y., Sun, H., & Chang, Y. (2023). A comprehensive tool for tumor precision medicine with pharmaco-omics data analysis. *Frontiers in Pharmacology*, 14, 1085765. DOI: 10.3389/fphar.2023.1085765 PMID: 36713829

Li, Z.-H., Wang, J., Xu, J.-P., Wang, J., & Yang, X. (2023). Recent advances in CRISPR-based genome editing technology and its applications in cardiovascular research. *Military Medical Research*, 10(1), 12. DOI: 10.1186/s40779-023-00447-x PMID: 36895064

Lo, C., Nguyen, S., Yang, C., Witt, L., Wen, A., Liao, T. V., Nguyen, J., Lin, B., Altman, R. B., & Palaniappan, L. (2020). Pharmacogenomics in Asian subpopulations and impacts on commonly prescribed medications. *Clinical and Translational Science*, 13(5), 861–870. DOI: 10.1111/cts.12771 PMID: 32100936

Lorell, J., Ivana, K., Tanoto, J. P. W., Michael, M., Jescika, S., Gautama, N. R. P., Tjoa, O., Alina, K., Parikesit, A. A., & Sutanto, F.LORELL. (2024). In silico testing of C9H12ClNO2 and C6H5Cl2NO as derivatives of acetaminophen using molecular docking method. *Notulae Scientia Biologicae*, 16(1), 11632–11632. DOI: 10.55779/nsb16111632

Love-Koh, J., Peel, A., Rejon-Parrilla, J. C., Ennis, K., Lovett, R., Manca, A., Chalkidou, A., Wood, H., & Taylor, M. (2018). The future of precision medicine: Potential impacts for health technology assessment. *PharmacoEconomics*, 36(12), 1439–1451. DOI: 10.1007/s40273-018-0686-6 PMID: 30003435

Luck, K. (2020). A proteome-scale map of the human interactome network. *Cell*, 184(10), 2745–2764.

Lu, M., Yin, J., Zhu, Q., Lin, G., Mou, M., Liu, F., Pan, Z., You, N., Lian, X., Li, F., Zhang, H., Zheng, L., Zhang, W., Zhang, H., Shen, Z., Gu, Z., Li, H., & Zhu, F. (2023). Smart Process Manufacturing toward Carbon Neutrality-Review Artificial Intelligence in Pharmaceutical Sciences. *Engineering (Beijing)*, 27, 37–69. DOI: 10.1016/j.eng.2023.01.014

Lunshof, J. E. (2008). Genetic data and its ethical implications. *Nature Reviews. Genetics*, 9(4), 295–305.

Luo, F., Das, A., Khetarpal, S. A., Fang, Z., Zelniker, T. A., Rosenson, R. S., & Qamar, A. (2024). ANGPTL3 inhibition, dyslipidemia, and cardiovascular diseases. In *Trends in Cardiovascular Medicine* (Vol. 34, Issue 4). DOI: 10.1016/j.tcm.2023.01.008

Lynch, C., Sakamuru, S., Ooka, M., Huang, R., Klumpp-Thomas, C., Shinn, P., Gerhold, D., Rossoshek, A., Michael, S., Casey, W., Santillo, M. F., Fitzpatrick, S., Thomas, R. S., Simeonov, A., & Xia, M. (2024). High-throughput screening to advance in vitro toxicology: Accomplishments, challenges, and future directions. *Annual Review of Pharmacology and Toxicology*, 64(1), 191–209. DOI: 10.1146/annurev-pharmtox-112122-104310 PMID: 37506331

Mackay, D., & Soares, G. (2020). Ethical and legal considerations in genomic testing. *Journal of Medical Ethics*, 46(3), 179–185.

Madan, A., Walker, C. R., Weinstein, B., & Fowler, J. C. (2015). Pharmacogenomics in practice: A case report of personalized inpatient psychiatric care. *Pharmacogenomics*, 16(5), 433–439. DOI: 10.2217/pgs.15.9 PMID: 25916515

Madejczyk, A. M., Canzian, F., Góra-Tybor, J., Campa, D., Sacha, T., Link-Lenczowska, D., Florek, I., Prejzner, W., Całbecka, M., Rymko, M., Dudziński, M., Orzechowska, M. J., & Jamroziak, K. (2022). Impact of genetic polymorphisms of drug transporters ABCB1 and ABCG2 and regulators of xenobiotic transport and metabolism PXR and CAR on clinical efficacy of dasatinib in chronic myeloid leukemia. *Frontiers in Oncology*, 12, 952640. DOI: 10.3389/fonc.2022.952640 PMID: 36212403

Magavern, E. F., Kaski, J. C., Turner, R. M., Drexel, H., Janmohamed, A., Scourfield, A., Burrage, D., Floyd, C. N., Adeyeye, E., Tamargo, J., Lewis, B. S., Kjeldsen, K. P., Niessner, A., Wassmann, S., Sulzgruber, P., Borry, P., Agewall, S., Semb, A. G., Savarese, G., & Caulfield, M. J. (2022). Challenges in cardiovascular pharmacogenomics implementation: A viewpoint from the European Society of Cardiology Working Group on Cardiovascular Pharmacotherapy. *European Heart Journal. Cardiovascular Pharmacotherapy*, 8(1), 100–103. DOI: 10.1093/ehjcvp/pvab063 PMID: 34463331

Magavern, E. F., Kaski, J. C., Turner, R. M., Drexel, H., Janmohamed, A., Scourfield, A., Burrage, D., Floyd, C. N., Adeyeye, E., Tamargo, J., Lewis, B. S., Kjeldsen, K. P., Niessner, A., Wassmann, S., Sulzgruber, P., Borry, P., Agewall, S., Semb, A. G., Savarese, G., & Caulfield, M. J. (2022). The role of pharmacogenomics in contemporary cardiovascular therapy: A position statement from the European Society of Cardiology Working Group on Cardiovascular Pharmacotherapy. *European Heart Journal. Cardiovascular Pharmacotherapy*, 8(1), 85–99. DOI: 10.1093/ehjcvp/pvab018 PMID: 33638977

Maguire, J. J., Kuc, R. E., & Davenport, A. P. 2012. Radioligand binding assays and their analysis. *Receptor binding techniques*, pp.31-77.

Mahajan, A., & Taliun, D. (2018). Fine-mapping of an additional 11,000 variants for association with risk of type 2 diabetes. *Nature Genetics*, 50(5), 1270–1282.

Mahler, M., Martinez-Prat, L., Sparks, J. A., & Deane, K. D. (2020). Precision medicine in the care of rheumatoid arthritis: Focus on prediction and prevention of future clinically-apparent disease. *Autoimmunity Reviews*, 19(5), 102506. DOI: 10.1016/j.autrev.2020.102506 PMID: 32173516

Makarova, A. V. (2020). CRISPR/Cas9 gene editing technology: A potential for understanding and treating neurological diseases. *Frontiers in Neurology*, 11, 572317.

Makarov, D. V., Loeb, S., Getzenberg, R. H., & Partin, A. W. (2009). Biomarkers for prostate cancer. *Annual Review of Medicine*, 60(1), 139–151. DOI: 10.1146/annurev.med.60.042307.110714 PMID: 18947298

Malik, S., Muhammad, K., & Waheed, Y. (2022). Emerging Applications of Nanotechnology in Healthcare and Medicine. *Molecules (Basel, Switzerland)*, 28(18), 6624. DOI: 10.3390/molecules28186624 PMID: 37764400

Malin, B. A., & Sweeney, L. G. (2010). Determining the identifiability of DNA database entries. *The Journal of Privacy and Confidentiality*, 2(1), 1–19.

Malkawi, R. (2024). Revolutionizing drug delivery innovation: Leveraging AI-driven Chabot's for enhanced efficiency. *International Journal of Applied Pharmaceutics*, 16(2), 52–56. DOI: 10.22159/ijap.2024v16i2.50182

Mallal, S. A., Phillips, E., Carosi, G., Molina, J.-M., Workman, C., Tomažič, J., Jägel-Guedes, E., Rugina, S., Kozyrev, O., Cid, J. F., Hay, P., Nolan, D., Hughes, S., Hughes, A., Ryan, S., Fitch, N., Thorborn, D., & Benbow, A. (2008). HLA-B*57:01 screening before starting abacavir. *The New England Journal of Medicine*, 358(6), 568–579. DOI: 10.1056/NEJMoa0706135 PMID: 18256392

Malo, N., Hanley, J. A., Cerquozzi, S., Pelletier, J., & Nadon, R. (2006). Statistical practice in high-throughput screening data analysis. *Nature Biotechnology*, 24(2), 167–175. DOI: 10.1038/nbt1186 PMID: 16465162

Mandelker, D. (2017). The role of exome sequencing in the identification of pathogenic variants in cancer. *Nature Reviews. Cancer*, 17(9), 618–630.

Manolio, T. A. (2013). Implementing genomic medicine in clinical practice: The opportunities and challenges. *Journal of the American Medical Association*, 310(11), 1161–1162.

Manolio, T. A., Collins, F. S., Cox, N. J., Goldstein, D. B., Hindorff, L. A., Hunter, D. J., McCarthy, M. I., Ramos, E. M., Cardon, L. R., Chakravarti, A., Cho, J. H., Guttmacher, A. E., Kong, A., Kruglyak, L., Mardis, E., Rotimi, C. N., Slatkin, M., Valle, D., Whittemore, A. S., & Visscher, P. M. (2019). Finding the missing heritability of complex diseases. *Nature*, 461(7265), 747–753. DOI: 10.1038/nature08494 PMID: 19812666

Mantere, T. (2019). Mutational landscape of BRCA1 and BRCA2 in hereditary breast and ovarian cancer. *Nature Reviews. Cancer*, 19(12), 723–735.

Mapes, B. M., Foster, C. S., Kusnoor, S. V., Epelbaum, M. I., AuYoung, M., Jenkins, G., Lopez-Class, M., Richardson-Heron, D., Elmi, A., Surkan, K., Cronin, R. M., Wilkins, C. H., Pérez-Stable, E. J., Dishman, E., Denny, J. C., & Rutter, J. L. (2020). Diversity and inclusion for the All of Us research program: A scoping review. *PLoS One*, 15(7), e0234962. DOI: 10.1371/journal.pone.0234962 PMID: 32609747

Mardis, E. R. (2008). Next-generation DNA sequencing methods. *Annual Review of Genomics and Human Genetics*, 9(1), 387–402. DOI: 10.1146/annurev.genom.9.081307.164359 PMID: 18576944

Mardis, E. R. (2017). DNA sequencing technologies: 2006–2016. *Nature Protocols*, 12(2), 213–218. DOI: 10.1038/nprot.2016.182 PMID: 28055035

Margaretha, F., Subagja, M., Putri, A. D., Salam, C. Q., Valenthenardo, L., Adhiwijaya, P. K., Kaitlyn, G. Z., & Parikesit, A. (2023). Comprehensive analysis of protein-protein interactions (PPIs) with structure prediction program for breast cancer determination. *Berkala Penelitian Hayati*, 29(3), 87–94. DOI: 10.23869/bphjbr.29.3.20232

Marques, L., Costa, B., Pereira, M., Silva, A., Santos, J., Saldanha, L., Silva, I., Magalhães, P., Schmidt, S., & Vale, N. (2024). Advancing Precision Medicine: A Review of Innovative In Silico Approaches for Drug Development, Clinical Pharmacology and Personalized Healthcare. *Pharmaceutics*, 16(3), 332. Advance online publication. DOI: 10.3390/pharmaceutics16030332 PMID: 38543226

Martis, E. A., Radhakrishnan, R., & Badve, R. R. (2011). High-throughput screening: The hits and leads of drug discovery—An overview. [Google Scholar]. *Journal of Applied Pharmaceutical Science*, 1, 2–10.

Marx, V. (2013). The big challenges of big data. *Nature*, 498(7453), 255–260. DOI: 10.1038/498255a PMID: 23765498

Masoudkabir, F., Mohammadifard, N., Mani, A., Ignaszewski, A., Davis, M. K., Vaseghi, G., Mansourian, M., Franco, C., Gotay, C., & Sarrafzadegan, N. (2023). Shared Lifestyle-Related Risk Factors of Cardiovascular Disease and Cancer: Evidence for Joint Prevention. *TheScientificWorldJournal*, 2023, 1–15. Advance online publication. DOI: 10.1155/2023/2404806 PMID: 37520844

Mathur, S., & Sutton, J. (2017). Personalized medicine could transform healthcare. *Biomedical Reports*, 7(1), 3–5. DOI: 10.3892/br.2017.922 PMID: 28685051

Matthew, A. (2006). Current biosensor technologies in drug discovery. Drug discovery, 69.

Mattick, J. S. (2005). The functional genomics of noncoding RNA. *Science*, 309(5740), 1527–1528. DOI: 10.1126/science.1117806 PMID: 16141063

Mavaddat, N., Michailidou, K., Dennis, J., Lush, M., Fachal, L., Lee, A., & MacInnis, R. J. (2019). Polygenic risk scores for prediction of breast cancer and breast cancer subtypes. *American Journal of Human Genetics*, 104(1), 21–34. DOI: 10.1016/j.ajhg.2018.11.002 PMID: 30554720

Mayer, I. A., & Arteaga, C. L. (2016). The PI3K/AKT/mTOR pathway in cancer therapeutics. *Current Opinion in Pharmacology*, 29, 7–12.

Mayoh, C., Mao, J., Xie, J., Tax, G., Chow, S.-O., Cadiz, R., Pazaky, K., Barahona, P., Ajuyah, P., Trebilcock, P., Malquori, A., Gunther, K., Avila, A., Yun, D. Y., Alfred, S., Gopalakrishnan, A., Kamili, A., Wong, M., Cowley, M. J., & Dolman, M. E. M. (2023). High-throughput drug screening of primary tumor cells identifies therapeutic strategies for treating children with high-risk cancer. *Cancer Research*, 83(16), 2716–2732. DOI: 10.1158/0008-5472.CAN-22-3702 PMID: 37523146

McCarthy, M. I. (2016). Genomics and drug discovery. *Science*, 352(6291), 78–83.

McDonough, C. W. (2021). Pharmacogenomics in cardiovascular diseases. *Current Protocols*, 1(7), e189. DOI: 10.1002/cpz1.189 PMID: 34232575

McKenna, A., Hanna, M., Banks, E., Sivachenko, A., Cibulskis, K., Kernytsky, A., Garimella, K., Altshuler, D., Gabriel, S., Daly, M., & DePristo, M. A. (2010). The Genome Analysis Toolkit: A MapReduce framework for analyzing next-generation DNA sequencing data. *Genome Research*, 20(9), 1297–1303. DOI: 10.1101/gr.107524.110 PMID: 20644199

McMurray, C. T. (2018). Mechanisms of disease: Dynamic mutations in human genetic disease. *Nature Reviews. Genetics*, 19(7), 440–450.

McShane, L. M. (2020). Clinical trial validation of biomarker-driven therapeutic strategies. *Nature Reviews. Drug Discovery*, 19(10), 616–632.

Medwid, S., & Kim, R. B. (2022). Implementation of pharmacogenomics: Where are we now? *British Journal of Clinical Pharmacology*. PMID: 36366858

Meenu, S. (2023). K., Yadav, S., & Yadav, D. (2023). Recent Advances in Pharmacogenomics and Clinical Pharmacology: A Systematic Review. *Recent Advances in Pharmacogenomics and Clinical Pharmacology: A Systematic Review. European Chemical Bulletin*, (12), 1776–1788. DOI: 10.48047/ecb/2023.12.si12.154

Membrive-Jiménez, C., Vieira-Maroun, S., Márquez-Pete, N., Cura, Y., Pérez-Ramírez, C., Tercedor-Sánchez, J., Jiménez-Morales, A., & Ramírez-Tortosa, M. C. (2023). ABCC1, ABCG2 and FOXP3: Predictive Biomarkers of Toxicity from Methotrexate Treatment in Patients Diagnosed with Moderate-to-Severe Psoriasis. *Biomedicines*, 11(9), 2567–2567. DOI: 10.3390/biomedicines11092567 PMID: 37761008

Metzker, M. L. (2010). Sequencing technologies—The next generation. *Nature Reviews. Genetics*, 11(1), 31–46. DOI: 10.1038/nrg2626 PMID: 19997069

Mezinska, S., Gallagher, L., Verbrugge, M., & Bunnik, E. M. (2021). Ethical issues in genomics research on neurodevelopmental disorders: A critical interpretive review. *Human Genomics*, 15(1), 16. DOI: 10.1186/s40246-021-00317-4 PMID: 33712057

Miao, F., Smith, D. D., Zhang, L., Min, A., Feng, W., & Natarajan, R. (2008). Lymphocytes from patients with type 1 diabetes display a distinct profile of chromatin histone H3 lysine 9 dimethylation: An epigenetic study in diabetes. *Diabetes*, 57(12), 3189–3198. DOI: 10.2337/db08-0645 PMID: 18776137

Michael, M., Liauw, W., McLachlan, S.-A., Link, E., Matera, A., Thompson, M., Jefford, M., Hicks, R. J., Cullinane, C., Hatzimihalis, A., Campbell, I. G., Rowley, S., Beale, P. J., Karapetis, C. S., Price, T., & Burge, M. E. (2021). Pharmacogenomics and functional imaging to predict irinotecan pharmacokinetics and pharmacodynamics: The predict IR study. *Cancer Chemotherapy and Pharmacology*, 88(1), 39–52. DOI: 10.1007/s00280-021-04264-8 PMID: 33755789

Mirsadeghi, S., & Larijani, B. (2017). Personalized Medicine: Pharmacogenomics and Drug Development. *Acta Medica Iranica*, 55(3), 150–165. PMID: 28282716

Mishra, S. K., & Georrge, J. J. In Silico investigation of Epigallocatechin-3-Gallate and its derivatives against the VEGF-A cancerous protein through molecular docking and MD simulation studies.

Mishra, S. K., Jeba Praba, J., & Georrge, J. J. (2024). An emerging trends of bioinformatics and big data analytics in healthcare. *Digital Transformation in Healthcare 5.0: Volume 2: Metaverse, Nanorobots and Machine Learning*, 159.

Mishra, S. K., Priya, P., Rai, G. P., Haque, R., & Shanker, A. (2023). Coevolution based immunoinformatics approach considering variability of epitopes to combat different strains: A case study using spike protein of SARS-CoV-2. *Computers in Biology and Medicine*, 163, 107233. DOI: 10.1016/j.compbiomed.2023.107233 PMID: 37422941

Miteva-Marcheva, N. N., Ivanov, H. Y., Dimitrov, D. K., & Stoyanova, V. K. (2020). Application of pharmacogenetics in oncology. *Biomarker Research*, 8(1), 32. DOI: 10.1186/s40364-020-00213-4 PMID: 32821392

Mitri, Z., Constantine, T., & O'Regan, R. (2012). The HER2 receptor in breast cancer: Pathophysiology, clinical use, and new advances in therapy. *Chemotherapy Research and Practice*, 2012(1), 743193. DOI: 10.1155/2012/743193 PMID: 23320171

Moc, C. (2020). Pharmacogenomics: An evolving clinical tool for precision medicine. *Cleveland Clinic Journal of Medicine*, 87(2), 91–99. DOI: 10.3949/ccjm.87a.19073 PMID: 32015062

Mok, T. S. (2009). Epidermal growth factor receptor (EGFR) mutation status in non-small cell lung cancer: A review and the importance of EGFR mutations in treatment. *The Lancet. Oncology*, 10(4), 321–329. PMID: 19350698

Molecular Devices. (n.d.). ImageXpress Micro Confocal High-Content Imaging System. Retrieved from Molecular Devices website

Moore, L. D., Le, T., & Wong, H. K. (2019). DNA methylation and its basic functions. *Nature Reviews. Genetics*, 10(2), 125–137.

Morgens, D. W. (2016). Genome-scale identification of CRISPR-Cas9 off-targets in human cells. *Nature Biotechnology*, 34(3), 368–371.

Morley, K. I., & Hall, W. D. (2004). Using pharmacogenetics and pharmacogenomics in the treatment of psychiatric disorders: Some ethical and economic considerations. *Journal of Molecular Medicine*, 82(1), 21–30. DOI: 10.1007/s00109-003-0496-x PMID: 14598045

Mostafa, S., Polasek, T. M., Bousman, C. A., Müeller, D. J., Sheffield, L. J., Rembach, J., & Kirkpatrick, C. M. (2022). Pharmacogenomics in psychiatry–the challenge of cytochrome P450 enzyme phenoconversion and solutions to assist precision dosing. *Pharmacogenomics*, 23(15), 857–867. DOI: 10.2217/pgs-2022-0104 PMID: 36169629

Mroz, P., Michel, S., Allen, J. D., Meyer, T., McGonagle, E. J., Carpentier, R., Vecchia, A., Schlichte, A., Bishop, J. R., Dunnenberger, H. M., Yohe, S., Thyagarajan, B., Jacobson, P. A., & Johnson, S. G. (2021). Development and Implementation of In-House Pharmacogenomic Testing Program at a Major Academic Health System. *Frontiers in Genetics*, 12, 712602. DOI: 10.3389/fgene.2021.712602 PMID: 34745204

Mulder, D. J., Khalouei, S., Warner, N., Gonzaga-Jauregui, C., Church, P. C., Walters, T. D., Ramani, A. K., Griffiths, A. M., Cohn, I., & Muise, A. M. (2020). Utilization of Whole Exome Sequencing Data to Identify Clinically Relevant Pharmacogenomic Variants in Pediatric Inflammatory Bowel Disease. *Clinical and Translational Gastroenterology*, 11(12), e00263. Advance online publication. DOI: 10.14309/ctg.0000000000000263 PMID: 33512800

Muteeb, G., Rehman, M. T., Shahwan, M., & Aatif, M. (2023). Origin of antibiotics and antibiotic resistance, and their impacts on drug development: A narrative review. *Pharmaceuticals (Basel, Switzerland)*, 16(11), 1615. DOI: 10.3390/ph16111615 PMID: 38004480

Naithani, N., Atal, A. T., Tilak, T. V. S. V. G. K., Vasudevan, B., Misra, P., & Sinha, S. (2021). Precision medicine: uses and challenges. *medical journal armed forces india, 77*(3), 258-265.

Naithani, N., Sinha, S., Misra, P., Vasudevan, B., & Sahu, R. (2021). Precision medicine: Concept and tools. In *Medical Journal Armed Forces India* (Vol. 77, Issue 3). DOI: 10.1016/j.mjafi.2021.06.021

Naithani, N., Sinha, S., Misra, P., Vasudevan, B., & Sahu, R. (2021). Precision medicine: Concept and tools. *medical journal armed forces india, 77*(3), 249-257.

Natarajan, P. (2018). Genetic mapping and mechanisms of coronary artery disease. *Circulation Research*, 122(2), 267–286. PMID: 29167274

Nath, A., Wang, J., & Huang, R. S. (2017). PHARMACOGENETICS AND PHARMACOGENOMICS OF TARGETED THERAPEUTICS IN CHRONIC MYELOID LEUKEMIA. *Molecular Diagnosis & Therapy*, 21(6), 621–631. DOI: 10.1007/s40291-017-0292-x PMID: 28698977

National Academies Press (US) (2017 Jul 10). Case Studies in Precision Drug Development. National Academies of Sciences, Engineering, and Medicine; Health and Medicine Division; Board on Health Sciences Policy; Roundtable on Genomics and Precision Health; Forum on Drug Discovery, Development, and Translation. Enabling Precision Medicine: The Role of Genetics in Clinical Drug Development: Proceedings of a Workshop. Washington (DC): Available from: https://www.ncbi.nlm.nih.gov/sites/books/NBK458878/

Ndadza, A., Muyambo, S., Mntla, P., Wonkam, A., Chimusa, E., Kengne, A. P., Ntsekhe, M., & Dandara, C. (2021). Profiling of warfarin pharmacokinetics-associated genetic variants: Black Africans portray unique genetic markers important for an African specific warfarin pharmacogenetics-dosing algorithm. Journal of thrombosis and haemostasis. *Journal of Thrombosis and Haemostasis*, 19(12), 2957–2973. DOI: 10.1111/jth.15494 PMID: 34382722

Nebert, D. W., Zhang, G., & Vesell, E. S. (2008). From human genetics and genomics to pharmacogenetics and pharmacogenomics: Past lessons, future directions. *Drug Metabolism Reviews*, 40(2), 187–224. DOI: 10.1080/03602530801952864 PMID: 18464043

Nge, P. N., Rogers, C. I., & Woolley, A. T. (2013). Advances in microfluidic materials, functions, integration, and applications. *Chemical Reviews*, 113(4), 2550–2583. DOI: 10.1021/cr300337x PMID: 23410114

Ngoepe, M., Choonara, Y. E., Tyagi, C., Tomar, L. K., du Toit, L. C., Kumar, P., Ndesendo, V. M., & Pillay, V. (2013, June 14). Integration of biosensors and drug delivery technologies for early detection and chronic management of illness. *Sensors (Basel)*, 13(6), 7680–7713. DOI: 10.3390/s130607680 PMID: 23771157

Ng, P. C., & Henikoff, S. (2003). SIFT: Predicting amino acid changes that affect protein function. *Nucleic Acids Research*, 31(13), 3812–3814. DOI: 10.1093/nar/gkg509 PMID: 12824425

NHS England. (n.d.). Personalised medicine. www.england.nhs.uk/ourwork/qual-clin-lead/personalisedmedicine/ [Accessed 23 July 2024].

Nicholson, W. T., Formea, C. M., Matey, E. T., Wright, J. A., Giri, J., & Moyer, A. M. (2021). Considerations When Applying Pharmacogenomics to Your Practice. *Mayo Clinic Proceedings*, 96(1), 218–230. DOI: 10.1016/j.mayocp.2020.03.011 PMID: 33308868

Nicholson, W. T., Formea, C. M., Matey, E. T., Wright, J. A., Giri, J., & Moyer, A. M. (2021). Considerations when applying pharmacogenomics to your practice. Mayo Clinic Proceedings, Pirmohamed, M. (2023). Pharmacogenomics: Current status and future perspectives. *Nature Reviews. Genetics*, 24(6), 350–362.

Nickischer, D., Elkin, L., Cloutier, N., O'Connell, J., Banks, M., & Weston, A. (2018). Challenges and opportunities in enabling high-throughput, miniaturized high content screening. *High Content Screening: A Powerful Approach to Systems Cell Biology and Phenotypic Drug Discovery*, 165-191.

Nierode, G., Kwon, P. S., Dordick, J. S., & Kwon, S. J. (2016). Cell-based assay design for high-content screening of drug candidates. *Journal of Microbiology and Biotechnology*, 26(2), 213–225. DOI: 10.4014/jmb.1508.08007 PMID: 26428732

Nisar, H., & Srivastava, R. (2018). Fundamental concept of psychosomatic disorders: a review. *International Journal of contemporary Medicine surgery and radiology, 3*(1), 12-18.

Nobile, C. (2023). Bioinformatics in drug discovery and development: Methods and applications. *Pre QC. Revised*, 12. Advance online publication. DOI: 10.35841/bmcr

Nogueiras-Álvarez, R. (2023). Pharmacogenomics in clinical trials: An overview. *Frontiers in Pharmacology*, 14, 1247088. DOI: 10.3389/fphar.2023.1247088 PMID: 37927590

Nordberg, G. F. (2010). Biomarkers of exposure, effects and susceptibility in humans and their application in studies of interactions among metals in China. *Toxicology Letters*, 192(1), 45–49. DOI: 10.1016/j.toxlet.2009.06.859 PMID: 19540908

Norinder, U., & Bergström, C. A. S. (2006). Prediction of ADMET properties. In *ChemMedChem* (Vol. 1, pp. 920–937). DOI: 10.1002/cmdc.200600155

Nosek, B. A., Alter, G., Banks, G. C., Borsboom, D., Bowman, S. D., Breckler, S. J., Buck, S., Chambers, C. D., Chin, G., Christensen, G., Contestabile, M., Dafoe, A., Eich, E., Freese, J., Glennerster, R., Goroff, D., Green, D. P., Hesse, B., Humphreys, M., & Yarkoni, T. (2015). Promoting an open research culture. *Science*, 348(6242), 1422–1425. DOI: 10.1126/science.aab2374 PMID: 26113702

Nwosu, S. N. (2024). Revolutionizing Drug Discovery: The Impact of Artificial Intelligence on Efficiency and Accuracy. *Journal of Advances in Medical and Pharmaceutical Sciences*, 26(6), 56–63. DOI: 10.9734/jamps/2024/v26i6695

O'Boyle, N. M., Banck, M., James, C. A., Morley, C., Vandermeersch, T., & Hutchison, G. R. (2011). Open Babel: An open chemical toolbox. *Journal of Cheminformatics*, 3(1), 1–14. DOI: 10.1186/1758-2946-3-33 PMID: 21982300

O'Sullivan, B. P., & Freedman, S. D. (2020). Cystic fibrosis. *Lancet*, 373(9678), 1572–1583. PMID: 19403164

Oakley, A. (2023). HIPAA, HIPPA, or HIPPO: What Really Is the Heath Insurance Portability and Accountability Act? *Biotechnology Law Report*, 42(6), 306–318. DOI: 10.1089/blr.2023.29329.aso

Olivier, C., & Williams-Jones, B. (2011). Pharmacogenomic technologies: A necessary" luxury" for better global public health? *Globalization and Health*, 7(1), 1–13. DOI: 10.1186/1744-8603-7-30 PMID: 21864366

Olson, M. V. (1993). The human genome project. *Proceedings of the National Academy of Sciences of the United States of America*, 90(10), 4338–4344. DOI: 10.1073/pnas.90.10.4338 PMID: 8506271

Ota, M., & Fujio, K. (2021). Multi-omics approach to precision medicine for immune-mediated diseases. *Inflammation and Regeneration*, 41(1), 23. DOI: 10.1186/s41232-021-00173-8 PMID: 34332645

Ouzounis, C. A. (2012). Rise and demise of bioinformatics? promise and progress. *PLoS Computational Biology*, 8(4), e1002487. Advance online publication. DOI: 10.1371/journal.pcbi.1002487 PMID: 22570600

Palmirotta, R., Lovero, D., Delacour, H., Le Roy, A., Cremades, S., & Silvestris, F. (2019). Rare Dihydropyrimidine Dehydrogenase Variants and Toxicity by Floropyrimidines: A Case Report. *Frontiers in Oncology*, 9, 139. DOI: 10.3389/fonc.2019.00139 PMID: 30915274

Pandi, M. T., Koromina, M., Tsafaridis, I., Patsilinakos, S., Christoforou, E., van der Spek, P. J., & Patrinos, G. P. (2021). A novel machine learning-based approach for the computational functional assessment of pharmacogenomic variants. *Human Genomics*, 15(1), 1–13. DOI: 10.1186/s40246-021-00352-1 PMID: 34372920

Pant, P., Chitme, H., Sircar, R., Prasad, R., & Prasad, H. O. (2023). Genome-wide association study for single nucleotide polymorphism associated with mural and cumulus granulosa cells of PCOS (polycystic ovary syndrome) and non-PCOS patients. *Future Journal of Pharmaceutical Sciences*, 9(1), 27. Advance online publication. DOI: 10.1186/s43094-023-00475-3

Pareek, V., Tuteja, L., Sharma, L., Kumar, S., & Verma, N. (2023). Revolutionizing Drug Design with Artificial Intelligence: A Comprehensive Review of Techniques, Applications, and Case Studies. *Journal of Pharmacy Research*, 22(3), 103–112.

Parikesit, A. A. (2018). Introductory Chapter: The Contribution of Bioinformatics as Blueprint Lead for Drug Design. In Glavic, I. (Ed.), *Molecular Insight of Drug Design* (p. 7). InTech., DOI: 10.5772/intechopen.79664

Parikesit, A. A. (2021). Introductory Chapter: The Emerging Corner of the Omics Studies for Rational Drug Design. In Parikesit, A. A. (Ed.), *Drug Design: Novel Advances in the Omics Field and Applications* (p. 4). IntechOpen., DOI: 10.5772/intechopen.95544

Parikesit, A. A., Hermantara, R., Kevin, G., & Sidhartha, E. (2023). Designing hybrid CRISPR-Cas12 and LAMP detection systems for treatment-resistant Plasmodium falciparum with in silico method. *Narra J*, 3(3), e301. DOI: 10.52225/narra.v3i3.301 PMID: 38455618

Parikesit, A. A., Nurdiansyah, R., Parikesit, A. A., Nurdiansyah, R., Aditya Parikesit, A., & Nurdiansyah, R. (2022). *Introductory Chapter: Current Perspective of COVID-19 Drugs. COVID-19 Drug Development - Recent Advances*. New Perspectives and Applications., DOI: 10.5772/intechopen.105537

Parker, L. S. (2024). Ethical considerations in precision medicine. In *The New Era of Precision Medicine* (pp. 143–172). Academic Press. DOI: 10.1016/B978-0-443-13963-5.00002-9

Pasmant, E., Sabbagh, A., Vidaud, M., & Bieche, I. (2011). ANRIL, a long, non-coding RNA, is an unexpected major hotspot in GWAS. *The FASEB Journal*, 25(2), 444–448. Advance online publication. DOI: 10.1096/fj.10-172452 PMID: 20956613

Paul, D., Sanap, G., Shenoy, S., Kalyane, D., Kalia, K., & Tekade, R. K. (2021, January). Artificial intelligence in drug discovery and development. *Drug Discovery Today*, 26(1), 80–93. DOI: 10.1016/j.drudis.2020.10.010 PMID: 33099022

Pedersen, A. J. T., Stage, T. B., Glintborg, D., Andersen, M., & Christensen, M. M. H. (2018). The Pharmacogenetics of Metformin in Women with Polycystic Ovary Syndrome: A Randomized Trial. *Basic & Clinical Pharmacology & Toxicology*, 122(2), 239–244. DOI: 10.1111/bcpt.12874 PMID: 28834135

Pence, H. E., and Williams, A. (2010). ChemSpider: an online chemical information resource.

Pereira, N. L., Rihal, C. S., So, D. Y. F., Rosenberg, Y., Lennon, R. J., Mathew, V., Goodman, S. G., Weinshilboum, R. M., Wang, L., Baudhuin, L. M., Lerman, A., Hasan, A., Iturriaga, E., Fu, Y. P., Geller, N., Bailey, K., & Farkouh, M. E. (2019). Clopidogrel Pharmacogenetics. *Circulation: Cardiovascular Interventions*, 12(4), e007811. DOI: 10.1161/CIRCINTERVENTIONS.119.007811 PMID: 30998396

Perez-Paramo, Y. X., Watson, C. J. W., Chen, G., Thomas, C. E., Adams-Haduch, J., Wang, R., Khor, C. C., Koh, W. P., Nelson, H. H., Yuan, J. M., & Lazarus, P. (2023). Impact of Genetic Variants in the Nicotine Metabolism Pathway on Nicotine Metabolite Levels in Smokers. *Cancer Epidemiology, Biomarkers & Prevention*, 32(1), 54–65. DOI: 10.1158/1055-9965.EPI-22-0868 PMID: 36252563

PerkinElmer. (n.d.). Operetta CLS High-Content Analysis System. Retrieved from PerkinElmer website

Peterson, B. S. (2020). Biomarkers in precision medicine for mental illnesses. *Journal of Child Psychology and Psychiatry, and Allied Disciplines*, 61(12), 1279–1281. DOI: 10.1111/jcpp.13357 PMID: 33252151

Phillips, S. (2018). Data governance in genomics: From ethical dilemmas to practical frameworks. *Genome Research*, 28(9), 1347–1355.

Piccart-Gebhart, M. J., Procter, M., Leyland-Jones, B., Goldhirsch, A., Untch, M., Smith, I., Gianni, L., Baselga, J., Bell, R., Jackisch, C., Cameron, D., Dowsett, M., Barrios, C. H., Steger, G., Huang, C.-S., Andersson, M., Inbar, M., Lichinitser, M., Láng, I., & Gelber, R. D. (2005). Trastuzumab after Adjuvant Chemotherapy in HER2-Positive Breast Cancer. *The New England Journal of Medicine*, 353(16), 1659–1672. DOI: 10.1056/NEJMoa052306 PMID: 16236737

Pinto, D., & Pagnamenta, A. T. (2020). Functional impact of copy number variations in neurodevelopmental disorders. *Nature*, 516(7529), 407–414.

Pinto, N., & Dolan, M. E. (2011). Clinically relevant genetic variations in drug metabolizing enzymes. *Current Drug Metabolism*, 12(5), 487–497. DOI: 10.2174/138920011795495321 PMID: 21453273

Pinu, F. R., Beale, D. J., Paten, A. M., Kouremenos, K., Swarup, S., Schirra, H. J., & Wishart, D. (2019). Systems biology and multi-omics integration: Viewpoints from the metabolomics research community. *Metabolites*, 9(4), 76. DOI: 10.3390/metabo9040076 PMID: 31003499

Pirmohamed, M. (2023). Pharmacogenomics: Current status and future perspectives. *Nature Reviews. Genetics*, 24(6), 350–362. DOI: 10.1038/s41576-022-00572-8 PMID: 36707729

Plans-Beriso, E., Petrova, D., Barahona-López, C., Diez-Echave, P., Hernández, O. R., Fernández-Martínez, N. F., Turner, H., García-Ovejero, E., Craciun, O., Fernández-Navarro, P., Fernández-Larrea, N., García-Esquinas, E., Kuhn, I., Jiménez-Planet, V., Moreno, V., Rodríguez-Artalejo, F., Sánchez, M. J., Pollan-Santamaria, M., Blackburn, L., & Pérez-Gómez, B. (2024). Biomarkers for personalised prevention of chronic diseases: A common protocol for three rapid scoping reviews. *Systematic Reviews*, 13(1), 147. Advance online publication. DOI: 10.1186/s13643-024-02554-9 PMID: 38824585

Poste, G. (2011). Bring on the biomarkers. *Nature*, 469(7329), 156–157. DOI: 10.1038/469156a PMID: 21228852

Prainsack, B., & Buyx, A. (2017). The value of biomedical research and its moral obligations. *Cambridge Quarterly of Healthcare Ethics*, 26(1), 15–27.

Pranay, O. V., Prajapati, A. P., Patel, B. S., Narkhede, S. B., & Luhar, S. (2016). A review on integrating artificial intelligence into drug development: Revolutionizing the pharmaceutical landscape. *I J R D.*, 8(16), 205–210.

Pratama, R. A., Astina, J., & Parikesit, A. A. (2023). In Silico Antidiabetic Study of Phenolic Compounds from Banana (Musa spp.) Peel Targeting PTP1B Protein. *Journal of Tropical Biodiversity and Biotechnology*, 8(3), 83124. DOI: 10.22146/jtbb.83124

Price, A. L., Patterson, N. J., Plenge, R. M., Weinblatt, M. E., Shadick, N. A., & Reich, D. (2006). Principal components analysis corrects for stratification in genome-wide association studies. *Nature Genetics*, 38(8), 904–909. DOI: 10.1038/ng1847 PMID: 16862161

Pritchard, D. E., Moeckel, F., Villa, M. S., Housman, L. T., McCarty, C. A., & McLeod, H. L. (2017). Strategies for integrating personalized medicine into healthcare practice. *Personalized Medicine*, 14(2), 141–152. DOI: 10.2217/pme-2016-0064 PMID: 29754553

Proper, S. P., Azouz, N. P., & Mersha, T. B. (2021). Achieving precision medicine in allergic disease: Progress and challenges. *Frontiers in Immunology*, 12, 720746. DOI: 10.3389/fimmu.2021.720746 PMID: 34484229

Qahwaji, R., Ashankyty, I., Sannan, N. S., Hazzazi, M. S., Basabrain, A. A., & Mobashir, M. (2024). Pharmacogenomics: A Genetic Approach to Drug Development and Therapy. *Pharmaceuticals (Basel, Switzerland)*, 17(7), 940. Advance online publication. DOI: 10.3390/ph17070940 PMID: 39065790

Raman, R., Utts, J., Cohen, A. I., & Hayat, M. J. (2023). Integrating Ethics into the Guidelines for Assessment and Instruction in Statistics Education (GAISE). *The American Statistician*, 77(3), 323–330. DOI: 10.1080/00031305.2022.2156612

Ramaswami, R., Bayer, R., & Galea, S. (2018). Precision medicine from a public health perspective. *Annual Review of Public Health*, 39(1), 153–168. DOI: 10.1146/annurev-publhealth-040617-014158 PMID: 29166244

Ramezankhani, R., Solhi, R., Chai, Y. C., Vosough, M., & Verfaillie, C. (2022, April). Organoid and microfluidics-based platforms for drug screening in COVID-19. *Drug Discovery Today*, 27(4), 1062–1076. DOI: 10.1016/j.drudis.2021.12.014 PMID: 34954328

Rao, S. J., Isath, A., Krishnan, P., Tangsrivimol, J. A., Virk, H. U. H., Wang, Z., Glicksberg, B. S., & Krittanawong, C. (2024). ChatGPT: A Conceptual Review of Applications and Utility in the Field of Medicine. *Journal of Medical Systems*, 48(1), 59. DOI: 10.1007/s10916-024-02075-x PMID: 38836893

Rastogi, P. (2008). *Bioinformatics Methods And Applications: Genomics Proteomics And Drug Discovery 3Rd Ed.* Prentice-Hall Of India Pvt. Limited. http://books.google.co.id/books?id=YP1rEFxFgDcC

Raza, M. A., Aziz, S., Noreen, M., Saeed, A., Anjum, I., Ahme, M., & Raza, S. M. (2022). AIin Pharmacy: An Overview of Innovations. *Pharmacy Practice and Practice-Based Research.*, 13(2), 1–13.

Redon, R., Ishikawa, S., Fitch, K. R., Feuk, L., Perry, G. H., Andrews, T. D., Fiegler, H., Shapero, M. H., Carson, A. R., Chen, W., Cho, E. K., Dallaire, S., Freeman, J. L., González, J. R., Gratacòs, M., Huang, J., Kalaitzopoulos, D., Komura, D., MacDonald, J. R., & Hurles, M. E. (2018). Global variation in copy number in the human genome. *Nature*, 444(7118), 444–454. DOI: 10.1038/nature05329 PMID: 17122850

Rehman, A. U., Li, M., Wu, B., Ali, Y., Rasheed, S., Shaheen, S., Liu, X., Luo, R., & Zhang, J. (2024). Role of Artificial Intelligence in Revolutionizing Drug Discovery. *Fundamental Research (Beijing)*. Advance online publication. DOI: 10.1016/j.fmre.2024.04.021

Reizine, N. M., & O'Donnell, P. H. (2022). Modern developments in germline pharmacogenomics for oncology prescribing. *CA: a Cancer Journal for Clinicians*, 72(4), 315–332. DOI: 10.3322/caac.21722 PMID: 35302652

Robillard, J. M., Roskams-Edris, D., Kuzeljevic, B., & Illes, J. (2014). Prevailing public perceptions of the ethics of gene therapy. *Human Gene Therapy*, 25(8), 740–746. DOI: 10.1089/hum.2014.030 PMID: 24773182

Roden, D. M. (2003). Cardiovascular pharmacogenomics. *Circulation*, 108(25), 3071–3074. DOI: 10.1161/01.CIR.0000110626.24310.18 PMID: 14691022

Roden, D. M., Wilke, R. A., Kroemer, H. K., & Stein, C. M. (2011). Pharmacogenomics: The genetics of variable drug responses. *Circulation*, 123(15), 1661–1670. DOI: 10.1161/CIRCULATIONAHA.109.914820 PMID: 21502584

Rodríguez-Vicente, A., Lumbreras, E., Hernández, J., Martín, M., Calles, A., Otín, C., Algarra, S., Páez, D., & Taron, M. (2016). Pharmacogenetics and pharmacogenomics as tools in cancer therapy. *Drug Metabolism and Personalized Therapy*, 31(1), 25–34. DOI: 10.1515/dmpt-2015-0042 PMID: 26863347

Roos, D. S. (2001). COMPUTATIONAL BIOLOGY: Bioinformatics—Trying to Swim in a Sea of Data. *Science*, 291(5507), 1260–1261. DOI: 10.1126/science.291.5507.1260 PMID: 11233452

Ross, J. S. (2009). HER2 testing in breast cancer: Current status and future directions. *Journal of Clinical Oncology*, 27(32), 5453–5462.

Ross, J. S., Symmans, W. F., Pusztai, L., & Hortobagyi, G. N. (2005). Pharmacogenomics and clinical biomarkers in drug discovery and development. *Pathology Patterns Reviews*, 124(suppl_1), S29–S41. DOI: 10.1309/XYQAFANAPYNC6X59 PMID: 16468416

Roudnitzky, N., Behrens, M., Engel, A., Kohl, S., Thalmann, S., Hübner, S., Lossow, K., Wooding, S. P., & Meyerhof, W. (2015). Receptor Polymorphism and Genomic Structure Interact to Shape Bitter Taste Perception. *PLOS Genetics*, 11(9), e1005530. DOI: 10.1371/journal.pgen.1005530 PMID: 26406243

Russell, L. E., Zhou, Y., Almousa, A. A., Sodhi, J. K., Nwabufo, C. K., & Lauschke, V. M. (2021). Pharmacogenomics in the era of next generation sequencing - from byte to bedside. *Drug Metabolism Reviews*, 53(2), 253–278. DOI: 10.1080/03602532.2021.1909613 PMID: 33820459

Ryszkiewicz, P., Malinowska, B., & Schlicker, E. (2023). Polypharmacology: promises and new drugs in 2022. In *Pharmacological Reports* (Vol. 75, Issue 4). DOI: 10.1007/s43440-023-00501-4

Sackmann, E. K., Fulton, A. L., & Beebe, D. J. (2014). The present and future role of microfluidics in biomedical research. *Nature*, 507(7491), 181–189. DOI: 10.1038/nature13118 PMID: 24622198

Sadee, W., Wang, D., Hartmann, K., & Toland, A. E. (2023). Pharmacogenomics: Driving personalized medicine. *Pharmacological Reviews*, 75(4), 789–814. DOI: 10.1124/pharmrev.122.000810 PMID: 36927888

Saha, D., Chan, A., Stacy, B., Javkar, K., Patkar, S., & Mazurek, M. L. (2020). User Attitudes On Direct-to-Consumer Genetic Testing. In *2020 IEEE European Symposium on Security and Privacy (EuroS&P)* (pp. 120-138). Genoa, Italy: IEEE. DOI: 10.1109/EuroSP48549.2020.00016

Sakaeda, T., Nakamura, T., & Okumura, K. (2002). MDR1 Genotype-Related Pharmacokinetics and Pharmacodynamics. *Biological & Pharmaceutical Bulletin*, 25(11), 1391–1400. DOI: 10.1248/bpb.25.1391 PMID: 12419946

Sakshi, V. (2023). Artificial Intelligence and Machine Learning in the Pharmaceutical Industry. *IJPPR*, 28(2), 111–131.

Saleh, S. H., Ismail, R., Ibrahim, Z., & Hussin, N. (2018). Issues, challenges and solutions of big data in information management: An overview. *International Journal of Academic Research in Business & Social Sciences*, 8(12), 1382–1393.

Salo-Ahen, O. M., Alanko, I., Bhadane, R., Bonvin, A. M., Honorato, R. V., Hossain, S., Juffer, A. H., Kabedev, A., Lahtela-Kakkonen, M., Larsen, A. S., Lescrinier, E., Marimuthu, P., Mirza, M. U., Mustafa, G., Nunes-Alves, A., Pantsar, T., Saadabadi, A., Singaravelu, K., & Vanmeert, M. (2020). Molecular dynamics simulations in drug discovery and pharmaceutical development. *Processes (Basel, Switzerland)*, 9(1), 71. DOI: 10.3390/pr9010071

Sampene, A. K., & Nyirenda, F. (2024). Evaluating the effect of artificial intelligence on pharmaceutical product and drug discovery in China. *Future Journal of Pharmaceutical Sciences*, 10(58), 1–11. DOI: 10.1186/s43094-024-00632-2

Sanders, S. J., & Campbell, A. J. (2020). Copy number variation in neurodevelopmental disorders. *Nature Reviews. Genetics*, 21(7), 407–418.

Sanjay, S. T., & Vishal, G. (2022). Artificial Intelligence in the Pharmaceutical Industry-An Overview of Innovations. *International Journal of Research Publication and Reviews.*, 3(7), 822–834.

Sankar, P., & Parker, L. (2017). The Precision Medicine Initiative's All of Us Research Program: An agenda for research on its ethical, legal, and social issues. *Genetics in Medicine*, 19(7), 743–750. DOI: 10.1038/gim.2016.183 PMID: 27929525

Sathyanarayanan, A., Mueller, T. T., Moni, M. A., Schueler, K., Baune, B. T., Lio, P., & Xicota, L. (2023). Multi-omics data integration methods and their applications in psychiatric disorders. *European Neuropsychopharmacology*, 69, 26–46. DOI: 10.1016/j.euroneuro.2023.01.001 PMID: 36706689

Saukko, P. (2018). Digital health–a new medical cosmology? The case of 23andMe online genetic testing platform. *Sociology of Health & Illness*, 40(8), 1312–1326. DOI: 10.1111/1467-9566.12774 PMID: 29998537

Saulsberry, L., & Olopade, O. I. (2021). Precision oncology: Directing genomics and pharmacogenomics toward reducing cancer inequities. *Cancer Cell*, 39(6), 730–733. DOI: 10.1016/j.ccell.2021.04.013 PMID: 34019805

Sauver, J. L. S., Bielinski, S. J., Olson, J. E., Bell, E. J., Mc Gree, M. E., Jacobson, D. J., McCormick, J. B., Caraballo, P. J., Takahashi, P. Y., & Roger, V. L. (2016). Integrating pharmacogenomics into clinical practice: promise vs reality. *The American journal of medicine, 129*(10), 1093-1099. e1091.

Sayam, A. G., Pradhan, M., & Choudhury, A. K. (2023). Artificial Intelligence the Futuristic Technology in the Drug Discovery Process: A Review. *Journal of Young Pharmacists : JYP*, 15(3), 390–396. DOI: 10.5530/jyp.2023.15.54

Sazonova, E. V., Chesnokov, M. S., Zhivotovsky, B., & Kopeina, G. S. (2022). Drug toxicity assessment: Cell proliferation versus cell death. *Cell Death Discovery*, 8(1), 417. DOI: 10.1038/s41420-022-01207-x PMID: 36241623

Schaefer, E. I. (2021). Network analysis of Alzheimer's disease. *Neurobiology of Aging*, 91, 39–53.

Schaid, D. J. (2018). Genomic annotations and functional analysis of genetic variants in complex diseases. *American Journal of Human Genetics*, 102(1), 9–22.

Schenone, M., Dančík, V., Wagner, B. K., & Clemons, P. A. (2013). Target identification and mechanism of action in chemical biology and drug discovery. *Nature Chemical Biology*, 9(4), 232–240. DOI: 10.1038/nchembio.1199 PMID: 23508189

Schindelin, J., Arganda-Carreras, I., Frise, E., Kaynig, V., Longair, M., Pietzsch, T., Preibisch, S., Rueden, C., Saalfeld, S., Schmid, B., Tinevez, J.-Y., White, D. J., Hartenstein, V., Eliceiri, K., Tomancak, P., & Cardona, A. (2012). Fiji: An open-source platform for biological-image analysis. *Nature Methods*, 9(7), 676–682. DOI: 10.1038/nmeth.2019 PMID: 22743772

Schneider, E. A. (2020). Clinical applications of genomic biomarkers in drug development. *The New England Journal of Medicine*, 383(1), 34–47.

Schomaker, S., Ramaiah, S., Khan, N., & Burkhardt, J. (2019). Safety biomarker applications in drug development. *The Journal of Toxicological Sciences*, 44(4), 225–235. DOI: 10.2131/jts.44.225 PMID: 30944276

Schork, N. J. (2018). Personalized medicine: Time for one-person trials. *Nature*, 520(7549), 609–611. DOI: 10.1038/520609a PMID: 25925459

Schwaederle, M. (2015). Impact of precision medicine in diverse cancers: A meta-analysis of 1,000 patients. *Journal of Clinical Oncology*, 33(31), 3329–3336.

Schwartz, T. S. (2021). In vitro and in vivo models for the study of genetic variants in disease. *The Journal of Experimental Medicine*, 218(7), e20210492.

Schwarz, J. M., Rödelsperger, C., Schuelke, M., & Seelow, D. (2010). MutationTaster evaluates disease-causing potential of sequence alterations. *Nature Methods*, 7(8), 575–576. DOI: 10.1038/nmeth0810-575 PMID: 20676075

Scionti, F., Di Martino, M. T., Caracciolo, D., Pensabene, L., Tagliaferri, P., & Arbitrio, M. (2022). Tools in Pharmacogenomics Biomarker Identification for Cancer Patients. *Methods in Molecular Biology (Clifton, N.J.)*, 2401, 1–12. DOI: 10.1007/978-1-0716-1839-4_1 PMID: 34902118

Selker, H. P., Eichler, H. G., Stockbridge, N. L., McElwee, N. E., Dere, W. H., Cohen, T., Erban, J. K., Seyfert-Margolis, V. L., Honig, P. K., Kaitin, K. I., Oye, K. A., & D'Agostino, R. B.Sr. (2019). Efficacy and Effectiveness Too Trials: Clinical Trial Designs to Generate Evidence on Efficacy and on Effectiveness in Wide Practice. *Clinical Pharmacology and Therapeutics*, 105(4), 857–866. DOI: 10.1002/cpt.1347 PMID: 30610746

Sendekie, A. K., Kasahun, A. E., Limenh, L. W., Dagnaw, A. D., & Belachew, E. A. (2023). Clinical and economic impact of adverse drug reactions in hospitalised patients: Prospective matched nested case-control study in Ethiopia. *BMJ Open*, 13(6), e073777. DOI: 10.1136/bmjopen-2023-073777 PMID: 37280017

Sequist, L. V. (2013). Phase II study of the EGFR inhibitor erlotinib for patients with advanced non-small cell lung cancer (NSCLC) and EGFR mutations. *The Lancet. Oncology*, 14(6), 603–610.

Serelli-Lee, V., Ito, K., Koibuchi, A., Tanigawa, T., Ueno, T., Matsushima, N., & Imai, Y. (2022). A State-of-the-Art Roadmap for Biomarker-Driven Drug Development in the Era of Personalized Therapies. In *Journal of Personalized Medicine* (Vol. 12, Issue 5). DOI: 10.3390/jpm12050669

Sethi, Y., Patel, N., Kaka, N., Kaiwan, O., Kar, J., Moinuddin, A., Goel, A., Chopra, H., & Cavalu, S. (2023). Precision Medicine and the future of Cardiovascular Diseases: A Clinically Oriented Comprehensive Review. *Journal of Clinical Medicine*, 12(5), 1799. Advance online publication. DOI: 10.3390/jcm12051799 PMID: 36902588

Shabaruddin, F. H., Fleeman, N. D., & Payne, K. (2015). Economic evaluations of personalized medicine: Existing challenges and current developments. *Pharmacogenomics and Personalized Medicine*, •••, 115–126. DOI: 10.2147/PGPM.S35063 PMID: 26309416

Shah, B. (2023). Revolutionizing Drug Discovery on the role of Artificial Intelligence. *International Journal of Scientific Research*, 12(12), 1848–1852.

Shalem, O., Sanjana, N. E., & Zhang, F. (2015, May). High-throughput functional genomics using CRISPR-Cas9. *Nature Reviews. Genetics*, 16(5), 299–311. DOI: 10.1038/nrg3899 PMID: 25854182

Shama, G., & Malik, D. J. (2013). The uses and abuses of rapid bioluminescence-based ATP assays. *International Journal of Hygiene and Environmental Health*, 216(2), 115–125. DOI: 10.1016/j.ijheh.2012.03.009 PMID: 22541898

Shao, F., Lee, P. W., Li, H., Hsieh, K., & Wang, T. H. (2023). Emerging platforms for high-throughput enzymatic bioassays. *Trends in Biotechnology*, 41(1), 120–133. DOI: 10.1016/j.tibtech.2022.06.006 PMID: 35863950

Sharma, P., & Hu-Lieskovan, S. (2018). Companion diagnostics in oncology. *Nature Reviews. Clinical Oncology*, 15(4), 228–237.

Shemuel, J., Angelique, P., Josephine, E., Fugaha, D. R., Gabriela, V., Alhusain, S., & Parikesit, A. A. (2023). ANALISIS GEN KOMPARATIF KARSINOMA SEL SKUAMOSA PARU-PARU ANTARA INDIVIDU MEROKOK DAN TIDAK MEROKOK [Comparative Gene Analysis of Squamous Cell Lung Carcinoma Between Smoking and Non-smoking Individuals]. *Berita Biologi*, 22(3), 291–304. DOI: 10.55981/beritabiologi.2023.908

Shendure, J., & Ji, H. (2008). Next-generation DNA sequencing. *Nature Biotechnology*, 26(10), 1135–1145. DOI: 10.1038/nbt1486 PMID: 18846087

Shinwari, Z. K., Tanveer, F., & Khalil, A. T. (2018). Ethical Issues Regarding CRISPR Mediated Genome Editing. *Current Issues in Molecular Biology*, 26, 103–110. DOI: 10.21775/cimb.026.103 PMID: 28879860

Shukla, R. (2020). *Pharmacogenomics: overview, applications, and recent developments*. Drug Design-Novel Adv Omics Field Appl.

Shukla, R. (2020). *Pharmacogenomics: Overview*. Applications, and Recent Developments., DOI: 10.5772/intechopen.93737

Shultz, L. D. (2020). Humanized mouse models for cancer immunotherapy. *Annual Review of Cancer Biology*, 4, 129–151.

Sia, S. K., & Whitesides, G. M. (2003). Microfluidic devices fabricated in poly (dimethylsiloxane) for biological studies. *Electrophoresis*, 24(21), 3563–3576. DOI: 10.1002/elps.200305584 PMID: 14613181

Simon, R., & Roychowdhury, S. (2013). Implementing personalized cancer genomics in clinical practice. *Nature Reviews. Clinical Oncology*, 10(7), 389–396.

Singh, D. B. (2020). The impact of pharmacogenomics in personalized medicine. *Current Applications of Pharmaceutical Biotechnology*, 369-394.

Singh, S., Singh, D. B., Gautam, B., Singh, A., & Yadav, N. (2022). Pharmacokinetics and pharmacodynamics analysis of drug candidates. In *Bioinformatics* (pp. 305–316). Elsevier. DOI: 10.1016/B978-0-323-89775-4.00001-8

Singla, A. (2024). Precision Medicine: Tailoring Treatment to Individual Genetic Profiles. *Shodh Sagar Journal for Medical Research Advancement*, 1(1), 27–37. DOI: 10.36676/ssjmra.v1.i1.04

Sisodiya, S. M. (2021). Precision medicine and therapies of the future. *Epilepsia*, 62(S2), S90–S105. DOI: 10.1111/epi.16539 PMID: 32776321

Sitapati, A., Kim, H., Berkovich, B., Marmor, R., Singh, S., El-Kareh, R., Clay, B., & Ohno-Machado, L. (2017). Integrated precision medicine: The role of electronic health records in delivering personalized treatment. *Wiley Interdisciplinary Reviews. Systems Biology and Medicine*, 9(3), e1378. DOI: 10.1002/wsbm.1378 PMID: 28207198

Slamon, D. J., Leyland-Jones, B., Shak, S., Fuchs, H., Paton, V., Bajamonde, A., Fleming, T., Eiermann, W., Wolter, J., Pegram, M., Baselga, J., & Norton, L. (2001). Use of chemotherapy plus a monoclonal antibody against HER2 for metastatic breast cancer that overexpresses HER2. *The New England Journal of Medicine*, 344(11), 783–792. DOI: 10.1056/NEJM200103153441101 PMID: 11248153

Slaughter, L. (2013). Genetic information non-discrimination act. *Harvard Journal on Legislation*, 50, 41.

Slezak, T. (2003). Comparative genomics tools applied to bioterrorism defence. *Briefings in Bioinformatics*, 4(2), 133–149. DOI: 10.1093/bib/4.2.133 PMID: 12846395

Slikker, W.Jr. (2018). Biomarkers and their impact on precision medicine. *Experimental Biology and Medicine*, 243(3), 211–212. DOI: 10.1177/1535370217733426 PMID: 28927290

Smokovski, I., Steinle, N., Behnke, A., Bhaskar, S. M. M., Grech, G., Richter, K., Niklewski, G., Birkenbihl, C., Parini, P., Andrews, R. J., Bauchner, H., & Golubnitschaja, O. (2024). Digital biomarkers: 3PM approach revolutionizing chronic disease management — EPMA 2024 position. *The EPMA Journal*, 15(2), 149–162. DOI: 10.1007/s13167-024-00364-6 PMID: 38841615

Smurnyy, Y., Cai, M., Wu, H., McWhinnie, E., Tallarico, J. A., Yang, Y., & Feng, Y. (2014). DNA sequencing and CRISPR-Cas9 gene editing for target validation in mammalian cells. *Nature Chemical Biology*, 10(8), 623–625. DOI: 10.1038/nchembio.1550 PMID: 24929529

Sonehara, K., & Okada, Y. (2021). Genomics-driven drug discovery based on disease-susceptibility genes. *Inflammation and Regeneration*, 41(1), 8. Advance online publication. DOI: 10.1186/s41232-021-00158-7 PMID: 33691789

Spreafico, R., Soriaga, L. B., Grosse, J., Virgin, H. W., & Telenti, A. (2020a). Advances in Genomics for Drug Development. *Genes*, 11(8), 1–19. DOI: 10.3390/genes11080942 PMID: 32824125

Stark, R. (2019). RNA sequencing: From technology to clinical applications. *Clinical Genetics*, 95(6), 557–564.

Stefanicka-Wojtas, D., & Kurpas, D. (2023). Barriers and Facilitators to the Implementation of Personalised Medicine across Europe. *Journal of Personalized Medicine*, 13(2), 203. Advance online publication. DOI: 10.3390/jpm13020203 PMID: 36836438

Stephens, Z. D., Lee, S. Y., Faghri, F., Campbell, R. H., Zhai, C., Efron, M. J., Iyer, R., Schatz, M. C., Sinha, S., & Robinson, G. E. (2015). Big data: Astronomical or genomical? *PLoS Biology*, 13(7), e1002195. DOI: 10.1371/journal.pbio.1002195 PMID: 26151137

Stoeckius, M. (2018). Simultaneous epitope and transcriptome measurement in single cells. *Nature Methods*, 15(5), 457–463. PMID: 28759029

Stolfi, F., Abreu, H., Sinella, R., Nembrini, S., Centonze, S., Landra, V., Brasso, C., Cappellano, G., Rocca, P., & Chiocchetti, A. (2024). Omics approaches open new horizons in major depressive disorder: From biomarkers to precision medicine. *Frontiers in Psychiatry*, 15, 1422939. DOI: 10.3389/fpsyt.2024.1422939 PMID: 38938457

Strianese, O., Rizzo, F., Ciccarelli, M., Galasso, G., Salvati, A., Giudice, C. D., Tesorio, P., & Rusciano, M. R. (2020). Precision and Personalized Medicine: How Genomic Approach Improves the Management of Cardiovascular and Neurodegenerative Disease. *Genes*, 11(7), 747. Advance online publication. DOI: 10.3390/genes11070747 PMID: 32640513

Stump, J. L. (2014). Henrietta Lacks and The HeLa Cell: Rights of Patients and Responsibilities of Medical Researchers. *The History Teacher*, 48(1), 127–180. https://www.jstor.org/stable/43264385

Subramanian, M., Wojtusciszyn, A., Favre, L., Boughorbel, S., Shan, J., Letaief, K. B., Pitteloud, N., & Chouchane, L. (2020). Precision medicine in the era of artificial intelligence: Implications in chronic disease management. *Journal of Translational Medicine*, 18(1), 1–12. DOI: 10.1186/s12967-020-02658-5 PMID: 33298113

Sugandh, F., Chandio, M., Raveena, F., Kumar, L., Karishma, F., Khuwaja, S., Memon, U. A., Bai, K., Kashif, M., Varrassi, G., Khatri, M., & Kumar, S. (2023). Advances in the Management of Diabetes Mellitus: A Focus on Personalized Medicine. *Cureus*, 15(8), e43697. DOI: 10.7759/cureus.43697 PMID: 37724233

Sukri, A., Salleh, M. Z., Masimirembwa, C., & Teh, L. K. (2022). A systematic review on the cost effectiveness of pharmacogenomics in developing countries: Implementation challenges. *The Pharmacogenomics Journal*, 22(3), 147–159. DOI: 10.1038/s41397-022-00272-w PMID: 35319010

Sullivan, P. F., & Grove, J. (2020). Statistical analysis in large-scale genetic association studies: Challenges and opportunities. *Nature Reviews. Genetics*, 21(6), 352–366.

Sultana, A., Maseera, R., Rahamanulla, A., & Misiriya, A. (2023). Emerging of artificial intelligence and technology in pharmaceuticals [review]. *Future Journal of Pharmaceutical Sciences*, 9(65), 1–9. DOI: 10.1186/s43094-023-00517-w

Swapna, B. V., Shetty, S., Shetty, M., & Shetty, S. S. (2024). Shibani Shetty, Manjunath Shetty, Smitha Sammith Shetty, Smart science: How artificial intelligence is revolutionizing pharmaceutical medicine. *Acta Marisiensis Seria Medica*, 70(1), 8–15. DOI: 10.2478/amma-2024-0002

Swinney, D. C., & Anthony, J. (2011, June 24). How were new medicines discovered? *Nature Reviews. Drug Discovery*, 10(7), 507–519. DOI: 10.1038/nrd3480 PMID: 21701501

Swinney, D. C., & Lee, J. A. (2020). Recent advances in phenotypic drug discovery. *F1000 Research*, •••, 9. PMID: 32850117

Syahruni, R., Umar, A. H., Rahman, H. N., & Kusuma, W. A. (2023). Exploration of Annona muricata (Annonaceae) in the Treatment of Hyperlipidemia through Network Pharmacology and Molecular Docking. *Sains Malaysiana*, 52(3), 899–939. DOI: 10.17576/jsm-2023-5203-17

Szymański, P., Markowicz, M., & Mikiciuk-Olasik, E. (2011). Adaptation of high-throughput screening in drug discovery—Toxicological screening tests. *International Journal of Molecular Sciences*, 13(1), 427–452. DOI: 10.3390/ijms13010427 PMID: 22312262

Tabor, H. (2011). Ethical challenges in genetic testing. *Journal of Genetic Counseling*, 20(4), 417–425.

Tade, R. S., Jain, S. N., Satyavijay, J. T., Shah, P. N., Bari, T. D., Patil, T. M., & Shah, R. P. (2024). Artificial Intelligence in the Paradigm Shift of Pharmaceutical Sciences: A Review. *Nano Biomedicine and Engineering*, 16(1), 64–77. DOI: 10.26599/NBE.2023.9290043

Tafazoli, A., Guchelaar, H.-J., Miltyk, W., Kretowski, A. J., & Swen, J. J. (2021). Applying next-generation sequencing platforms for pharmacogenomic testing in clinical practice. *Frontiers in Pharmacology*, 12, 693453. DOI: 10.3389/fphar.2021.693453 PMID: 34512329

Tafazoli, A., Wawrusiewicz-Kurylonek, N., Posmyk, R., & Miltyk, W. (2020). Pharmacogenomics, how to deal with different types of variants in next generation sequencing data in the personalized medicine area. *Journal of Clinical Medicine*, 10(1), 34. DOI: 10.3390/jcm10010034 PMID: 33374421

Tam, S. (2019). Genome-wide association studies in cardiovascular disease: Current status and future directions. *Lancet*, 393(10183), 1254–1264.

Tan, T. W., Tong, J. C., Khan, A. M., de Silva, M., Lim, K. S., & Ranganathan, S. (2010). Advancing standards for bioinformatics activities: Persistence, reproducibility, disambiguation and Minimum Information About a Bioinformatics investigation (MIABi). *BMC Genomics*, 11(SUPPL. 4), S27. DOI: 10.1186/1471-2164-11-S4-S27 PMID: 21143811

Tapper, W., Carneiro, G., Mikropoulos, C., Thomas, S. A., Evans, P. M., & Boussios, S. (2024). The Application of Radiomics and AI to Molecular Imaging for Prostate Cancer. *Journal of Personalized Medicine*, 14(3), 287. DOI: 10.3390/jpm14030287 PMID: 38541029

Tatsuno, S. Y., & Tatsuno, E. M. (2014). Does ethnicity play a role in the dosing of warfarin in Hawai'i? *Hawai'i Journal of Medicine & Public Health : a Journal of Asia Pacific Medicine & Public Health*, 73(3), 76–79. PMID: 24660123

Taylor, C. J., Pomberger, A., Felton, K. C., Grainger, R., Barecka, M., Chamberlain, T. W., Bourne, R. A., Johnson, C. N., & Lapkin, A. A. (2023). A brief introduction to chemical reaction optimization. *Chemical Reviews*, 123(6), 3089–3126. DOI: 10.1021/acs.chemrev.2c00798 PMID: 36820880

Taylor, C., Crosby, I., Yip, V., Maguire, P., Pirmohamed, M., & Turner, R. M. (2020). A Review of the Important Role of CYP2D6 in Pharmacogenomics. *Genes*, 11(11), 1295. DOI: 10.3390/genes11111295 PMID: 33143137

Teh, C.-K., Ong, A.-L., Kwong, Q.-B., Apparow, S., Chew, F.-T., Mayes, S., Mohamed, M., Appleton, D., & Kulaveerasingam, H. (2016). Genome-wide association study identifies three key loci for high mesocarp oil content in perennial crop oil palm. *Scientific Reports*, 6(1), 19075. DOI: 10.1038/srep19075 PMID: 26743827

The International HapMap Project. (2003). The International HapMap Project. *Nature*, 426(6968), 789–796. DOI: 10.1038/nature02168 PMID: 14685227

Thermo Fisher Scientific. (n.d.). [High-Content Screening] [HCS] [Platform. Retrieved from Thermo Fisher Scientific website]. *Cell Insight*, •••, CX7.

Thorn, C. F., Klein, T. E., & Altman, R. B. (2010). Pharmacogenomics and bioinformatics: PharmGKB. *Pharmacogenomics*, 11(4), 501–505. DOI: 10.2217/pgs.10.15 PMID: 20350130

Thorn, C. F., Klein, T. E., & Altman, R. B. (2013). PharmGKB: The Pharmacogenomics Knowledge Base. *Methods in Molecular Biology (Clifton, N.J.)*, 1015, 311–320. DOI: 10.1007/978-1-62703-435-7_20 PMID: 23824865

Thorsen, T., Maerkl, S. J., & Quake, S. R. (2002). Microfluidic large-scale integration. *Science*, 298(5593), 580–584. DOI: 10.1126/science.1076996 PMID: 12351675

Tiwari, P. C., Pal, R., Chaudhary, M. J., & Nath, R. (n.d.). Artificial intelligence revolutionizing drug development: Exploring opportunities and challenges, https://doi.org/DOI: 10.1002/ddr.22115

Tong, H., Phan, N. V., Nguyen, T. T., Nguyen, D. V., Vo, N. S., & Le, L. (2021). Review on databases and bioinformatic approaches on pharmacogenomics of adverse drug reactions. *Pharmacogenomics and Personalized Medicine*, 14, 61–75. DOI: 10.2147/PGPM.S290781 PMID: 33469342

Topol, E. J. (2019). High-performance medicine: The convergence of human and artificial intelligence. *Nature Medicine*, 25(1), 44–56. DOI: 10.1038/s41591-018-0300-7 PMID: 30617339

Tourish, D., & Willmott, H. (2023). Despotic Leadership and Ideological Manipulation at Theranos: Towards a theory of hegemonic totalism in the workplace. *Organization Studies*, 44(11), 1801–1824. DOI: 10.1177/01708406231171801

Tripathi, P., Singh, J., Lal, J. A., & Tripathi, V. (2019). Next-Generation Sequencing: An Emerging Tool for Drug Designing. *Current Pharmaceutical Design*, 25(31), 3350–3357. DOI: 10.2174/1381612825666190911155508 PMID: 31544713

Tristan, M. (2022). Pharmacogenomics Testing in Phase I Oncology Clinical Trials: Constructive Criticism Is Warranted. *Cancers (Basel)*, 14(5), 1131–1131. DOI: 10.3390/cancers14051131 PMID: 35267440

Trost, B. (2020). Targeted sequencing approaches in disease gene identification: A review of recent advancements. *Nature Reviews. Genetics*, 21(7), 431–444.

Trump, B., Cummings, C., Klasa, K., Galaitsi, S., & Linkov, I. (2023). Governing biotechnology to provide safety and security and address ethical, legal, and social implications. *Frontiers in Genetics*, 13, 1052371. DOI: 10.3389/fgene.2022.1052371 PMID: 36712887

Tsurusawa, N., Chang, J., Namba, M., Makioka, D., Yamura, S., Iha, K., Kyosei, Y., Watabe, S., Yoshimura, T., & Ito, E. (2021). Modified ELISA for ultrasensitive diagnosis. *Journal of Clinical Medicine*, 10(21), 5197. DOI: 10.3390/jcm10215197 PMID: 34768717

Tu, R., Zhang, Y., Hua, E., Bai, L., Huang, H., Yun, K., & Wang, M. (2021). Droplet-based microfluidic platform for high-throughput screening of Streptomyces. *Communications Biology*, 4(1), 647. DOI: 10.1038/s42003-021-02186-y PMID: 34059751

Turner, R. M., Magavern, E. F., & Pirmohamed, M. (2022). Pharmacogenomics: Relevance and opportunities for clinical pharmacology. *British Journal of Clinical Pharmacology*, 88(9), 3943–3946. DOI: 10.1111/bcp.15329 PMID: 35347744

Tyumentseva, M., Tyumentsev, A., & Akimkin, V. (2023). CRISPR/Cas9 landscape: Current state and future perspectives. *International Journal of Molecular Sciences*, 24(22), 16077. DOI: 10.3390/ijms242216077 PMID: 38003266

Udegbe, F. C., Ebulue, O. R., Ebulue, C. C., & Ekesiobi, C. S.Francisca Chibugo UdegbeOgochukwu Roseline EbulueCharles Chukwudalu EbulueChukwunonso Sylvester Ekesiobi. (2024). Precision Medicine and Genomics: A comprehensive review of IT-enabled approaches. *International Medical Science Research Journal*, 4(4), 509–520. DOI: 10.51594/imsrj.v4i4.1053

Uhlén, M., Fagerberg, L., Hallström, B. M., Lindskog, C., Oksvold, P., Mardinoglu, A., Sivertsson, Å., Kampf, C., Sjöstedt, E., Asplund, A., Olsson, I. M., Edlund, K., Lundberg, E., Navani, S., Szigyarto, C. A.-K., Odeberg, J., Djureinovic, D., Takanen, J. O., Hober, S., & Pontén, F. (2015). Tissue-based map of the human proteome. *Science*, 347(6220), 1260419. DOI: 10.1126/science.1260419 PMID: 25613900

Unertl, K. M., Jaffa, H., Field, J. R., Price, L., & Peterson, J. F. (2015). Clinician perspectives on using pharmacogenomics in clinical practice. *Personalized Medicine*, 12(4), 339–347. DOI: 10.2217/PME.15.10 PMID: 26635887

Urdinguio, R. G., Sanchez-Mut, J. V., & Esteller, M. (2009). Epigenetic mechanisms in neurological diseases: Genes, syndromes, and therapies. *Lancet Neurology*, 8(11), 1056–1072. DOI: 10.1016/S1474-4422(09)70262-5 PMID: 19833297

Vaghasia, V. V., Sharma, K., Mishra, S. K., & Georrge, J. J. (2024). In silico identification of natural product inhibitor for multidrug resistance proteins from selected gram-positive bacteria. In *Nanotechnology and In Silico Tools* (pp. 309–317). Elsevier. DOI: 10.1016/B978-0-443-15457-7.00015-0

Vakhariya Sakina, S., Mishra, S. K., Sharma, K., & Georrge, J. J. (2023). Designing of a novel curcumin analogue to inhibit mitogen-activated protein kinase: A cheminformatics approach. *Journal of Phytonanotechnology and Pharmaceutical Sciences*, 3(1), 37–47.

Vakhariya, S., Kumar Mishra, S., & Georrge, J. J. (2023). Identification of Novel Curcumin Analogues to Inhibit Mitogen-Activated Protein Kinase 1 Using in Silico Combinatorial Library Design and Molecular Docking Approach. *Available atSSRN* 4649193.

Valenti, A., & Burke, L. A. (2010). Post-Burlington: What Employers and Employees Need to Know About Retaliation. *Employee Responsibilities and Rights Journal*, 22(3), 235–251. DOI: 10.1007/s10672-009-9137-z

Valeska, M. D., Adisurja, G. P., Bernard, S., Wijaya, R., Aldino, M., & Parikesit, A. A. (2019). The Role of Bioinformatics in Personalized Medicine: Your Future Medical Treatment. *Cermin Dunia Kedokteran*, 46(12), 785–788. DOI: 10.5281/zenodo.4460835

Vamathevan, J., Clark, D., Czodrowski, P., Dunham, I., Ferran, E., Lee, G., Li, B., Madabhushi, A., Shah, P., Spitzer, M., & Zhao, S. (2019). Applications of machine learning in drug discovery and development. *Nature Reviews. Drug Discovery*, 18(6), 463–477. DOI: 10.1038/s41573-019-0024-5 PMID: 30976107

Van Cutsem, E., Köhne, C.-H., Hitre, E., Zaluski, J., Chang Chien, C.-R., Makhson, A., D'Haens, G., Pintér, T., Lim, R., Bodoky, G., Roh, J. K., Folprecht, G., Ruff, P., Stroh, C., Tejpar, S., Schlichting, M., Nippgen, J., & Rougier, P. (2009). Cetuximab and chemotherapy as initial treatment for metastatic colorectal cancer. *The New England Journal ofMedicine*, 360(14), 1408–1417. DOI: 10.1056/NEJMoa0805019 PMID: 19339720

van der Lee, M., Kriek, M., Guchelaar, H.-J., & Swen, J. J. (2020). Technologies for pharmacogenomics: A review. *Genes*, 11(12), 1456. DOI: 10.3390/genes11121456 PMID: 33291630

van der Lee, M., Rowell, W. J., Menafra, R., Guchelaar, H. J., Swen, J. J., & Anvar, S. Y. (2022). Application of long-read sequencing to elucidate complex pharmacogenomic regions: A proof of principle. *The Pharmacogenomics Journal*, 22(1), 75–81. DOI: 10.1038/s41397-021-00259-z PMID: 34741133

Van Driest, S. L., & Cascorbi, I. (2021). Progress and Challenges in Pharmacogenomics. *Clinical Pharmacology and Therapeutics*, 110(3), 529–532. DOI: 10.1002/cpt.2359 PMID: 34412159

Vandenberg, L. N., Colborn, T., Hayes, T. B., Heindel, J. J., Lee, H., Shioda, T., Soto, A. M., Welshons, W. V., Zoeller, R. T., & Myers, J. P. (2012). Hormones and Endocrine-Disrupting Chemicals: Low-Dose Effects and Nonmonotonic Dose Responses. *Endocrine Reviews*, 33(3), 378–455. DOI: 10.1210/er.2011-1050 PMID: 22419778

Vargas, A. J., & Harris, C. C. (2016). Biomarker development in the precision medicine era: Lung cancer as a case study. *Nature Reviews. Cancer*, 16(8), 525–537. DOI: 10.1038/nrc.2016.56 PMID: 27388699

Venkatachalapathy, P., Padhilahouse, S., Sellappan, M., Subramanian, T., Kurian, S. J., Miraj, S. S., Rao, M., Raut, A. A., Kanwar, R. K., Singh, J., Khadanga, S., Mondithoka, S., & Munisamy, M. (2021). Pharmacogenomics and Personalized Medicine in Type 2 Diabetes Mellitus: Potential Implications for Clinical Practice. *Pharmacogenomics and Personalized Medicine*, 14, 1441–1455. DOI: 10.2147/PGPM.S329787 PMID: 34803393

Ventola, C. L. (2011). Pharmacogenomics in Clinical Practice: Reality and Expectations. *P&T*, 36(7), 412. https://www.ncbi.nlm.nih.gov/pmc/articles/PMC3171815/ PMID: 21931473

Verma, A., & Awasthi, A. (2024). Revolutionizing Drug Discovery: The Role of Artificial Intelligence and Machine Learning. *Current Pharmaceutical Design*, 30(1), 807–810. DOI: 10.2174/0113816128298691240222054120 PMID: 38409722

Villeneuve, D. L., Coady, K., Escher, B. I., Mihaich, E., Murphy, C. A., Schlekat, T., & Garcia-Reyero, N. (2019). High-throughput screening and environmental risk assessment: State of the science and emerging applications. *Environmental Toxicology and Chemistry*, 38(1), 12–26. DOI: 10.1002/etc.4315 PMID: 30570782

Vincent, F., Nueda, A., Lee, J., Schenone, M., Prunotto, M., & Mercola, M. (2022). Phenotypic drug discovery: Recent successes, lessons learned and new directions. *Nature Reviews. Drug Discovery*, 21(12), 899–914. DOI: 10.1038/s41573-022-00472-w PMID: 35637317

Visscher, P. M., Wray, N. R., Zhang, Q., Sklar, P., McCarthy, M. I., Brown, M. A., & Yang, J. (2017). 10 years of GWAS discovery: Biology, function, and translation. *American Journal of Human Genetics*, 101(1), 5–22. DOI: 10.1016/j.ajhg.2017.06.005 PMID: 28686856

Visvikis-Siest, S., Theodoridou, D., Kontoe, M. S., Kumar, S., & Marschler, M. (2020). Milestones in personalized medicine: From the ancient time to nowadays—the provocation of COVID-19. *Frontiers in Genetics*, 11, 569175. DOI: 10.3389/fgene.2020.569175 PMID: 33424917

Vogel, F. (1959). Moderne probleme der humangenetik. In *Ergebnisse der inneren medizin und kinderheilkunde* (pp. 52–125). Springer Berlin Heidelberg. DOI: 10.1007/978-3-642-94744-5_2

Vogelstein, B., Papadopoulos, N., Velculescu, V. E., Zhou, S., Diaz, L. A.Jr, & Kinzler, K. W. (2013). Cancer genome landscapes. *Science*, 339(6127), 1546–1558. DOI: 10.1126/science.1235122 PMID: 23539594

Vora, L. K., Gholap, A. D., Jetha, K., Raghu, R. S. T., & Hetvi, K. (1916). Artificial Intelligence in Pharmaceutical Technology and Drug Delivery Design. *Pharmaceutics*, 2023(15), 1–46.

Wallace, D. C. (2019). Mitochondrial DNA and disease. *Human Genetics*, 138(1), 1–18. PMID: 30390160

Wang, C. W., Preclaro, I. A. C., Lin, W. H., & Chung, W. H. (2022). An Updated Review of Genetic Associations With Severe Adverse Drug Reactions: Translation and Implementation of Pharmacogenomic Testing in Clinical Practice. *Frontiers in Pharmacology*, 13, 886377. DOI: 10.3389/fphar.2022.886377 PMID: 35548363

Wang, D., Bolleddula, J., Coenen-Stass, A., Grombacher, T., Dong, J. Q., Scheuenpflug, J., Locatelli, G., & Feng, Z. (2024). Implementation of whole-exome sequencing for pharmacogenomics profiling and exploring its potential clinical utilities. *Pharmacogenomics*, 25(4), 197–206. DOI: 10.2217/pgs-2023-0243 PMID: 38511470

Wang, E., Cho, W. C., Wong, S. C., & Liu, S. (2017). Disease biomarkers for precision medicine: Challenges and future opportunities. *Genomics, Proteomics & Bioinformatics*, 15(2), 57–58. DOI: 10.1016/j.gpb.2017.04.001 PMID: 28392478

Wang, J., Wu, M. Y., Tan, J. Q., Li, M., & Lu, J. H. (2019, February 28). High content screening for drug discovery from traditional Chinese medicine. *Chinese Medicine*, 14(1), 5. DOI: 10.1186/s13020-019-0228-y PMID: 30858873

Wang, K. (2018). Statistical methods for genome-wide association studies. *Nature Reviews. Genetics*, 19(10), 597–610. PMID: 29727703

Wang, R. C., & Wang, Z. (2023). Precision Medicine: Disease Subtyping and Tailored Treatment. *Cancers (Basel)*, 15(15), 3837. DOI: 10.3390/cancers15153837 PMID: 37568653

Wang, T., Wei, J. J., Sabatini, D. M., & Lander, E. S. (2014). Genetic screens in human cells using the CRISPR-Cas9 system. *Science*, 343(6166), 80–84. DOI: 10.1126/science.1246981 PMID: 24336569

Wang, Y., Yang, Y., Chen, S., & Wang, J. (2021). DeepDRK: A deep learning framework for drug repurposing through kernel-based multi-omics integration. *Briefings in Bioinformatics*, 22(5), bbab048. DOI: 10.1093/bib/bbab048 PMID: 33822890

Weibel, D. B., & Whitesides, G. M. (2006). Applications of microfluidics in chemical biology. *Current Opinion in Chemical Biology*, 10(6), 584–591. DOI: 10.1016/j.cbpa.2006.10.016 PMID: 17056296

Weinshilboum, R. M., & Wang, L. (2017). Pharmacogenomics: Precision medicine and drug response. Mayo Clinic Proceedings, Whirl-Carrillo, M., McDonagh, E. M., Hebert, J. M., Gong, L., Sangkuhl, K., Thorn, C. F., Altman, R. B., & Klein, T. E. (2012). Pharmacogenomics knowledge for personalized medicine. *Clinical Pharmacology and Therapeutics*, 92(4), 414–417. DOI: 10.1038/clpt.2012.96

Weinstein, D., Leininger, J., Hamby, C., & Safai, B. (2014). Diagnostic and prognostic biomarkers in melanoma. *The Journal of Clinical and Aesthetic Dermatology*, 7(6), 13. PMID: 25013535

Whitesides, G. M. (2006). The origins and the future of microfluidics. nature, 442(7101), 368-373.

Wilkinson, S. (2008). "Eugenics talk" and the language of bioethics. *Journal of Medical Ethics*, 34(6), 467–471. DOI: 10.1136/jme.2007.021592 PMID: 18511622

Williams, T. D. (2008). Individual variation in endocrine systems: Moving beyond the 'tyranny of the Golden Mean'. *Philosophical Transactions of the Royal Society of London. Series B, Biological Sciences*, 363(1497), 1687–1698. DOI: 10.1098/rstb.2007.0003 PMID: 18048295

Wohlers, I., Garg, S., & Hehir-Kwa, J. Y. (2022). Editorial: Long-read sequencing-Pitfalls, benefits and success stories. *Frontiers in Genetics*, 13, 1114542. DOI: 10.3389/fgene.2022.1114542 PMID: 36685894

Woodcock, J., & LaVange, L. M. (2017). Master protocols to study multiple therapies, multiple diseases, or both. *The New England Journal of Medicine*, 377(1), 62–70. DOI: 10.1056/NEJMra1510062 PMID: 28679092

Woodcock, J., & LaVange, L. M. (2017). The regulatory approval of genomic biomarkers: A guide. *The New England Journal of Medicine*, 377(24), 2319–2321. PMID: 28402237

Wooller, S. K., Benstead-Hume, G., Chen, X., Ali, Y., & Pearl, F. M. G. (2017). Bioinformatics in translational drug discovery. *Bioscience Reports*, 37(4), 20160180. DOI: 10.1042/BSR20160180 PMID: 28487472

Xia, X. (2017). Bioinformatics and Drug Discovery. *Current Topics in Medicinal Chemistry*, 17(15), 1709–1726. DOI: 10.2174/1568026617666161116143440 PMID: 27848897

Xi, X., Li, T., Huang, Y., Sun, J., Zhu, Y., Yang, Y., & Lu, Z. J. (2017). RNA biomarkers: Frontier of precision medicine for cancer. *Non-Coding RNA*, 3(1), 9. DOI: 10.3390/ncrna3010009 PMID: 29657281

Xu, F., Zheng, C., Xu, W., Zhang, S., Liu, S., Chen, X., & Yao, K. (2024). Breaking genetic shackles: The advance of base editing in genetic disorder treatment. *Frontiers in Pharmacology*, 15, 1364135. DOI: 10.3389/fphar.2024.1364135 PMID: 38510648

Xu, Y., Jiang, Z., Wei Ting, D. S., Chieh Kow, A. W., Bello, F., Car, J., Tham, C., & Wong, T. Y. (2024). Medical education and physician training in the era of artificial intelligence. *Singapore Medical Journal*, 65(3), 159–166. DOI: 10.4103/singaporemedj.SMJ-2023-203 PMID: 38527300

Yadav, S., Singh, A., Singhal, R., & Yadav, J. P. (2024). *Revolutionizing drug discovery: The impact of artificial intelligence on advancements in pharmacology and the pharmaceutical industry*. Intelligent Pharmacy.

Yadav, S., Singh, A., Singhal, R., & Yadav, J. P. (2024). Revolutionizing drug discovery: The impact of artificial intelligence on advancements in pharmacology and the pharmaceutical industry. *Intelligent Pharmacy.*, 2(3), 367–380. DOI: 10.1016/j.ipha.2024.02.009

Yakubu, A. M., & Chen, Y.-P. P. (2020). Ensuring privacy and security of genomic data and functionalities. *Briefings in Bioinformatics*, 21(2), 511–526. DOI: 10.1093/bib/bbz013 PMID: 30759195

Yang, Z., Zhao, J. V., Qi, Y., Deng, X., Ji, Z., & Liu, J. (2024). A translational framework of genoproteomic studies for cardiovascular drug discovery. *Npj Cardiovascular Health 2024 1:1, 1*(1), 1–12. DOI: 10.1038/s44325-024-00015-9

Yang, H., Lou, C., Sun, L., Li, J., Cai, Y., Wang, Z., Li, W., Liu, G., & Tang, Y. (2019). AdmetSAR 2.0: Web-service for prediction and optimization of chemical ADMET properties. *Bioinformatics (Oxford, England)*, 35(6), 1067–1069. DOI: 10.1093/bioinformatics/bty707 PMID: 30165565

Yan, W., Cui, J., Wang, Y., Zhang, X., Yang, F., Chen, E. F., Wang, M., & Yang, J. (2012). Comparative Analysis of *UGT1A9* Genetic Polymorphisms between Chinese Han and Tibetan Populations. *American Medical Journal*, 3(2), 194–202. DOI: 10.3844/amjsp.2012.194.202

Yehya, A. (2022). Interplay between Pharmacokinetics and Pharmacogenomics. In *Dosage Forms-Innovation and Future Perspectives*. IntechOpen.

Yogeshappa, V. G. (2024). AI-driven precision medicine: Revolutionizing personalized treatment plans. [IJCET]. *International Journal of Computer Engineering and Technology*, 15(5), 455–474.

Yuan, B. (2022). What personalized medicine humans need and way to it——also on the practical significance and Scientific Limitations of Precision Medicine. *Pharmacogenomics and Personalized Medicine*, 15, 927–942. DOI: 10.2147/PGPM.S380767 PMID: 36353711

Yu, H. B., Li, M., Wang, W. P., & Wang, X. L. (2016). High throughput screening technologies for ion channels. *Acta Pharmacologica Sinica*, 37(1), 34–43. DOI: 10.1038/aps.2015.108 PMID: 26657056

Yuliwulandari, R., Prayuni, K., Susilowati, R. W., Subagyo, S., Soedarsono, S. M., Sofro, A. S., Tokunaga, K., & Shin, J. G. (2019). NAT2 slow acetylator is associated with anti-tuberculosis drug-induced liver injury severity in Indonesian population. *Pharmacogenomics*, 20(18), 1303–1310. DOI: 10.2217/pgs-2019-0131 PMID: 31699005

Yumiceba, V., López-Cortés, A., Pérez-Villa, A., Yumiseba, I., Guerrero, S., García-Cárdenas, J. M., Armendáriz-Castillo, I., Guevara-Ramírez, P., Leone, P. E., Zambrano, A. K., & Paz-y-Miño, C. (2020). Oncology and Pharmacogenomics Insights in Polycystic Ovary Syndrome: An Integrative Analysis. *Frontiers in Endocrinology*, 11, 585130. Advance online publication. DOI: 10.3389/fendo.2020.585130 PMID: 33329391

Yu-Wai-Man, P., & Votruba, M. (2020). Leber hereditary optic neuropathy. *Archives of Neurology*, 67(9), 1167–1173. PMID: 33298565

Zampatti, S., Fabrizio, C., Ragazzo, M., Campoli, G., Caputo, V., Strafella, C., Pellicano, C., Cascella, R., Spalletta, G., Petrosini, L., Caltagirone, C., Termine, A., & Giardina, E. (2021). Precision Medicine into Clinical Practice: A Web-Based Tool Enables Real-Time Pharmacogenetic Assessment of Tailored Treatments in Psychiatric Disorders. *Journal of Personalized Medicine*, 11(9), 851. Advance online publication. DOI: 10.3390/jpm11090851 PMID: 34575628

Zanger, U. M., & Schwab, M. (2013). Cytochrome P450 enzymes in drug metabolism: Regulation of gene expression, enzyme activities, and impact of genetic variation. *Pharmacology & Therapeutics*, 138(1), 103–141. DOI: 10.1016/j.pharmthera.2012.12.007 PMID: 23333322

Zeggini, E., Gloyn, A. L., Barton, A. C., & Wain, L. V. (2019). Translational genomics and precision medicine: Moving from the lab to the clinic. *Science*, 365(6460), 1409–1413. DOI: 10.1126/science.aax4588 PMID: 31604268

Zhang, L., Zhang, X., Ma, Q., Ma, F., & Zhou, H. (2010). Transcriptomics and Proteomics in the Study of H1N1 2009. In *Genomics, Proteomics and Bioinformatics* (Vol. 8, Issue 3, pp. 139–144). Elsevier. DOI: 10.1016/S1672-0229(10)60016-2

Zhang, C., Yang, J., Chen, S., Sun, L., Li, K., Lai, G., Peng, B., Zhong, X., & Xie, B. (2024). Artificial intelligence in ovarian cancer drug resistance advanced 3PM approach: Subtype classification and prognostic modeling. *The EPMA Journal*, 15(3), 525–544. Advance online publication. DOI: 10.1007/s13167-024-00374-4 PMID: 39239109

Zhang, J. H., Chung, T. D., & Oldenburg, K. R. (1999). A simple statistical parameter for use in evaluation and validation of high throughput screening assays. *Journal of Biomolecular Screening*, 4(2), 67–73. DOI: 10.1177/108705719900400206 PMID: 10838414

Zhang, L., Xia, Y., & Gui, Y. (2023). Neuronal ApoE4 in Alzheimer's disease and potential therapeutic targets. *Frontiers in Aging Neuroscience*, 15, 1199434. DOI: 10.3389/fnagi.2023.1199434 PMID: 37333457

Zhang, Y., Liu, B., & Tong, Z. (2022). Adenosine triphosphate (ATP) bioluminescence-based strategies for monitoring atmospheric bioaerosols. *Journal of the Air & Waste Management Association*, 72(12), 1327–1340. DOI: 10.1080/10962247.2022.2101566 PMID: 36226866

Zhang, Z. (2019). KRAS and EGFR mutations in non-small cell lung cancer: Targeted treatment approach and mechanisms. *Oncotarget*, 10(45), 4485–4500.

Zhao, X., Iqbal, S., Valdes, I. L., Dresser, M., & Girish, S. (2022). Integrating real-world data to accelerate and guide drug development: A clinical pharmacology perspective. *Clinical and Translational Science*, 15(10), 2293–2302. DOI: 10.1111/cts.13379 PMID: 35912537

Zhou, Y., & Lauschke, V. M. (2023). Challenges Related to the Use of Next-Generation Sequencing for the Optimization of Drug Therapy. *Handbook of Experimental Pharmacology*, 280, 237–260. DOI: 10.1007/164_2022_596 PMID: 35792943

Zhou, Y., & Lauschke, V. M. (2024). Next-generation sequencing in pharmacogenomics - fit for clinical decision support? *Expert Review of Clinical Pharmacology*, 17(3), 213–223. DOI: 10.1080/17512433.2024.2307418 PMID: 38247431

Zhou, Y., Peng, S., Wang, H., Cai, X., & Wang, Q. (2024). Review of Personalized Medicine and Pharmacogenomics of Anti-Cancer Compounds and Natural Products. *Genes*, 15(4), 468. Advance online publication. DOI: 10.3390/genes15040468 PMID: 38674402

Zhou, Y., Tremmel, R., Schaeffeler, E., Schwab, M., & Lauschke, V. M. (2022). Challenges and opportunities associated with rare-variant pharmacogenomics. *Trends in Pharmacological Sciences*, 43(10), 852–865. DOI: 10.1016/j.tips.2022.07.002 PMID: 36008164

Zhou, Y., Zhu, S., Cai, C., Yuan, P., Li, C., Huang, Y., & Wei, W. (2014). High-throughput screening of a CRISPR/Cas9 library for functional genomics in human cells. *Nature*, 509(7501), 487–491. DOI: 10.1038/nature13166 PMID: 24717434

Zhu, Y. (2017). Emerging technologies in genomic medicine: Single-cell sequencing and spatial transcriptomics. *Nature Reviews. Genetics*, 18(7), 485–497.

Zhu, Y., Swanson, K. M., Rojas, R. L., Wang, Z., St Sauver, J. L., Visscher, S. L., Prokop, L. J., Bielinski, S. J., Wang, L., Weinshilboum, R., & Borah, B. J. (2020). Systematic review of the evidence on the cost-effectiveness of pharmacogenomics-guided treatment for cardiovascular diseases. *Genetics in Medicine*, 22(3), 475–486. DOI: 10.1038/s41436-019-0667-y PMID: 31591509

Zock, J. M. (2009, November). Applications of high content screening in life science research. *Combinatorial Chemistry & High Throughput Screening*, 12(9), 870–876. DOI: 10.2174/138620709789383277 PMID: 19938341

Zook, J. M. (2020). Genomic data security and privacy: A global perspective. *Nature Reviews. Genetics*, 21(8), 516–530.

About the Contributors

Arli Aditya Parikesit received his doctoral degree PhD (Informatics) from Computational EvoDevo, Institute of Computer Science, University of Leipzig, Germany in 2012 under the supervision of Prof Dr. Sonja Prohaska. Presently he is holding the position of Vice-Rector (R&D) and Professor in the Department of Bioinformatics, School of Life Sciences, Indonesia International Institute for Life Sciences (i3L), Indonesia. His selected works- Evolutionary Analysis of the Protein Domain Distribution in Eukaryotes: Computing genome-wide Protein Domain Annotation of RefSeq and Gene Prediction. Investigating domain co-occurrences tendency of the Selected Genomes. In Silico Design of Therapeutic and Preventive agents for Virus-based disease. Ongoing Project: Drugs and Vaccines design on Influenza A, Dengue, and HPV. Breast Cancer RNA drug design, COVID-19 Herbal Medicine, and Fish Transcriptomics.

Anupam Singh received her doctoral degree PhD (Biotechnology) from Shri Ram Swaroop Memorial University, India in 2020. Presently he is holding the position of Head of Department (HoD) and Associate-Professor in Bansal Institute of Engineering and Technology, Lucknow (U.P.), India. His selected works- A Method for Precision Genome Editing for Human Disease. Development of curcumin loaded hybrid nanomaterials as a potential therapeutic agent in cancer and study of its harmful effects using Drosophila melanogaster as an In-vivo model. Synthesis of Silver Nanoparticles from V. Serpens and their Effect on MDR Strains. A Portable Assembly for Providing Treatment of Hazardous Materials in Oxygen Rich Environment and Method Thereof. A method of preparing light sequence from microscopy to transcriptomics. Eco-friendly slow-release urea fertilizer based on starch-g-poly composite with natural char nanoparticles. Identification of anti-Covid 19 agent Somniferine as an inhibitor of Mpro & ACE2-RBD Interaction. Identification of Withafastuosin E as an Inhibitor of RHO-Associated Kinase-2 Catalytic Domain: Potential to be Developed as Neurotherapeutic Agent for Ischemic

Stroke. Density Gradient-Based Nucleic Acid Isolation. In: DNA and RNA Isolation Techniques for Non-Experts. Double Gene Targeting Multiplex Polymerase Chain Reaction of Cow, Buffalo, Goat and Sheep Species in Raw Meat. Cytochrome B gene targeting multiplex polymerase chain reaction of Chicken (Gallus gallus), Horse (Equus caballus), Pig (Sus scrofa) and Rat (Rattus norvegicus) species in raw meat. Genetic Diversity Analysis in Turmeric (Curcuma Longa L.) Based on SSR Markers. Studies on Genetic Variability in Rice Using Molecular Markers. Identification and Molecular Characterization of Phytoplasma Associated with Peach Disease in Western Uttar Pradesh. In Vitro Antibacterial and Antifungal Activity of Medicinal Plants. Antibacterial Effectivity Of Tulsi (Ocimum Sanctum) And Peppermint (Mentha Piperita) Oils With Standards Antibiotics Against Selected Bacterial Strain. Virtual Screening Analysis for Drug Designing against Precursor Protein of Alzheimer's disease. In vitro antimicrobial activity of medicinal plants Ashwagandha (Withania Somnifera) and Papaya (Carica Papaya).

Rajesh G. Konnur received her doctoral degree PhD (Biotechnology) from Trinity College Seychelles in 2010. Presently he is holding the position of Director in Eudoxia Research Centre, India. His selected works- Book: Psychology for Graduate Nurses. Book: Text book of Psychiatric Nursing. Book: Clinical Record Book for Mental Health Nursing. Book: Advanced Nursing Practice. Book: Nursing Management. National Conference on Child Psychiatric Nursing. National Conference on Women's Mental Health: Nursing Perceptive. National Conference on drug Abuse. International Conference on Adolescent Mental Health Issues and Problems: Nursing perspectives. National Conference on Geriatric Mental Health Nursing. World Summit on Medical Sciences, Pharmaceutical sciences and Nursing sciences.

Saptarshi Mukherjee received her master degree M.Tech. (Bioinformatic) and M.Sc. (Computer Science and Engineering) from Maulana Abul Kalam Azad University of Technology, West Bengal, India. Presently he is holding the position of Research Administrator in GoldSmith Institute of Advanced Research (GIAR), India. His selected works- Precision Medicine in Lung Cancer Treatment. Precision Medicine in Acute myeloid leukemia. Biologically Synthesized Silver Nanoparticles: Characterization, Antimicrobial Efficacy, And Mechanistic Insights Using Leaf Extract. AI Integrated predictive inventory management system for manufacturing supply chains. Neonatal Diabetes Mellitus: Unravelling the clinical, genetic, and nutritional dimensions for enhanced management. Diabetes Mellitus and its role in the devolopment and progression of atrial fibrillation -A review and Recommendations. Computational Analysis for Dementia Drug Target Identification and Optimization. NGS Analysis Approach for Neurodegenerative

Disease with Biopython. Review on Identification and distribution of heat shock proteins in wheat genome. Bacteria with the correct plasmid are used to make more plasmid DNA or, in some cases, induced to express the gene and make protein. DNA cloning is a molecular biology technique that makes many identical copies of a piece of DNA, such as a gene. In a typical cloning experiment, a target gene is inserted into a circular piece of DNA called a plasmid. The plasmid is introduced into bacteria via a process called transformation, and bacteria carrying the plasmid are selected using antibiotics. Silver Nano Particles on antimicrobial plant and finding their physicochemical properties of the antimicrobial plant. Medical Laboratory Technologists perform analyses by preparing slides of human tissues and other specimens using lab equipment. Genetic mutations has led to various rare and prevalent neurological diseases. Parkinson's disease is a neurodegenerative disorder with a heterogeneous genetic etiology. An early and accurate molecular diagnosis are very much important for neurodegenerative patients to undergo proper therapeutic regimens. The next-generation sequencing (NGS) method is used for examining millions of sequences at a time. Finally, we discuss the NGS challenges and opportunities in neurodegenerative diseases.

***

Ch. Hima Bindu finished her master's degree and doctoral research at JNTUK in Kakinada. presently employed as a professor at the QIS College of Engineering and Technology in Ongole, Andhra Pradesh, India, in the Department of Electronics and Communication Engineering. She also attended numerous international conferences and produced a number of research papers in reputable journals, textbooks, and patents. Several organizations, including AICTE, DST, UGC, and others, awarded her research grants. Under her guidance, over five students are pursuing doctorates.

Devadasu Ch. has been working as an Associate Professor at QIS college of Pharmacy and he has 50 research papers of national and international repute to his credit. He has a teaching experience of Pharmaceutical analysis and chemistry for the past 14 years.

Tabsum Chhetri is currently involved in the drug target identification, transcriptomics data analysis, and novel inhibitor identification for the pathogens related research. Apart from that she also strongly interested in genomics profile and their impact on the disease understating via advance computational, bioinformatics approaches.

Gudimitla Raveendra Babu finished his master's degree and doctoral research at JNTUK in Kakinada and JNTUH, Hyderabad respectively. in the department of pharmaceutical analysis. presently employed as a professor at the QIS College of Pharmacy in Ongole, Andhra Pradesh, India, in the Department of pharmaceutical analysis. He also attended numerous international conferences and produced a number of research papers in reputable journals, textbooks, and patents.

Kavinaya Shri is a final year undergraduate student of biotechnology at Vels Institute of Science, Technology and Advanced Studies. Her field of interest focuses on drug discovery and delivery and global sustainable development. She has authored and co-authored few papers in top-tier journals, that are yet to be published.

Jalaiah Marri is a Professor and HOD in the department of pharmacy Practice at QIS College of Pharmacy, Vengamukkapalem,Ongole, Andhra Pradesh. With 14 years of experience he has made substantial contribution to his field, publishing 26 research papers in various national and international journals, authoring three books and holding five patents. His primary area of interest is neuropharmacology, Where he continues to innovate and lead in both research and education.

Index

A

B

C

D

E

G

www.ingramcontent.com/pod-product-compliance
Lightning Source LLC
LaVergne TN
LVHW081537300926
849161LV00006B/1463
* 9 7 9 8 3 6 9 3 6 5 9 7 7 *